THE LION

THE LION

BEHAVIOR, ECOLOGY, AND CONSERVATION
OF AN ICONIC SPECIES

CRAIG PACKER

WITH COLOR PHOTOGRAPHS BY
DANIEL ROSENGREN

PRINCETON UNIVERSITY PRESS

PRINCETON AND OXFORD

Copyright © 2023 by Princeton University Press

Princeton University Press is committed to the protection of copyright and the intellectual property our authors entrust to us. Copyright promotes the progress and integrity of knowledge. Thank you for supporting free speech and the global exchange of ideas by purchasing an authorized edition of this book. If you wish to reproduce or distribute any part of it in any form, please obtain permission.

Requests for permission to reproduce material from this work should be sent to permissions@press.princeton.edu

Published by Princeton University Press
41 William Street, Princeton, New Jersey 08540
99 Banbury Road, Oxford OX2 6JX

press.princeton.edu

All Rights Reserved

Library of Congress Cataloging-in-Publication Data

Names: Packer, Craig, author. | Rosengren, Daniel, 1978– photographer.
Title: The lion : behavior, ecology, and conservation of an iconic species / Craig Packer ; with color photographs by Daniel Rosengren.
Description: Princeton : Princeton University Press, 2023. | Includes bibliographical references and index.
Identifiers: LCCN 2022013103 (print) | LCCN 2022013104 (ebook) | ISBN 9780691215297 (hardback) | ISBN 9780691235950 (ebook)
Subjects: LCSH: Lion—Behavior—Africa, East.
Classification: LCC QL737.C23 P315 2023 (print) | LCC QL737.C23 (ebook) | DDC 599.75709676—dc23/eng/20220316
LC record available at https://lccn.loc.gov/2022013103
LC ebook record available at https://lccn.loc.gov/2022013104

British Library Cataloging-in-Publication Data is available

Editorial: Alison Kalett and Hallie Schaeffer
Production Editorial: Natalie Baan
Production: Jacquie Poirier
Publicity: Charlotte Coyne and Matthew Taylor
Copyeditor: Dana Henricks

Jacket photograph by Daniel Rosengren

This book has been composed in Spectral, Nocturne Serif, and Futura PT

Printed on acid-free paper. ∞

Printed in the United States of America

10 9 8 7 6 5 4 3 2 1

CONTENTS

	Preface	ix
1	**INTRODUCTION**	1
	A Lion Primer	5
	Key Points	28
2	**LION SOCIAL STRUCTURE**	32
	The Basic System	32
	Social Dominance	35
	Reproductive Success	40
	Pride Persistence versus Coalition Longevity	53
	Key Points	57
3	**INFANTICIDE AND EXPULSION**	59
	After the Storm	63
	Synchronizing the Reproductive Clock	68
	The Academic Politics of Infanticide	73
	Key Points	75
4	**CUB REARING**	76
	From the Den to the Crèche	76
	Nursing Behavior	80
	Non-offspring Nursing across Mammals	86
	Babysitting	88
	What about Grandma?	93
	What Keeps the Crèche Honest?	96
	Littering	96
	It's a Girl! It's a Boy!! It's Three Boys and a Girl . . .	101
	Key Points	104
5	**MATING COMPETITION AND PATERNAL CARE**	107
	Within-Coalition Competition	108
	One for All and All for One?	114
	Why Are Large Coalitions So Rare?	116
	Father(s) of the Pride	118
	I Love You, Dad, but . . .	128
	Key Points	130

6	**THE LION'S MANE**	**133**
	Manes and Wounding	137
	Manes versus Male Quality	140
	What Do the Lions Think?	144
	Why Lions? Why Manes?	145
	Key Points	148
7	**FORAGING BEHAVIOR**	**149**
	Background	150
	Hunting in Groups—or Not . . .	154
	But Do They Cooperate When They *Are* Together?	160
	Key Points	164
8	**INTERGROUP COMPETITION**	**166**
	Intergroup Conflict	167
	Blood on the Tracks	176
	War Games of the Sexes	180
	Intergenerational-Intergroup Competition	184
	Groups Begetting Groups—of Both Sexes	192
	Key Points	192
9	**THE EVOLUTION OF LION SOCIALITY**	**195**
	The Three Rules of Lion Real Estate	195
	The Landscape of Female Sociality	202
	Pride Size Revisited	208
	The Origins of Male Sociality	212
	The Lion: The Paragon of Group Living	219
	Key Points	222
10	**POPULATION REGULATION**	**224**
	Serengeti: Tracking Lions in a Changing Environment	224
	Top-Down Population Regulation: Predation, Infanticide, and Disease	230
	Ngorongoro Crater: Disease versus Density	240
	Keeping It in the Family	245
	Key Points	249
11	**INTERSPECIFIC INTERACTIONS**	**250**
	Lions and Their Prey	251
	Aren't Predators and Prey Supposed to Show Population Cycles?	255
	The Savanna Landscape of Fear	261
	Lions versus the Other Large Carnivores	265
	Lions, Hyenas, and Possible "Trophic Cascades"	275
	Lions versus Hyenas and the Evolution of Lion Sociality	277
	The Lion's Place in the Web	278
	Key Points	281

12 LION CONSERVATION — 282
Infectious Disease — 282
Human-Lion Conflict — 294
Sport Hunting — 309
Bushmeat Hunting — 315
The Future — 316
Key Points — 326

AFTERWORD — 327

Acknowledgments — 333
Appendix — 335
References — 337
Index — 353

PREFACE

> The world is not created once and for all time for each of
> us individually. There are added to it in the course of our
> life things of which we have never had any suspicion.
> —MARCEL PROUST

Whether as a symbol of courage or as a source of fear, few animals on earth have ever held the same grip on the human imagination as the African lion, and, in a world of shrinking wilderness, nowhere evokes such a breathtaking sense of nature as the sight of a million wildebeest migrating across the Serengeti with its wide-open spaces and tree-lined river crossings. For many, George Schaller was our first encounter with the ultimate predator and its iconic home with his landmark book, *The Serengeti Lion*. I hadn't yet started high school when Schaller first arrived in the Serengeti in March 1966. In those days, I spent each school year with my parents in Fort Worth, Texas, but my summers were always on my grandmother's farm a few miles from the town of Noodle, Texas (with a purported population of seven, though we didn't know any of the inhabitants). Mammaw's farm was only a few hundred acres of poor soil with modest harvests of wheat and sorghum that were just enough to get by from one year to the next. I helped my uncle with the chores, clearing brush and occasionally driving the tractor, but I otherwise took every opportunity to go fishing or look for arrowheads or wander around outdoors until it got too hot in the afternoon. I had little idea what I wanted to do with my life, except I was determined to be a scientist of some sort—I was always curious about everything and reckoned that scientific research would be the closest I'd ever get to becoming a twentieth-century explorer. Oddly, though, the white-coated stereotypical scientist of the time invariably worked in a sterile laboratory with test tubes or caged rodents, and I had a profound aversion to being trapped indoors.

By the time Schaller's book came out in 1972, I was in my final days as an undergraduate. I happened to be in Tanzania at the time, working as a field assistant for Jane Goodall at Gombe National Park and helping to set up a long-term study on three troops of olive baboons that paralleled Jane's famous chimp project. I went out every day noting down all the individuals I saw, who was mating with whom, who had given birth to which baby. We gave each animal a name; all the members of one troop were named after plants/vegetables (Cauliflower, Magnolia, Spud), another after famous characters (Hamlet, Persephone, Samson) and the third after place names (Amazon, Fargo, Santa Fe). I also helped bridge the gap back to several earlier baboon projects at Gombe, lining up their prior IDs with the current set of individuals. I can still remember the thrill of realizing that Asparagus had two older sisters, Apricot and Apple, thus making Azalea the grandmother of Algae. They not only bore a family resemblance, but their social relationships were different with each other than with anyone else. I spent most of my mornings in the forest just up the hill from the shores of Lake Tanganyika, following a particular individual for two hours at a stretch, collecting data on young male baboons before and after they left the troops of their birth to start breeding in neighboring troops. The work required long hours outside, surrounded by nature, absorbed by the lives of another species—and my teenaged fantasies of somehow doing science outdoors materialized almost too neatly in ways that could hardly have suited me better.

I read Schaller's lion book during my first year in graduate school at the University of Sussex in 1974—and though I was fascinated by his descriptions of the lions and their world, it was the simultaneous publication of another book, *The Spotted Hyena* by Hans Kruuk, that truly captured my imagination. Kruuk had been a student of Niko Tinbergen, the Nobel Prize–winning ethologist who founded the field of behavioral ecology—which seeks to understand the evolutionary underpinnings of animal behavior—and Kruuk's approach was more question-oriented than Schaller's; he collected quantitative data to test specific hypotheses about the survival value of specific aspects of hyena behavior. Hyenas may not be much to look at compared to lions, but their lives are every bit as interesting—even when reduced to a series of precise measurements of grouping patterns and hunting tactics.

Inspired by Kruuk's hyena study and motivated by my interactions with John Maynard Smith, Paul Harvey, and Tim Clutton-Brock at the University of Sussex, I went back to Gombe to complete my doctoral research with the goal of addressing specific hypotheses about the risks and benefits to a male baboon from leaving his mother's troop and entering a neighboring troop. Unlike any other primate study at the time, the three troops at Gombe made it possible to obtain a relatively large sample of individuals of just the right age to watch them leave home and/or establish themselves elsewhere, and, by 1974, the long-term records were long enough to start contemplating the impacts of dispersal on each male's survival and reproduction. But I had also learned an important lesson from Jane: sometimes the most valuable insights come from just watching the animals without any preconceived notions; don't forget to allow yourself to be surprised by things that you'd never contemplated before.

By the time I had completed my thesis in 1977, I knew this was the sort of work I wanted to do for the rest of my career. See something in nature that no one has ever seen before, assemble the observations into a larger picture, then write it all up. Watch and repeat. I was so deeply immersed in primate behavior that I next went to Japan to study snow monkeys, whose dispersal patterns showed certain parallels and contrasts with the baboons'. Many of the research sites in Japan had been monitoring their study animals for over a decade, so it would again be possible to measure survival and reproduction. But after six weeks of intensive Japanese lessons in California and an extended tour of Honshu, Kyushu, Shikoku, and Yakushima, the opportunities to collaborate with Japanese scientists turned out to be illusory—so I retreated back to the UK and rethought my future.

While at Gombe, I had become intrigued by the way that pairs of adult male baboons would sometimes team up to form a coalition and steal away the mating partner of a stronger, higher-ranking male. Two against one gave a powerful advantage, and the reward immediately improved the winner's mating opportunities. Charles Darwin had inspired generations of biologists to contemplate two evolutionary puzzles, and one concerned the origins of cooperative behavior.[1] The struggle for existence should usually reward selfishness, and most animals do seem pretty selfish most of the time—except for the notable exception of the social insects—and cooperation was a topic that transcended study species, being equally fascinating in birds, mammals, fish, or even slime mold. So, when the opportunity arose to take over the long-term study of the Serengeti lions, I was all in. What better species to study cooperation?

By the time I returned to Tanzania in 1978, the lion study already extended over three generations in a dozen Serengeti "prides." As Schaller was leaving the Serengeti in 1969, he was immediately followed by Brian Bertram, who extended the project to 1973 and then recruited Jeannette Hanby and David Bygott, who remained in the Serengeti until 1978, and Brian also facilitated the next handover to Anne Pusey (1978–1995) and me (1978–2015). David and Jeannette had also expanded the project to include all the lions living on the floor of the nearby Ngorongoro Crater—yet another iconic natural wonder ("the largest unflooded, unbroken caldera in the world" [Fosbrooke 1972]) and its extraordinary density of large mammals. Schaller's book had provided such rich descriptions of the lions' behavior and ecology—their interactions with each other, their vocalizations, their hunting techniques, their interactions with other carnivores—that there is still no better source for the basics of lion life. But Schaller had only spent three and a half years in the Serengeti, and it wasn't until Bertram, Hanby, and Bygott completed their stints that the lions' reproductive strategies could be characterized (Bertram 1975a) or quantified (Bygott, et al. 1979). Bertram (1976) had also written a provocative paper suggesting that the lions' most fascinating behaviors—communal nursing, group territoriality—might be ideal for testing ideas about the evolutionary pressures favoring cooperation between close genetic relatives.

Thus, Anne Pusey and I spent our first dozen years in the Serengeti collecting detailed data on lion behavior. We spent hundreds of hours with the lions as they fed at kills, competed for mates, and nursed their cubs, and we also took advantage of the ever-expanding long-term demographic data from the first few decades of the project to tackle larger questions about group living and conflicts between the sexes. But in 1994 a disease outbreak killed a third of the Serengeti lions, and my research interests shifted toward basic questions in population ecology: what regulates lion numbers, whether lions limit the abundance of their prey, what are the lions' impacts on other carnivore species. Meanwhile, lion populations throughout Africa had declined dramatically by the early 2000s, so I became increasingly focused on lion conservation: the impacts of poorly regulated sport hunting, the factors that prompt lions to become man-eaters, and the management strategies that successfully protect vulnerable lion populations.

My thirty-seven years in the Serengeti were first funded by the Harry Frank Guggenheim Foundation, followed by the National Geographic Society, the National Institute for Mental Health (NIMH) (through a training grant to the University of Chicago), and

---

1 The other topic was sexual selection, as exemplified by the peacock's tail—a trait that seemed to run directly counter to the precepts of natural selection, as it must surely render the male peacock more vulnerable to predation. Little did I know in the late 1970s that we would tackle the lion's mane with such alacrity at the turn of the millennium.

thirty-one years of grants from the National Science Foundation (NSF). The final NSF grant came from a program called "Opportunities for Promoting Understanding through Synthesis," or OPUS, that initiated the long process of writing this book. In some ways this is a long overdue sequel to Schaller, but it has also allowed a reassessment of our 140-plus scientific papers. The expectation of our funding agencies was always to address a limited set of questions, assess any insights generated by our data from a theoretical perspective and then add our findings to the overall body of knowledge by subjecting our papers to peer review. Thus, our published papers were scattered across dozens of learned journals and technical books, with one paper being largely disconnected from the next and available to only the most avid researcher.

In writing this synthesis, just trotting out the results from so many published articles and chaining them together in book form would have been pure torture for everyone involved, so I have instead tried to take a more encompassing approach: a scientific compendium that gives as much feel for the animal as I can. Thus, I have relied on three complementary approaches: quantitative, qualitative, and visual. First are the admittedly reductive and simplistic numerical data that are summarized in graphs and charts on virtually every page. Second are brief vignettes from field observations that often illuminate a topic by showing that we aren't just dealing with abstractions but with real-life, flesh-and-blood individuals. Third are the color photographs taken by Daniel Rosengren, who spent nearly four years working on the lion project before becoming a full-time photographer for Frankfurt Zoological Society; I have done my best to give center stage to Daniel's stunning portrayals of the lions and their world by placing them within the relevant context of our scientific measurements. Unless otherwise noted, all color photos are by Daniel Rosengren and all black and white photos are by me (CP).

Some of our earlier findings have stood the test of time, some have had to be reconsidered, and I have learned a lot not only from linking many of these findings together for the first time but by also addressing a number of neglected topics with new analyses from the long-term data. These new analyses help to connect the dots between related topics to provide a more coherent narrative, and I have tried to write the text in as user-friendly manner as possible. Above all, I want the lions themselves to emerge as the stars of the show so that anyone who comes into this thinking that lions are pretty darned cool will hopefully come out the other side thinking, damn, lions are downright *amazing*.

THE LION

CHAPTER 1

INTRODUCTION

You know you are truly alive when you're living among lions.
—KAREN BLIXEN

In some ways the lion needs no introduction as our relationship with the species extends back for millennia. Our ancestors drew paintings of archaic lions on the walls of their caves, ancient civilizations portrayed lions as sphinxes (human-headed lions); griffins (half-lions, half-eagles); servants of the goddesses Ishtar and Parvati; and the vanquished foes of exalted Assyrian kings. Many of us grew up with the lion as a character in children's literature (the Cowardly Lion from the *Wizard of Oz*, Aslan from *The Lion, the Witch and the Wardrobe*, Simba from *The Lion King*), and we often describe someone we admire as being brave as a lion, as having a leonine grace, as being lionhearted. But none of these portrayals tell us much about what it's really like to be a lion. What it's like to belong to an extended family of multiple mothers, aunts, cousins, and grandmothers, as well as fathers and uncles and sometimes even dad's best friend. What it's like to raise cubs in a world where the next big meal may not arrive in time. What it's like to be surrounded by neighboring lions who would be perfectly happy to cripple you or kill your young. But most importantly, this long history doesn't prepare us for the very real possibility that the lion may not survive in a world that is increasingly filled with humans.

The reality of the beast is far more engaging than the historical myths and the literary characterizations. Despite an ancient feline ancestry of habitual distrust toward others of their own kind, lions tackle most challenges as a group: they hunt together, raise cubs together, and defend a joint territory. Female pridemates are a sisterhood not only in terms of their genetic relationships but also in the idealized sense of a nonhierarchical feminism; males are comrades in arms to an almost gladiatorial extent.[1] But are any of these behaviors truly cooperative, or do slackers parasitize the generosity of the rest? And if any of these behaviors is truly cooperative, what is its basis? Close genetic kinship? Some cold calculating form of scorekeeping? Or the warm comfort of mutual dependency? And, if all this cooperation is such a good idea, why haven't tigers and leopards followed a similar path in their own evolution?

But not all aspects of lion behavior are admirable; lions also have their dark side. Family life involves prolonged periods of parental care—mostly by the mothers, but fathers matter, too—and the very fact of extended care provides incentives for outsiders

---

[1] The Cowardly Lion needed a companion more than he needed a medal; Aslan could never have ruled Narnia alone; Scar would never have contrived to kill Mufasa.

to try to eliminate any obstacles so as to speed up their own chances for reproduction. How do families protect themselves against this constant threat of disruption and infanticide? And how do the potential consequences of infanticide extend to seemingly distant aspects of lion life, including the choices females make when seeking new mates or even to the defining characteristic of the mature male, the conspicuousness of his mane?

As impressed as we may be with the lions' strength and size, they face the fundamental challenge of trying to capture an animal that doesn't want to be eaten, and some prey species are more difficult to catch than others. How do the lions overcome these challenges? Is it always better to hunt together or should they sometimes revert to the ways of their solitary ancestors? What happens when the herds are out of reach for months at a stretch? Beyond the daily drama of catching their next meal, the back-and-forth between predator and prey plays out over a longer time scale in ways that may determine the number of lions that can live in a particular area. Hence, we might ask how many wildebeest does it take to feed a lion? Conversely, do too many lions ever threaten the future of their own food supply? Lions can be terrible, horrible, awful animals when it comes to their interactions with smaller carnivores. What happens to these species in a landscape filled with lions?

If we take a wide-angle view of the lion across the broad landscape of twenty-first-century Africa, we cannot ignore the dark clouds on the horizon. Successful lion conservation is in doubt even in many of the best-protected parks in Africa, and solutions are urgently needed. Geneticists have long warned about the consequences of close inbreeding, and many of the smaller parks and reserves host lion populations that are unlikely to remain viable in the long term. But how does this actually manifest itself in a species as robust and resilient as the lion, and is there any practical way to address the problem? Human populations have grown to such an extent that lion habitat now directly abuts livestock pastures and agricultural fields across much of their remaining range. What happens when lions leave the confines of a national park and come face-to-face with livestock herders? Is there any way to promote human-lion coexistence? Even worse is when lions become man-eaters—yes, man-eating lions still exist even in the twenty-first century. What leads to this sort of behavior? Can we predict when and where people will be most at risk? And at what point should we just give up on coexistence and erect physical barriers to protect people from lions and vice versa?

These are just a few of the questions that we addressed during our intensive studies of the Serengeti and Ngorongoro Crater lions. In the following pages, the initial chapters will focus on various aspects of lion behavior before broadening out to include ecological issues and then expanding to cover lion conservation. But in first introducing the lions, I want to provide a picture of how we were able to keep track of everyone on a day-to-day basis. Though I will often describe lions in general terms such as "males," "females," or "cubs," we recognized every animal as an *individual*. Some lions had telltale markings like conspicuous scars or broken tails, and almost every adult had a diverse collection of ear notches acquired from squabbles with its companions at kills. But most lions are difficult to distinguish as they lack the conspicuous black stripes or rosettes of a tiger or a leopard. We therefore relied on Pennycuick and Rudnai's (1970) method of identifying individuals by the pattern of their whisker spots.

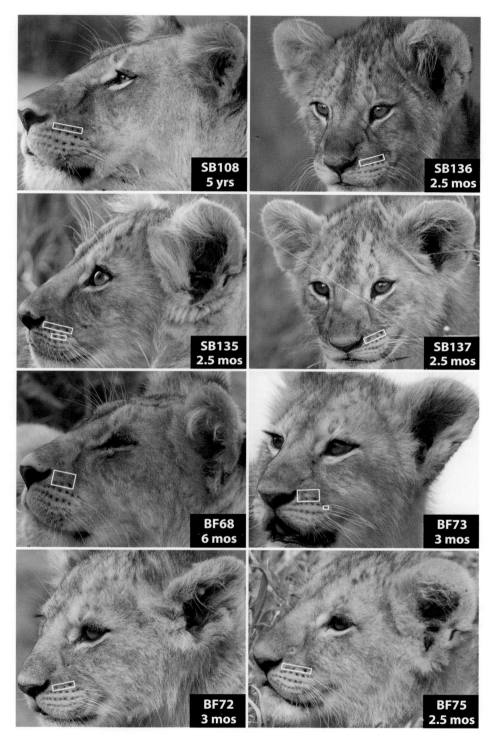

FIGURE 1.1. Identification photos for eight Serengeti lions in the Simba (SB) and Barafu (BF) prides. Large boxes indicate the primary whisker spots used to recognize each individual; small boxes highlight the "between-row spots" of SB135 and BF73. SB108 is the mother of littermates SB135–137; BF72 and BF73 are also littermates.

The well-defined rows of vibrissae on either side of each lion's face almost always include a few extra dots above the top row (figure 1.1), and these can be easily noted on an ID card. Whisker spots are distinct even in the smallest cubs and though these may become obscured by scars or age, the steady accumulation of additional markings keeps each animal unique throughout its entire life.

The daily monitoring required an inordinate amount of patience. The initial pulse of excitement at finding a group of lions was soon tempered by the fact that they have usually melted for the day and were unlikely to rise again until the cool cover of darkness (see figure 1.16a). But we needed to know who was there, who was mating with whom, who had just had babies, what species of prey they had obtained. Often, we found only two or three adults in a group, and the chores were fairly simple, but a bigger group with a large pile of cubs could be exhausting. Grant Hopcraft's description of one particular sighting provides a sense of the challenge:

> **4-Dec-1999 09:00.** I identified each one of a group of 17 lions from the Masai Kopje (MK) pride as they meandered toward a nearby kopje [rocky outcrop]. The four adult females were easy. Each marking helped to tell them apart—a notched ear, scarred nose, or distinctive dilated pupil with a brown dot in the iris. The 13 cubs were the hardest to identify, yowling behind their mothers or ambushing their unsuspecting sisters, and only differentiated from each other by a subtle whisker spot pattern. I spent nearly two hours, sometimes feeling like a rock climber clinging to an identification by my fingertips. An extra spot between the ordered rows of whisker spots is like a Christmas present as it distinguishes one cub from the next. The trick is to start with the easy ones and whittle down to the final few. Deductive logic, note taking, and a series of sketches helps keep track of each one before they disappear into a gully or kopje; any leftovers are listed as "unknowns" that at least provide an idea of the number of individuals in the group that day.

Though often tedious and laborious, these observations were the bread and butter of the long-term lion project, and, with 56,640 separate sightings like Grant's encounter with the MK pride, we were able to assemble mosaic portraits of 5,309 known individuals in 111 prides.

To assure an accurate accounting of so many individuals over the fifty years of the long-term study, we gave each animal a unique name. Schaller had mostly used numbers (Female-49, Male-107) except for a few distinctive individuals (Brownmane, Blackmane, One-eye). Bertram also used numbers (S-79, M-22), whereas the Bygotts preferred Swahili words: Mbili (two), Miwani (eyeglasses), and Mwindaji (hunter), in one pride; Shida (trouble), Shika (hold), and Safi (clean), in another. Anne Pusey and I kept the names of animals that had been christened by our predecessors but then christened each new cub with a code beginning with the two-letter abbreviation of its natal pride (SM for the Sametu Pride; LL for Loliondo) either followed by the letters of the alphabet or by numbers. Hence, SB108 in figure 1.1 was the 108th cub born to the Simba (SB) pride and BF68 was the 68th cub in the Barafu (BF) pride after we joined the project. We weren't averse to giving them longer names, but each name could only be used once, and so many cubs die before reaching maturity that there aren't enough names to go around. On the other hand, the relatively small number of older animals that immigrated into the study areas from elsewhere were far more likely to survive, so these received names that ranged from the sublime (Dorian and Gray) to the ridiculous (Twirp-1 and Twirp-2), but generally with a common theme indicating who had first been seen with whom.

A LION PRIMER

Before taking a deep dive into specific research topics, I first want to give a sense of the lions' food supply in the Serengeti and Ngorongoro, introduce more lion-monitoring methodology, provide a few basics about the lion's day (and night), and briefly describe their overall life course.

All Ecology Is Local

The Serengeti National Park is located two degrees south of the equator, just west of the Ngorongoro Crater highlands and east of Lake Victoria, and the Serengeti study area lies in the approximate center of the park (figure 1.2). Rain clouds moving off the Indian Ocean only surmount the Crater highlands for a few months each year, as the mountains produce a "rain shadow" that limits total rainfall over the open plains of the Ngorongoro Conservation Area (NCA) and the southern Serengeti, whereas Lake Victoria generates its own weather in the far west of the park (Sinclair and Norton-Griffiths 1979), thus, the Serengeti experiences a pronounced rainfall gradient from north to south and from east to west (figure 1.3). The seasonal rainfall patterns drive the annual migration of wildebeest, zebra, and Thomson's gazelle as they seek out the nutritious grasses on the volcanic soils of the eastern plains each wet season then retreat to the north and west each dry season (figure 1.4). Rainfall is heavier in the woodlands portion of the lion study area, allowing grazing species, such as Cape buffalo, hartebeest, and warthog, to remain throughout the year, and the woody vegetation also supports resident browsers, such as impala and giraffe. In contrast, the 250-km^2 floor of the Ngorongoro Crater is an island of open grasslands interspersed with swamps and marshes. The Crater receives substantially more rainfall than the Serengeti (figure 1.3), and as the floods recede each dry season, the emerging vegetation remains green at the margins, providing a continuous supply of forage (Estes and Small 1981). Thus, the larger herbivores reside on the Crater floor all year round, except for a proportion of wildebeest that move to the Serengeti plains each wet season and about a third of the Cape buffalo that move up to the Crater rim each dry season (figure 1.5).

The Crater lions subsist almost entirely on three species: wildebeest, zebra, and buffalo, with wildebeest being the most common prey throughout the year (figure 1.6a); zebra and buffalo are both taken somewhat more often in the dry season compared to their seasonal availability, possibly reflecting lower body condition in the driest months. During the wet season, the lions on the Serengeti plains are regularly able to "feast" on the migratory wildebeest, zebra, and Thomson's gazelle, but they are forced to subsist on a broader range of prey during the "famine" of the dry season (figure 1.6b). The lions in the woodlands study area mostly take wildebeest, zebra, and

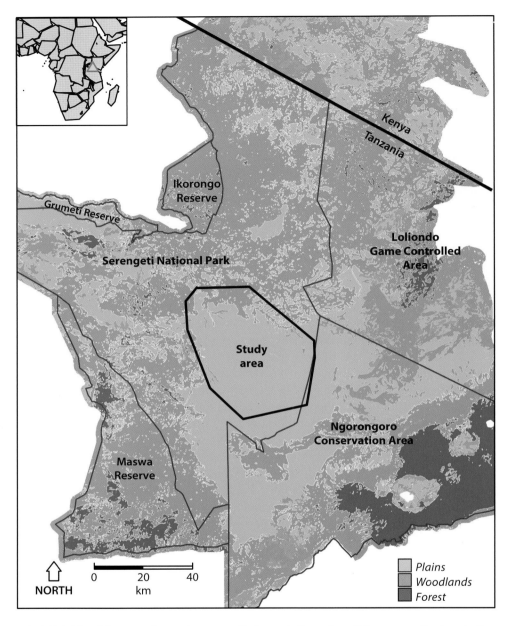

FIGURE 1.2. Map of the greater Serengeti ecosystem. Dark green indicates forest, light green woodlands, and yellow open grasslands. The long-term Serengeti study area covers about 2,000 km² and extends from the central woodlands to cover much of the open plains. The Ngorongoro Crater is located inside the Ngorongoro Conservation Area.

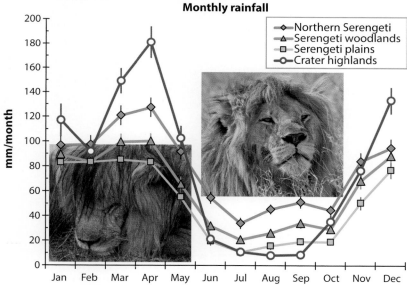

FIGURE 1.3. Monthly rainfall in the greater Serengeti ecosystem. Data from the Serengeti were collected between 1966 and 2000 from three to eighteen rain gauges in each part of the park; data from the Crater highlands were collected from a single gauge between 1963 and 2014; see Sinclair et al. (2013) for details. Vertical bars are standard errors.

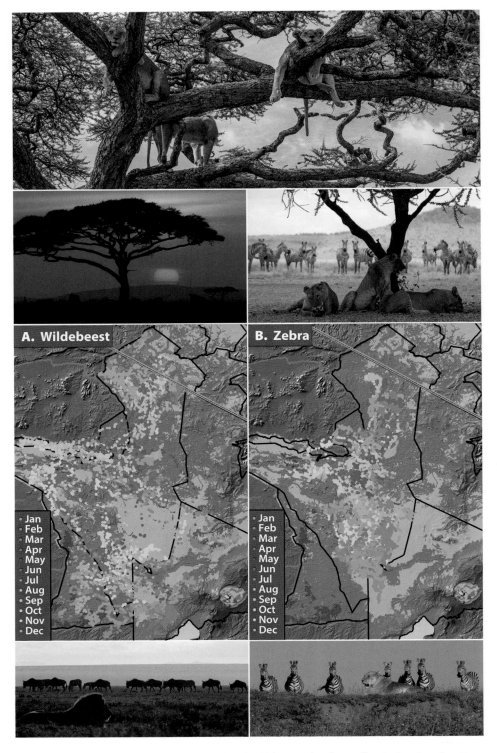

FIGURE 1.4. Daily movements of **A.** three GPS-collared wildebeest and **B.** three collared zebra. Data from Boone et al. (2006) and Hopcraft et al. (2014).

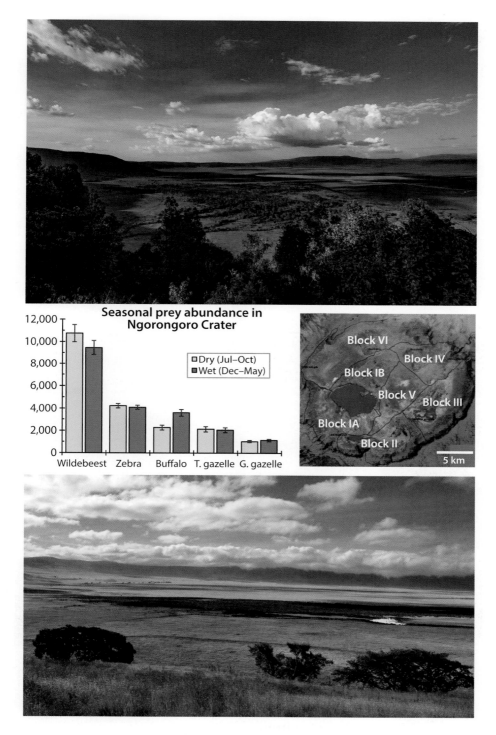

FIGURE 1.5. Seasonal prey counts in Ngorongoro Crater 1969–2012. Data were restricted to the thirty-three years when animals were counted in both the wet and dry seasons of the same year. All ground counts were conducted by MWEKA wildlife college; see Estes et al. (2006) for details. **Inset:** infrared satellite photo outlining the blocks surveyed during the ground surveys.

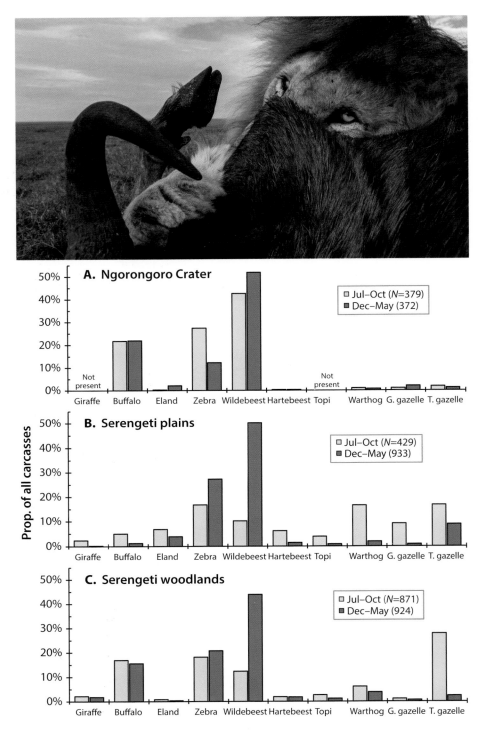

FIGURE 1.6. Species composition of the prey consumed by the lions in each habitat. Data are divided between the driest and wettest months of the year. Sample sizes are given in brackets and include all carcasses, regardless of whether they were scavenged or killed by the feeding lions. Data from the Ngorongoro Crater are from 1974 to 2015 and the two Serengeti study areas are from 1966 to 2015.

buffalo in the wet season, whereas wildebeest are replaced by Thomson's gazelle in the dry season (figure 1.6c). Thus, the plains lions endure the greatest seasonal variation, while the woodlands lions' diet is more similar to the Crater lions' in the wet season and less harsh than the plains lions' during the dry season.

Tabulating the number of prey animals in the lions' diet ignores important differences in food intake, as larger individual prey animals provide far larger meals.[2] A freshly caught male buffalo, for example, provides over 400 kg of edible biomass—enough food to sustain a large pride for several days—compared to only 10 kg from a female gazelle (see appendix). The Crater lions thus subsist almost entirely on buffalo, wildebeest, and zebra meat all year round (figure 1.7a). In contrast, lions on the Serengeti plains largely consume wildebeest, zebra, and eland during the wet season, and a more diverse diet during the dry season (figure 1.7b), whereas the woodlands lions rely heavily on buffalo, wildebeest, and zebra with the occasional giraffe during the wet season versus a continued high intake of buffalo in the dry season along with the three major migratory species and eland (figure 1.7c).

Tracking the Lions

The vast majority of the Serengeti and Ngorongoro lions live in stable pride territories that persist for decades. Figure 1.8 shows the movements of a GPS-collared female, K82, over a two-year period.[3] K82 belonged to a pride that occupied a small stretch of the Ngare Nanyuki River at the northern edge of the Serengeti plains, and she generally centered her activities along the course of the riverbed in both the wettest and driest months of the year, returning repeatedly to the confluences of two small tributaries. As will be discussed in chapter 9, river confluences are the most important "real estate" within a lion pride's territory, as they provide a relatively consistent supply of food and water as well as greater vegetative cover.

Despite their overall stability, pride territories in the Serengeti generally shift to the south and east each wet season as the lions respond to the seasonal migration out to the open grasslands: whereas prides in the largely nonmigratory Crater only move 1 to 2 km (and in no consistent direction) between seasons, the prides in the Serengeti woodlands shift 2 to 5 km to the southeast versus 5-to-10-km shifts by the prides on the Serengeti plains (figure 1.9). The influence of the Serengeti migration is even more apparent in the case of nomadic males. After leaving their natal prides, most

---

2 We almost exclusively monitored the lions during the daytime, so we undoubtedly underestimated the number of smaller prey animals captured during the hours of darkness. We did occasionally see the lions feed on animals as small as hares (2 kg) or spring hares (3 kg), but these were rare even during nighttime observations. The lion's diet throughout Africa is dominated by species weighing 200 kg or more (Hayward and Kerley 2005), thus the contribution of these smaller species to the lions' overall food intake is insignificant.

3 Beginning in 1984, we attached VHF radio collars to one or two females per Serengeti study pride. Each transmitter had a range of 2 to 5 km on the ground (depending on topography) and 5 to 20 km from the air. We fixed a directional antenna to the roof of each research vehicle and drove to the location of the signal during day-to-day monitoring. We also located every collared animal once each month from a light aircraft. Whereas we typically only tracked the VHF collars during daylight hours and usually obtained only a single data point per day, GPS collars transmitted multiple locations per day without our having to track the animals ourselves; however, our modest research budget only permitted the use of a limited number of GPS collars each year.

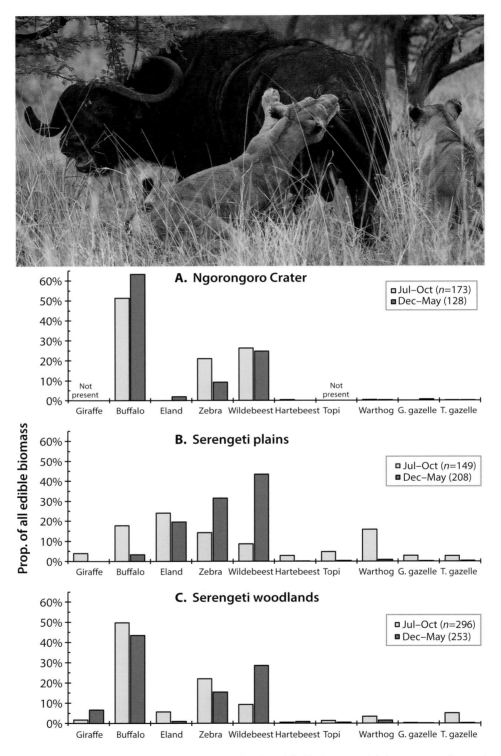

FIGURE 1.7. Edible biomass of all carcasses known to have been killed by lions in each habitat. Biomass from each carcass is taken from the age-sex data in appendix; sample periods as in figure 1.6.

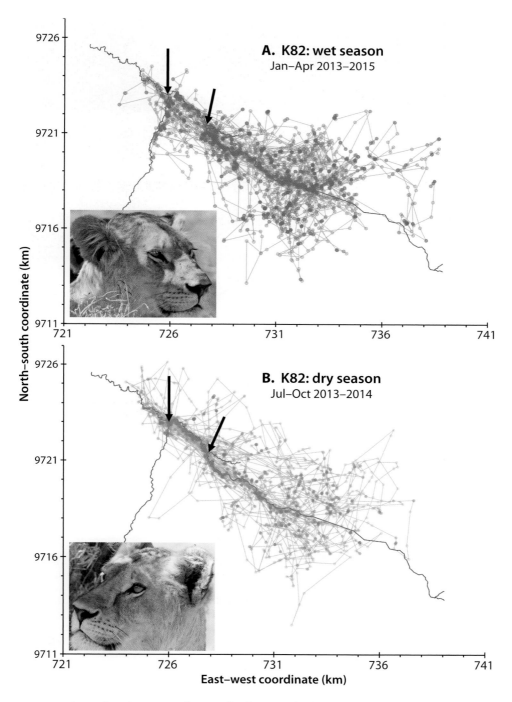

FIGURE 1.8. Seasonal ranging patterns of a GPS-collared Serengeti female named K82. Dots indicate hourly locations from 19:00 to 07:00 and at noon each day. Arrows highlight confluences along the Ngare Nanyuki River. Note that each map includes the same number of sightings (n = 3,588). Insets: ID photos of K82.

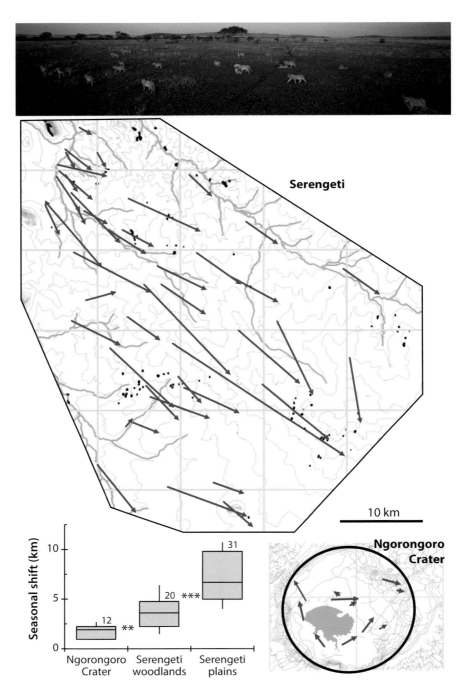

FIGURE 1.9. Seasonal shifts in midpoints of home ranges. In the wet season, all fifty-one Serengeti prides shifted eastwards and forty-nine of fifty-one shifted southward; Crater prides moved into more open areas, but otherwise showed no consistent directional change. Shifts by Crater prides were significantly smaller than in woodlands prides, which in turn moved shorter distances than plains prides. Box plots indicate medians, quartiles, and 10th and 90th percentiles. n = number of prides. Data include all prides observed at least ten times each season (median = 371 dry-season and 240 wet-season sightings/pride); prides were not all contemporaneous.

In this and all subsequent figures: *** $p < 0.001$, ** $p < 0.01$, * $p < 0.05$.

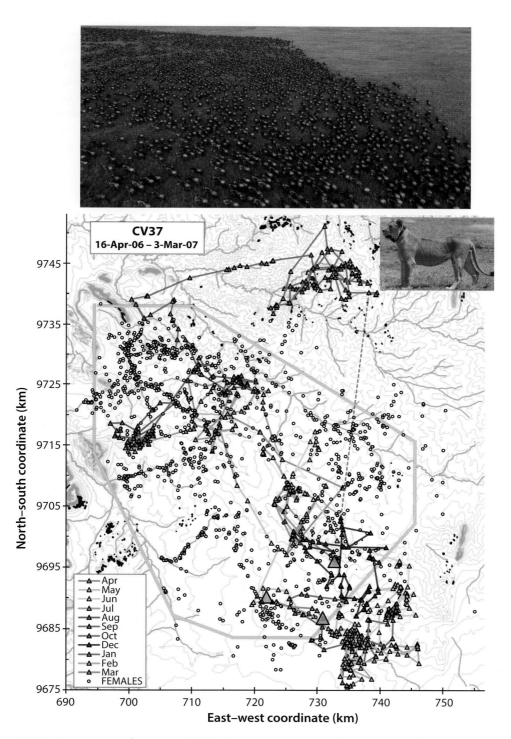

FIGURE 1.10. Movements of a two-year-old GPS-collared nomadic male in the Serengeti. Colored lines connect locations of CV37, who was born in March 2004; solid lines indicate approximately six-hour intervals; dotted line indicates a fifty-six-day gap between October 16 and December 12. Large green triangle marks site where CV37 was collared, the collar was removed at the large turquoise triangle, and the large gray triangle marks his last-ever sighting on March 9. CV37 was alone when collared, but accompanied by his littermate, CV36, at almost every sighting from June 2006 onward. Black circles indicate concurrent sightings of every VHF-collared female in the Serengeti study prides.

subadult males move across large areas, as seen in figure 1.10, where the two-year-old male, CV37, traveled around a substantial swath of the Serengeti, remaining in the open plains during two successive wet seasons and moving deep into the central woodlands in the intervening dry season (figure 1.10). Because year-round residency is impossible on the short-grass plains to the south and east of our long-term study area, this featureless landscape provides a seasonal refuge among the migratory herds of wildebeest, zebra, and gazelle, thus allowing young nomads like CV37 a short-term reprieve before following the migration into the densely occupied woodlands during the dry season.

Fueling the Beast

Feeding lions engorge themselves to such an extent that their bellies often become obviously distended; conversely, their bellies become deeply concave if they have not fed in several days (figure 1.11). Bertram (1975b) developed a "belly scale" that ranges from 1.0 for maximum distension to 4.0 for hungry lions with protuberant ribs. We subdivided these scores in 0.25 increments and assessed their belly sizes whenever we saw their standing profiles. Female lions generally keep their belly size at around 2.50–2.85, and recent food intake has a strong effect on the lions' subsequent feeding and ranging behavior: thinner lions are more likely to have fed by the next day, whereas the belly sizes of "fuller" lions typically shrink (figure 1.12a), and these patterns are still apparent four days later (figure 1.12b). Although females are similarly successful in maintaining consistent short-term food intakes in all three habitats, the lions on the Serengeti plains travel considerably further to do so: plains females move an average of about 4 km overnight when they are very thin versus 2 km when they are "full," whereas Crater and woodlands females only move about 1 km per day regardless of belly size (figure 1.12c). After four days, plains females have typically shifted 5 km when thin versus 3 km when full, whereas Crater and woodlands females have only moved about 2 km (figure 1.12d).

At monthly time scales, food intake rates are virtually constant throughout the year in the Crater and the Serengeti woodlands, whereas the lions on the Serengeti plains enjoy the feast of abundant migratory prey during the wet season and suffer from the famine of prey scarcity in the dry season (figure 1.13a). The seasonal variation in food intake rates profoundly influences female reproductive rates on the Serengeti plains. Although lions give birth every month of the year in the Ngorongoro Crater and Serengeti woodlands, females in the plains show a striking birth peak in March and April and rarely give birth during the heart of the dry season (figure 1.13b). This peak occurs about two to three months after the plains females attain their largest average belly sizes, and the trough in reproduction follows a few months later (figure 1.13c). Gestation in lions is about 3.5 months, so fertility of the plains females is presumably triggered by the return of the migration in November or December (when their average belly size first reaches ~2.5 before peaking at ~2.4 in March–May) whereas their nutritional plane falls too low (~2.85) to maintain reproduction during the hardest months of the dry season.

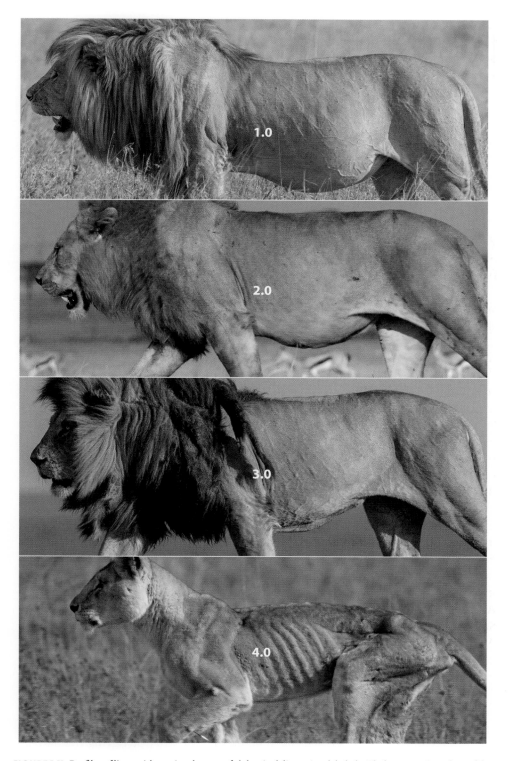

FIGURE 1.11. Profiles of lions with varying degrees of abdominal distension, labeled with the respective values of the "belly scale." Bottom photo by Ingela Jansson.

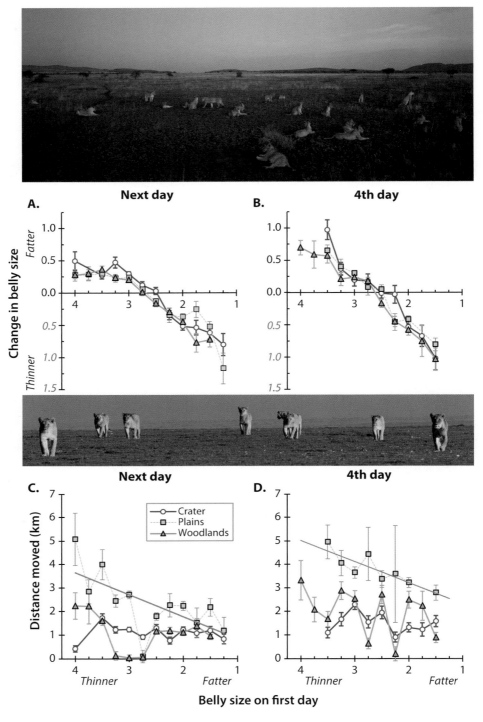

FIGURE 1.12. Changes in female belly sizes over (**A**) one- and (**B**) four-day time spans. "Full" females are thinner the next day whereas thin females tend to have eaten by the next day. Straight-line distances between locations at (**C**) one- and (**D**) four-day time spans. Plains females traveled farthest, especially when they were thinnest. Each point is based on at least six observations.

FIGURE 1.13. Seasonal differences in food intake and movement patterns of adult females in each habitat. **A.** Average belly sizes. **B.** Food intake rates versus birth rates in the plains prides. Food intake is highest from November through June; births mostly occur in March and April. White polygon indicates values if births were evenly distributed across the year; belly sizes have been rescaled to show the same range as births. **C.** Slopes of regressions between belly size and birth rate in the plains lions. The largest belly sizes were strongly correlated with birth rates two to three months later (i.e., the height of the rainy season) and negatively correlated with birth rates six to eight months later (i.e., the following dry season).

Across all three habitats, individual females that maintain higher belly sizes produce more milk (figure 1.14a),[4] enjoy higher cub survival (figure 1.14b), and shorter interbirth intervals for mothers whose prior cubs reached their second birthdays (figure 1.14c). Brian Bertram (1975b) found that chest circumference was closely related to overall bodyweight (excluding current stomach contents) (figure 1.15a), so we used chest girth as our standard for body size,[5] as lions are too difficult to weigh under most circumstances. Our belly size measures suggest that females do not translate better nutrition into more body mass: females that maintained larger belly sizes over the course of a year were no larger the following year than were poorly fed females (figure 1.15b). In contrast, well-fed males had significantly larger chest measurements the following year (figure 1.15b). Taken together, these data suggest that females translate more food into greater fertility, more frequent litters, and better cub survival, whereas males translate more food into more bulk.

A Day in the Life

Lions have the reputation of being profoundly lazy, but most people encounter lions during the day, and lions are primarily active at night, as illustrated by the movements of K82, the female that wore a GPS collar for two years (figure 1.16a). K82 traveled consistently farther each hour through the night than during the daytime, although her movements were highly sporadic: on a quarter of all days, she moved less than 30 m on average even during the most active hours of the night and less than 5 m during each hour of daylight. During the 1980s, we followed lions for ninety-six consecutive hours just before or after the day of the full moon (Packer et al. 1990), and most prey captures occurred during the hours of darkness; the few daytime kills were mostly juveniles (e.g., calves, fawns, foals, etc.) (figure 1.16b). Similarly, lions scavenged more carcasses during the night (figure 1.16c), stealing daytime carcasses mostly from diurnal species like cheetahs and vultures. Lions have far better night vision than their prey, but this advantage is reduced during nights of the full moon with a concomitant loss in hunting success (Van Orsdol 1984; Funston et al. 2001). Figure 1.17a shows that lions were thinnest in the days closest to the full moon—and this apparently forced them to forage more during the daytime, with a greater incidence of daytime kills and scavenging in the days closest to the full moon (figure 1.17b). Given that our extended nocturnal observations were made so close to the full moon, the number of *daytime* feeding events illustrated in figure 1.16b and figure 1.16c was no doubt higher than would have been observed during other times of the lunar cycle.

---

4 We immobilized eleven lactating females and injected them with 1 cc of oxytocin while they were apart from their cubs then extracted the milk by hand from a single anterior mammary gland (Pusey and Packer 1994a).

5 From Bertram's data, $y = 5.6554 e^{0.0287 x}$ where y = weight in kg and x = chest girth in cm, but I will report chest girths rather than bodyweights throughout the remainder of the book.

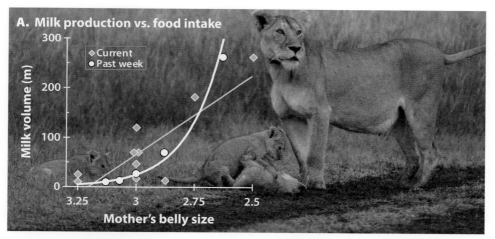
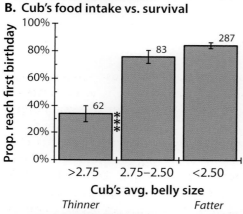

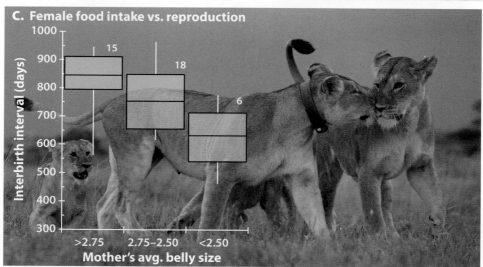

FIGURE 1.14. **A.** Milk production as a function of the mother's current and average belly size over the past week; $n = 11$ and 5 lactating females, respectively. Data replotted from Pusey and Packer (1994a). **B.** Cub survival versus average belly size prior to first birthday. Data restricted to cubs whose belly sizes were measured at least three times. The difference between cubs with average belly sizes that were thinner than 2.75 were significantly less likely to survive than those that maintained more moderate belly sizes ($p < 0.0001$). n = number of cubs. **C.** The interval between the birth of a female's surviving cubs and her next birth declined significantly with increasing food intake; Spearman rank test across all intervals: $t = 2.28$, rs = 0.351, $p = 0.028$, two-tailed. n = number of females.

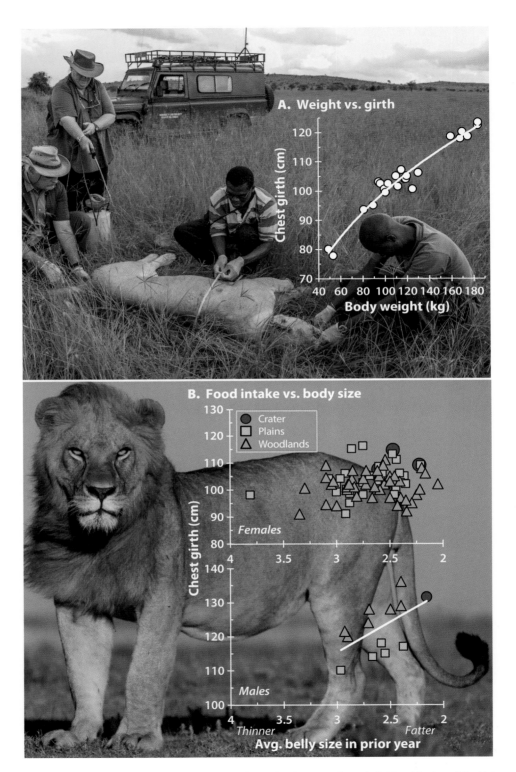

FIGURE 1.15. Body-size measurements. **A.** The relationship between chest girth (y) and body weight (x) is $y = 21.853x^{0.3296}$. Data are replotted from Bertram (1975b). **B.** Chest girths versus average belly sizes of females (top) and males (bottom). Data are restricted to individuals with at least five belly size measurements in the twelve months prior to immobilization. The relationship is significant for males ($p < 0.05$, including all data) and highly significant after controlling for habitat ($p < 0.0001$).

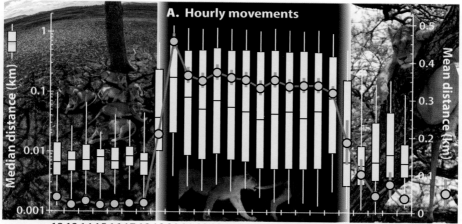

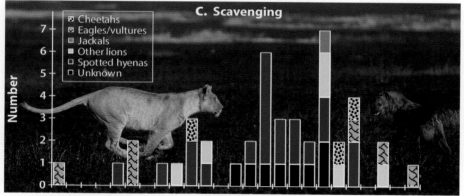

FIGURE 1.16. Daily activity patterns. **A.** Hourly distances moved by a GPS-collared female in the Serengeti plains study area in 2013–2014. Measurements are straight-line distances between successive GPS coordinates. **B.** Number of prey animals captured by female lions each hour. Data are taken from 194 complete days of observation in the first 4 days preceding and the first 4 days following the full moon each month in 1984–1987. Adult prey were mostly captured at night ($p < 0.001$); juveniles were more likely to be caught in the day than were adults ($p < 0.02$). **C.** Time of day that lions scavenged from other species during the 194 days of twenty-four-hour observation. Note: over the course of the year, sunrise and sunset in the Serengeti vary between 06:18–06:49 and 18:30–19:01 respectively.

FIGURE 1.17. Relationships between moon phase and food intake. **A.** Effect of moon phase on belly size. **B.** The number of daytime kills and scavenged carcasses across the lunar cycle. Redrawn from Packer, Swanson, et al. (2011).

A Pride Is Not a Pack—the Fission-Fusion Nature of Lion Society

Lions are social, but they are mostly found in "subgroups," and pride composition is only obvious after repeated observations (Schaller 1972). For example, figure 1.18a illustrates the daily locations and associations of an adult female, Glossie, over a ten-day period. Glossie was a member of the Gol Pride, which at the time consisted of six adult females and a "coalition" of seven adult males. On the first day, Glossie was found with three of her five pridemates, and the group had moved 3 km by the following morning. On the third day, Glossie was alone with three of the resident males, one of which "guarded" her as if she were coming into estrus. She had reunited with her three initial companions by the fourth day, and the quartet was joined by a fifth female on the fifth day. Over the course of the ten days of observation, Glossie was eventually seen with every member of her pride, but they were never found in the same place at the same time. Thus, subgroups generally vary in size and composition from one day to the next, with each individual spending time alone and entire prides rarely being all together.

We defined two lions as being together in the same subgroup whenever they were found within 200 m of each other, but pridemates can spread out over considerable distances. We usually radio collared only one female per pride but sometimes collared multiple pridemates (either two females or one female and one male) and could thereby measure the extent of their spatial separation. When separate, collared females were typically located ≈2 km apart and even spent about 10 percent of their time >5–6 km apart, depending on season and habitat (figure 1.18b) while females in neighboring prides typically remain 5–9 km apart (figure 1.18b).

Keeping in Touch

Given that the lions' fission-fusion social system often involves being separated from their companions, how do the pridemates communicate when they're so far apart? Lions rely on long-range vocalizations, "roars," to keep track of distant pridemates and members of neighboring prides (Schaller 1972). The lion's roar (figure 1.19a) carries up to ≈8 km,[6] and these vocalizations are individually distinctive: as shown in chapter 8, lions distinguish between the roars of friends and foes, and they can accurately count the number of unseen roaring individuals—an essential skill in a group-territorial species. Suffice it to say here that lions are far more likely to roar in response to distant roars when they are separated from most of their companions (figure 1.18c) and that lions mostly roar at night (figure 1.19b), when they are most active. Thus, roaring plays an important role in maintaining social cohesion despite being physically separated for varying periods of time—and it is noteworthy that pridemates spend the great majority of their time less than 6 km apart and, hence, well within earshot of each other, whereas neighbors generally remain close to the limits of the audible range of their rivals (figure 1.18b). Note, too, that while single-sexed groups of males most often roar in the hours before dawn, when sound is likely to carry farthest (see Larom et al.

---

[6] Schaller reported hearing roars from 3 to 4 km but suggested that they might be audible at 8 km, a distance we were able to confirm by sitting a known distance apart in two separate vehicles.

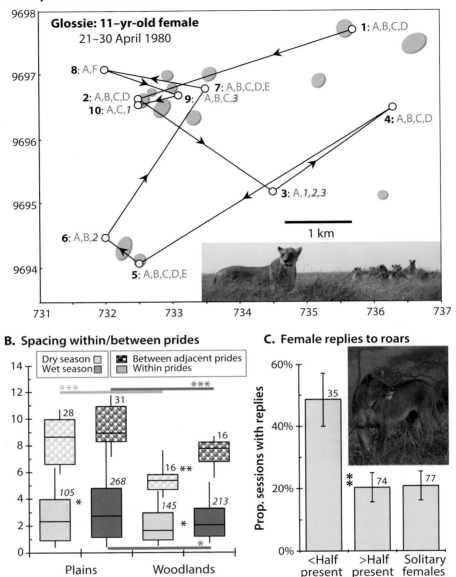

FIGURE 1.18. Fission-fusion grouping patterns and long-distance communication. **A.** Locations and group compositions of an adult female over ten consecutive days. Red letters refer to individual females (Glossie is represented as "A"); blue numbers refer to individual males. Gray ovals represent rocky outcrops. Modified from Packer (1986). **B.** Spatial separation of radio-collared pridemates and neighboring prides. Within-pride data are restricted to collared females that were observed at separate locations (more than 200 m apart) on the same day; n = number of paired sightings. Between-pride data are the average distance (in kilometers) between a collared female in each focal pride and her four closest neighbors; distances are from pairwise observations made on the same days (including interpride encounters where prides were less than 200 m apart); n = number of prides (overall, the four seasonal/habitat comparisons involved a total of more than eleven thousand pairwise measurements). Female pridemates in separate subgroups were spaced farther apart during the wet season in both habitats, and plains subgroups were more widely spaced than woodlands subgroups during the wet season. Females in adjacent woodlands prides were farther apart during the wet season than in the dry season; and plains prides were farther apart than woodlands prides in both seasons. **C.** Females in small subgroups roared significantly more often in response to the roars of distant lions than were groupings containing the majority of female pridemates ($p = 0.0035$); small subgroups also responded more often than did solitary females (who have no companions) ($p = 0.0013$). Sample sizes indicate the total number of roar sequences that were separated by gaps of at least thirty minutes; "replies" were defined as responses within fifteen minutes of the distant roars.

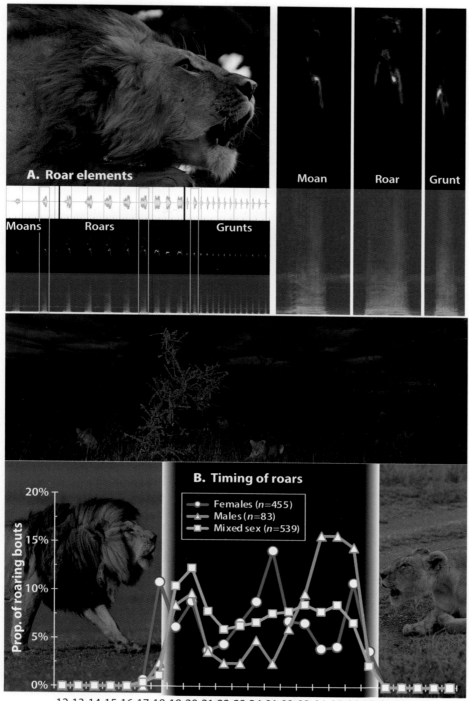

FIGURE 1.19. The lion's roar. **A.** Waveform and melodic range spectrogram of the complete roar sequence of an adult male. *Bottom left*: The sequence involves an initial pair of moans followed by a series of eight distinct roars and a sequence of fourteen staccato grunts. The vertical lines are spaced at three-second intervals. The three highlighted elements are presented in more detail on the right. **B.** Timing of roaring bouts by single-sexed and mixed-sexed groups. All data were collected during the first four days preceding and the first four days following the full moon. Each roaring "bout" consists of a connected series of roars followed by a sequence of grunts.

1997, Wijers et al. 2021), female groups also show peaks at dusk and midnight, whereas mixed-sex groups roar most in the first hours after dusk (figure 1.19b).

A Life in Full

An animal's body size, its age at maturation, its lifespan—these fundamental features of the lion's natural history are summarized in figure 1.20. As is typical for mammals, female lions are smaller and live longer than males: females reach full body size between their second and third birthday, whereas males reach full size about a year or two later (figure 1.20a). Mortality is much higher in males than females at all ages, resulting in an average life expectancy of approximately 4.5 years for females versus about 2.5 years for males (figure 1.20b), and the oldest-known-aged female in our study reached 19.75 years versus 17.59 years for the oldest male. For most purposes, we classify females as "adult" once they reach their second birthday: since the typical interbirth interval is about 730 days (figure 1.14c), two-year-old females are capable of capturing their own prey, and they fully participate in territorial defense (chapter 8). However, successful reproduction is highest in mothers between the ages of 3 to 14 years (figure 1.20c), so I will sometimes consider only the reproductive performance of older females. We generally classified males as adults after their fourth birthdays, allowing for the extra time required to reach full body size as well as to complete mane development. DNA fingerprinting confirmed that resident males are the fathers of all the cubs conceived during their tenure (chapter 2), and the age structure of resident coalitions indicates that male reproductive rates are highest between the ages of four to fourteen years (figure 1.20c). Note that because so few males gain residence, per capita reproduction is higher for *pride* males than for females. Only a few individuals survive to advanced ages, so age-specific measures exaggerate the apparent contribution of older individuals, thus figure 1.20d shows the proportion of surviving cubs produced by adults of each age. At any one time, most cubs in the population have mothers that are three to eleven years old and fathers that are four to nine years old.

Both sexes show clear signs of physiological aging (senescence) (figures 1.20 and 1.21). Cub survival follows an inverted U-shaped relationship with maternal age (figure 1.21a), presumably because of the lack of experience in very young mothers and reduced physical capabilities at advanced maternal age (chapter 4). Litter size also declines to a single cub in the final years before the female lions' equivalent of "menopause" (figure 1.21b). Because males must protect their prides from incursions by rival coalitions (chapter 3), cub survival declines linearly with paternal age as the fathers' physical prowess fades to the point that they can no longer maintain pride residence (figure 1.21a).

KEY POINTS

1. Though lions lack conspicuous coat patterns, their "whisker spots" persist until they are sufficiently tattered and scarred to become more readily recognizable. This made it possible to track thousands of lions in the Serengeti and Ngorongoro Crater throughout their entire lifespans over a dozen generations.

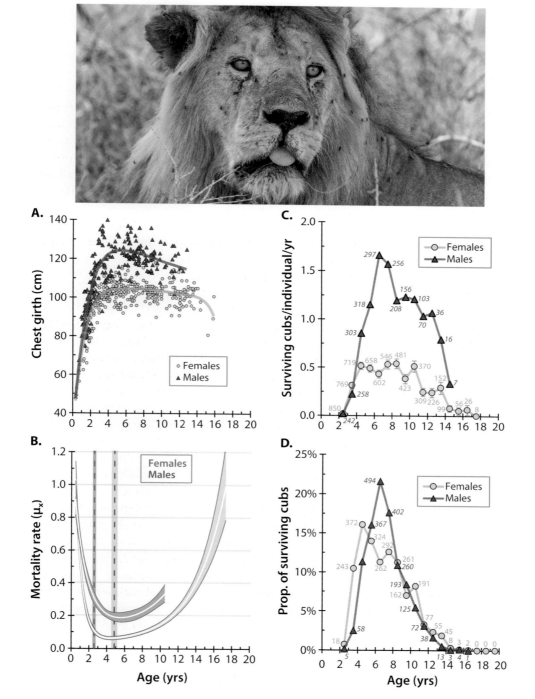

FIGURE 1.20. Age-specific rates of growth, mortality, and reproduction. **A.** Chest girth versus age. Data include occasional re-sampling of the same individuals. **B.** Age-specific mortality of males and females. The male curve treats disappearance of potential dispersers as right-censored; all other records are considered deaths. Polygons represent 95 percent credible intervals of age-specific mortality; curves extend to the age where 95 percent of each sex are assumed to have died. Vertical dashed lines give the mean life expectancy of cubs when first seen (typically three months of age); widths indicate 95 percent confidence intervals. Redrawn from Barthold et al. (2016). **C.** Age-specific production of surviving cubs. Cubs with multiple "candidate" mothers or fathers are partitioned as described in chapter 4; error bars cannot be estimated for males. n = number of males/females at each age. **D.** Proportion of surviving cubs with parents of each age. n = cubs born to mothers/fathers of each age.

FIGURE 1.21. Age-specific reproductive performance. **A.** First-year cub survival versus parental age at birth. Mean parental age was calculated for cubs with multiple "candidate" mothers/fathers. The rise and fall with maternal age are both highly significant ($p < 0.0001$, $n = 5020$ cubs); the decline with paternal age is also significant ($p < 0.0001$, $n = 4819$ cubs). **B.** Age-specific litter size of females. Litter size data are restricted to cubs with known maternity; redrawn from Packer et al. (1998).

2. The Serengeti and Ngorongoro Crater both support remarkably high concentrations of large herbivores, but while the Crater lions experience relatively constant access to their major prey, lions on the Serengeti plains endure an annual cycle of feast and famine, and seasonal prey abundance in the Serengeti woodlands is intermediate between these two extremes.
3. Lion prides defend territories that remain relatively stable for generations, but whereas Crater prides occupy similar areas throughout the year, Serengeti prides shift their home ranges 2 to 10 km toward the migratory herds each wet season, and nomadic Serengeti lions follow the migration for hundreds of kilometers over the course of the year.
4. Food intake affects lion behavior and demography. Thinner lions move farther in search of prey; lions with seasonal prey abundance breed more when food intake is highest; mothers produce more milk when they have recently fed, and females translate higher food intake rates into more rapid reproduction, while males appear to convert food to greater body mass.
5. Lions are primarily nocturnal, moving greater distances and hunting, scavenging, and roaring more at night. However, nights are not all the same, as lions suffer lower food intake during the brightest nights of the lunar cycle and thus compensate by foraging more during the day around the full moon.
6. Lion prides are fission-fusion social units; pridemates spend time in subgroups of various sizes, but distant pridemates are able to communicate over long distances via roaring. Separated pridemates remain close enough to be within earshot of each other's roars and are more likely to respond to distant roars when they are in smaller subgroups.
7. Females reach full size by about 2.5 years of age and typically breed for the first time shortly after their third birthday; males are not fully mature until they are closer to four years of age. Female reproductive rates are relatively constant from three to fourteen years of age but decline rapidly thereafter. Male reproduction is highest between five and twelve years.

CHAPTER 2

LION SOCIAL STRUCTURE

> Society does not consist of individuals, but expresses the sum of interrelations, the relations within which these individuals stand.
> —KARL MARX

Lion sociality shows many parallels with our own. Individuals are free to spend time alone, individuals may also enjoy the company of one or more members of their family, and, as we shall see, the whole is often greater than the sum of its parts, as the pride responds to the challenges of rearing cubs, catching the next meal, or defending its patch of turf. Females and males form lifelong relationships with same-sexed partners, and these partnerships have profound impacts on individual survival and reproduction, as do interactions between the sexes. Understanding how lions developed such a complex social system requires insights from multiple perspectives, but in providing an initial overview of lion society, I first want to provide basic descriptions of their social structure and the reproductive consequences of group living for each sex and set the stage for our detailed behavioral and ecological studies in the following chapters.

THE BASIC SYSTEM

Figure 2.1 summarizes the typical reproductive cycle within a lion pride. A male coalition enters the pride from elsewhere, the adult females soon breed, and the mothers raise their similarly aged cubs together in a nursery group or *crèche*. Spatial associations between mothers and offspring start to decline at about sixteen months of age with a notable sex difference appearing by twenty-four months (figure 2.2a)—the age when mothers typically give birth to their next litters. After reaching their second birthday, about a quarter of female cohorts disperse to form new prides, while the remainder are integrated into their mothers' pride. In contrast, all of the young males disperse, and male cohorts spend several years as nomads before gaining residence in a new pride (Schaller 1972). Two-to-four-year-old males transition to a more peripheral or nomadic existence (figure 2.2a), and nomadic males take over their first pride by about their fourth birthdays (median age: 3.8 yrs, $n = 193$ males in 83 coalitions) regardless of the age when they left their natal pride, thus their age at departure closely predicts the duration of their nomadic period: for each additional year in his natal pride, a subadult male typically spends about a year less as a nomad (figure 2.2b).

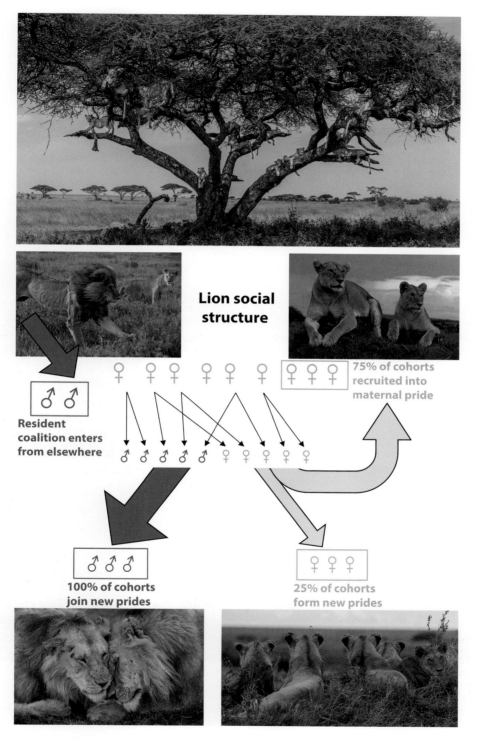

FIGURE 2.1. Schematic representation of lion social structure. A *coalition* of adult males enters a preexisting *pride* of adult females; multiple females give birth within a few weeks of each other. Surviving sons disperse and become resident elsewhere. Most surviving daughters remain in their natal pride, while the remainder disperse to form new prides.

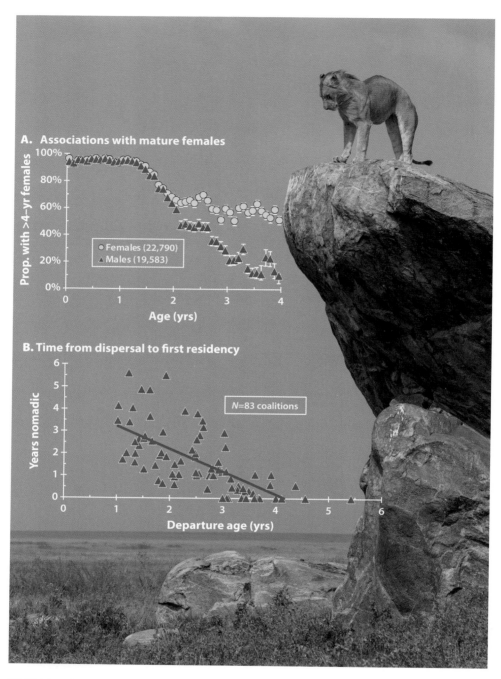

FIGURE 2.2. A. Associations of young animals with mature pride females. Data based one point per observation, taking the age of the youngest individual in each group. This includes all individuals that had not yet formed or joined another pride; new prides originated as soon as females stopped associating with their mothers, whereas males were nomadic until gaining residence elsewhere. Numbers refer to group observations of each sex. **B.** Time spent nomadic by young males. "Departure" is either defined as the date when the males' fathers were replaced by a new coalition or when the males directly entered a new pride from their natal pride. The nomadic phase decreased by about one year for every additional year spent in the natal pride ($p < 0.0001$) but did not vary significantly with the size of the male cohort. Sample size is the number of coalitions that both originated *and* resided in one of the long-term study prides.

Brian Bertram (1976) first pointed out that female pridemates are typically close kin, as established prides are maintained by the repeated recruitment of daughters with their mothers, aunts, and grandmothers, and new prides originate as cohorts of dispersing siblings and cousins. Male coalitions generally originate in the same manner as new prides, but dispersing solitaries and pairs often team up with an unrelated partner during their nomadic phase (Packer and Pusey 1982). We confirmed these genetic relationships using DNA fingerprinting: female pridemates show far greater genetic similarity to each other than to females from distant prides (figure 2.3a), and male coalition partners can either be classified as close relatives or unrelated to each other (figure 2.3b). The resident coalition fathers all of the cubs born during its tenure (figure 2.3d), and because the resident males are almost always unrelated to the pride females (figure 2.3c), kinship between prides declines with time after prides split (figure 2.3e) (Packer, Gilbert, et al. 1991). Note that each female is usually monopolized by a single "consorting" male for the duration of her estrous period, and thus littermates are almost always full siblings (figure 2.3.d). Also note that the band-sharing data in figure 2.3 are from the Serengeti, as the long history of close inbreeding in the small, isolated Crater lion population (see figs. 2.14 and 2.16 and chapter 10) rendered the genetic band-sharing patterns less clear-cut.

Although kinship often appears to serve as the glue that binds like-sexed pridemates together, associations between first-order kin contrasts strikingly with that of most other group-living mammals. After reaching adulthood, female lions do not maintain especially close relationships with their mothers: two-to-nine-year-old daughters are no more likely to associate with their mothers than with any other older females of the same age (figure 2.4a). Even though two-to-three-year-old females do associate significantly more with littermates (which are likely to be full siblings) than with other same-aged females, this pattern fades thereafter (figure 2.4b). Thus, female subgroupings in lions do not consist of the distinct matrilines that are commonly seen in primates (Silk et al. 2010), elephants (Moss et al. 2010) or spotted hyenas (Smith et al. 2010). Instead, female lions mostly associate with pridemates that have given birth to a new set of cubs at about the same time (see chapter 4)—hence their subgroupings are more strictly utilitarian than kin-based. Similarly, resident males do not preferentially associate with their male littermates (figure 2.4c), in contrast to male chimpanzees, which similarly form male-bonded societies but preferentially associate with their maternal siblings (Langergraber et al. 2007; Bray and Gilby 2020).

SOCIAL DOMINANCE

Not only are social relationships relatively undifferentiated in terms of their associations with close kin, but the Serengeti and Ngorongoro lions do not display obvious dominance hierarchies. This is most striking in females, as female pridemates do not harass each other or assert themselves during social interactions, in marked contrast to females in species with well-developed dominance relationships (e.g., canids: Macdonald and Moehlman 1982; meerkats: Clutton-Brock et al. 2001; spotted hyenas: Frank 1986; primates: Walters and Seyfarth 1987), Dominance behavior certainly exists

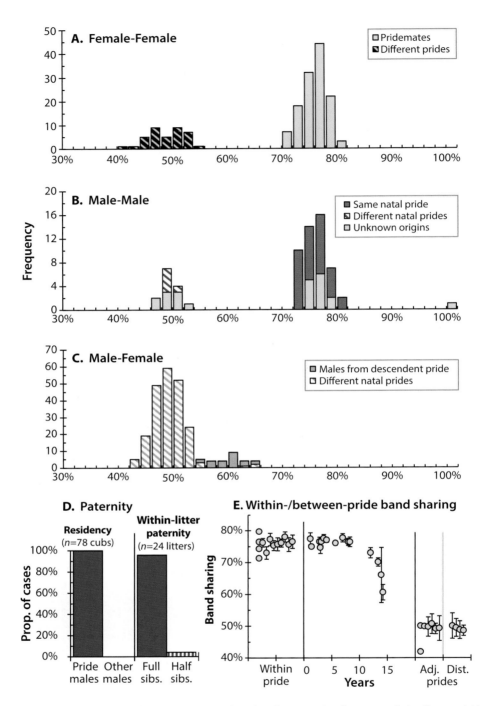

FIGURE 2.3. Genetic analysis of kinship in Serengeti lions based on restriction-fragment analysis of hypervariable minisatellite sequences. DNA fragments ("bands") are separated by electrophoresis, and band sharing calculated for each pair (see Gilbert et al. 1991). **A.** Female pridemates showed characteristically high degree of band sharing compared to females in different prides. **B.** Male coalition partners born in the same pride showed similar band sharing as female pridemates. Partners known to have been born in different prides were as divergent as females from different prides, while males of unknown origins fell into one category or the other. Note pair of identical twins. **C.** Resident males were almost always genetically divergent from their pride females, except for one coalition born in a descendant of their resident pride. **D.** Resident males fathered all cubs born during their tenure in each pride; littermates were almost always full siblings. **E.** Relatedness between prides declines through time after dispersal from the maternal pride. **A., B., C., E.** redrawn from Packer, Gilbert, et al. (1991).

A. Assoc. with mother vs. nonmother

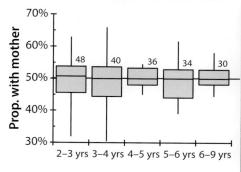

B. Assoc. with sister vs. nonsister

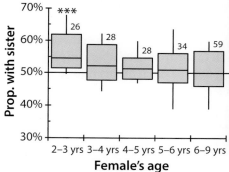

C. Resident males assoc. with brother vs. nonbrother

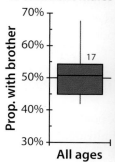

FIGURE 2.4. Comparative associations of females with their **A.** mothers and **B.** sisters, and **C.** of resident males with their brothers. **A.** only includes daughters of known maternity that were seen at least twenty times with their mother and/or a second female that closely matched the mother's age and longevity. **B.** and **C.** only include known littermates compared with a matched nonlittermate. Plots show medians, quartiles, and 10th/90th percentiles; sample size is the number of triads. Between their second and third birthdays, females associated significantly more often with littermates than with nonlittermates ($p < 0.0001$).

within lion prides: adult males easily dominate all other age-sex classes when feeding at kills (Schaller 1972). However, feeding competition *between* males—and *between* females—is largely regulated by a well-defined respect of "ownership" of specific feeding sites (Packer et al. 2001).

Lions rely on their scissoring carnassial teeth to cut openings in the hide of each carcass so as to gain access to the internal organs and muscle tissue. Only a few access points may be available at any given time, and a feeding lion (the "owner") will prevent other lions from feeding at the same opening by clamping his/her claws onto the carcass and growling, snarling, and/or lunging toward any animal that approaches too closely, forcing the "rival" to either halt or shift to a different part of the carcass. Figure 2.5 shows the outcomes of aggressive interactions at carcasses. Males routinely displace smaller age-sex classes from feeding sites, but adults of both sexes show a clear ownership "rule" whereby lions that are actively feeding are almost never supplanted from the kill (figure 2.5a). Females even respect the ownership of subadults, yearlings, and cubs (although females are significantly more likely to supplant cubs from feeding sites than vice versa) (figure 2.5b), and females never supplanted each other from a substantial quantity of meat: a feeding female was supplanted by another adult female in only 5 of 112 interactions. In three of these cases, the two females were feeding side by side, and the loser moved to a different part of the carcass. In the other two cases, the feeding female inadvertently or coincidentally lost a small scrap of bone when approached by the rival.

Respect of ownership defines lion etiquette, and lions learn the rules at an early age. After being kept apart from the rest of the pride for their first six weeks of life, mothers merge their litters to form a persistent nursery group or crèche, where they nurse each other's cubs to varying degrees (chapter 4). Nursing cubs in these communal litters frequently attempt to displace each other from a mother's nipple: larger cubs win most encounters with smaller cubs, but the owner wins most encounters between same-sized cubs, with exceptions usually occurring when the nursing cub had apparently fallen asleep at the nipple (figure 2.5c). Young lions also respect ownership at carcasses, with owners winning most cub-cub encounters and every encounter between same-aged subadults (figure 2.5c).

Respect of ownership at carcasses arises from two factors. First, the owner of a feeding site (whether meat or milk) possesses a clear positional advantage, latching on to its prey or to its mother's abdomen with powerful claws. Second, and perhaps most important, ownership rules are most likely to develop when the costs of fighting are high and contestants possess similar fighting abilities (Hammerstein 1981; Hammerstein and Riechert 1988). The lions' extensive weaponry carries a greater risk of "mutually assured destruction" than in other social species (Packer and Pusey 1985): squabbling pridemates sometimes tatter each other's ears with their claws and can even risk blinding.

However, the absence of overt aggression does not preclude the possibility of more subtle forms of feeding competition. For example, a particular individual might repeatedly arrive first at a carcass and thereby be the owner of a feeding site more often than its pridemates. To test whether companions might experience consistently different food intake rates, I compared the average belly size of all possible combinations of like-sexed companions who had been measured in the same subgroup on the

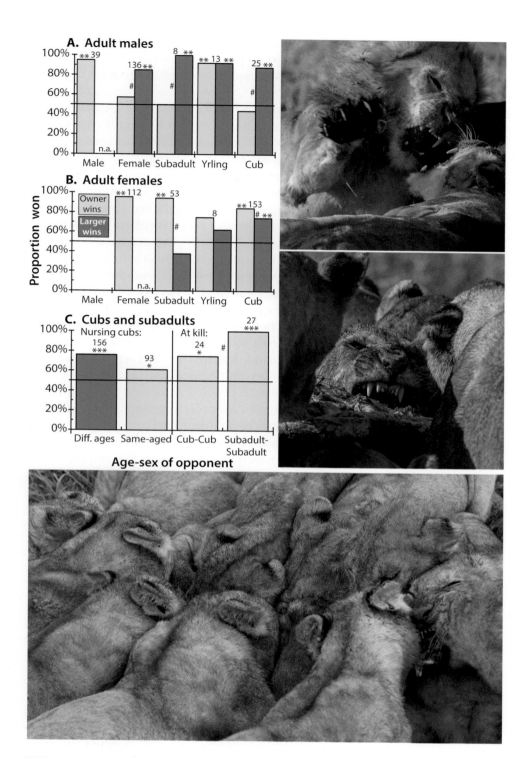

FIGURE 2.5. Outcomes of pairwise contests for access to food. **A.** Males dominate at carcasses, supplanting all smaller age-sex classes, but respecting each other's "ownership" of a feeding site. **B.** Females respect each other's ownership as well as that of smaller age classes, though they are more likely to supplant cubs than vice versa. **C.** Larger cubs supplant smaller cubs during nursing, but same-sized cubs enjoy an ownership advantage; ownership is also advantageous at carcasses in interactions between cubs and subadults. Asterisks highlight significant deviations from random for body size and/or ownership; pound signs indicate significant differences between size versus ownership in **A.** and **B.** and in the advantages of ownership between cubs and between subadults in **C.** Both **A.** and **B.** are redrawn from Packer et al. (2001).

same day.[1] Differences were remarkably small, as pridemates and coalition partners typically had the same belly size in the great majority of observations, and the average difference in belly size between companions was usually less than 0.125 on the scale that ranges from 1.0 to 4.0 (figure 2.6a). Consistent, statistically significant (at the 5% level) differences in belly size were only found in 16 (5.9%) of 271 female dyads and none of the 22 male dyads (figure 2.6b). Notably, about half of the significantly divergent cases involved pairs of females with pronounced differences in age: elderly females are often conspicuously thin (figure 2.6c), and older females were often notably thinner than their prime-aged companion (figure 2.6d). The remaining significant asymmetries involved one female that had lost the use of a hind leg and always arrived last at the carcass, two females that may have suffered from illness over the months of measurements, and two females that could have potentially been pregnant. But across all statistical probabilities (p values) associated with pairwise differences in belly sizes, there was a notable tendency for companions to have belly sizes that were somewhat more *similar* than expected by chance ($p > 0.50$). Thus, female pridemates and male coalition partners generally achieve comparable food intake rates, and, even in the few cases where females consistently differed in belly size, it is by no means certain that the thinner of the two was subordinate to her companion. Older females, for example, have conspicuously worn teeth, so they may have been unable to eat as quickly when feeding with pridemates. Additionally, like older wolves (MacNulty et al. 2009), older female lions may be less successful hunters and hence be thinner after spending time apart from a foraging group.

Note that although we could not find any obvious feeding hierarchies within the male coalitions in Serengeti and Ngorongoro, disparities in food consumption between coalition partners have been reported in the lions of India's Gir Forest (Chakrabarti and Jhala 2017), although their sample sizes only included three to four separate observations per coalition. As described in chapter 5, coalition partners compete with each other for mating opportunities with individual females within their prides. Pride size is smaller in Gir than in the Serengeti and Ngorongoro Crater (chapter 9), and male-male competition for mates appears to be more overt in the Gir lions.

REPRODUCTIVE SUCCESS

Differential survival and reproduction are fundamental to evolution by natural selection, as certain individuals will inevitably be more successful in the inescapable competition of life. Like every other organism, lions produce far more offspring than can possibly survive to reproduce: only 48.1 percent of 5,138 cubs survived to their first birthday. Of the individuals that did manage to survive to adulthood and gain breeding status, there was a wide range in reproductive performance: the most successful 10 percent of 625 breeding females each produced eight to thirteen surviving offspring, whereas the most successful 10 percent of 418 breeding males produced an estimated twelve to forty-one surviving offspring. But even among breeding animals, a

---

1 These data include 268 different females and 19 different males and exclude observations of females that were visibly pregnant as well as of mating pairs as they do not feed for four to five days at a time (see figure 5.3d).

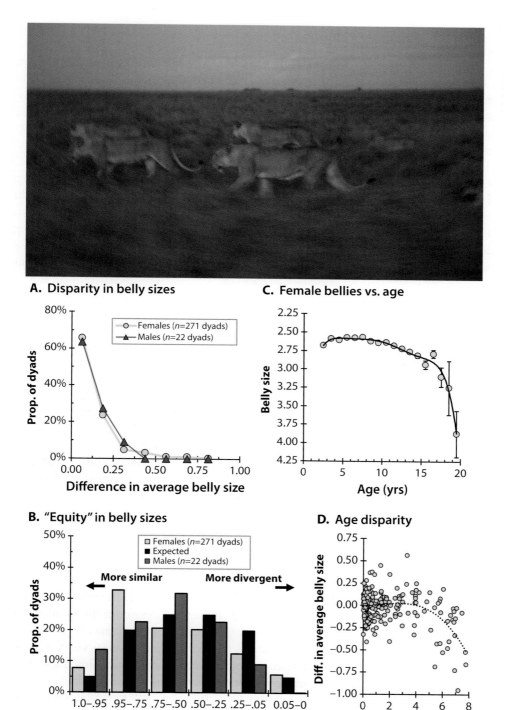

FIGURE 2.6. Within-group variation in belly size. **A.** Differences in the average belly size of like-sexed companions on days when they were in the same subgroups. Dyads include a total of 268 females and 29 males; each dyad was measured on at least ten different days. **B.** Overall distribution of probabilities associated with the differences in belly size in each dyad. Probabilities are based on t-tests for each pairwise comparison. **C.** Average female belly size peaks at about 6.5 ± 2.0 years of age and declines rapidly at the oldest ages. Data based on 20,043 measurements. **D.** Differences in belly size for each female dyad in **A.** and **B.** plotted against their respective deviations from 5.5 years of age; negative numbers indicate that the older female was thinner than the younger.

substantial proportion of individuals—12 percent of females and 13 percent of males—only managed to produce a single surviving cub.

What are the factors that produce such a wide range in performance? While much of the year-to-year variation in the lions' reproductive success is driven by large-scale ecological factors (e.g., population density in the Crater and rainfall patterns in the Serengeti; see chapter 10), a substantial proportion results from variations in pride size and coalition size. The individual-level impacts of group living differ between the two sexes, so I will consider females separately from males.

Females

The lack of dominance relationships in female lions contrasts with virtually every other vertebrate species (Hager and Jones 2009) and underlies the egalitarian nature of lion society: all females reproduce at the same rate, there is no "alpha" female, no division between breeders and helpers. Cubs are almost always pooled together and raised in a crèche: out of 3,436 cubs that were observed at least six times, 94.4 percent belonged to a crèche, and, as discussed in chapter 4, cubs are able to nurse from females other than their own mothers, thus we could not always be certain of maternity. Out of the 5,138 cubs in the long-term study, we knew the mother for 3,311 (64.4%), but there were two candidate mothers for 17.9 percent, three for 7.8 percent, and four or more for the remaining 9.9 percent. Whereas many of our statistical analyses are restricted to the 3,311 cubs of known maternity, others apportion fractional "shares" of each cub to each candidate mother (e.g., 100 percent to a known mother, 25 percent apiece to four possible mothers).

Figure 2.7a summarizes the relative reproductive performance of individual females within an illustrative set of prides. Over their lifetimes, some females produced more surviving litters per year than did others, but some females died relatively young and may have lost their one or two sets of cubs to demographic events (e.g., infanticide, chapter 3) while others may have been lucky (e.g., years with exceptional prey availability, chapter 10). But even ignoring these stochastic events, individual reproductive success is only skewed more than expected by chance in half of the study prides, while reproduction is more *similar* than expected in the remaining half (figure 2.7b). Across all study prides, the degree of reproductive skew is almost exactly what would be expected by chance (i.e., about one pride in twenty was skewed at the 5% level), regardless of whether the analysis is restricted to cubs of known maternity or also includes the summed shares of every cub (figure 2.7b). Slight variations in reproductive success exist in every pride, but pridemates all have the same opportunity to breed. If one female happens to rear more offspring over her lifetime, she is just lucky.[2] But if lions were like spotted hyenas (where the top-ranking females are disproportionately suc-

---

2 Ironically, the most successful female in the MS pride (figure 2.7a), owed her success to being *unlucky*: one of Mapema's litters was born during a six-month period when twelve females gave birth, and she was reluctant to pool them in such a large crèche. But when she left her cubs untended one day, another mother apparently took them to the crèche. Mapema returned to where she last left her cubs, calling and clearly distressed, but she continued to avoid the crèche. She gave up after a week, resumed mating, and gave birth again while her prior litter was still with the mothers in the crèche. Both litters survived, and Mapema's interbirth interval was the shortest on record (eleven months).

cessful; Holekamp et al. 1996) or any species where the alpha female is the sole breeder (e.g., wolves: Mech 1999; dwarf mongooses: Creel et al. 1992; meerkats: Clutton-Brock et al. 2006), the histogram in figure 2.7b would look entirely different, with nearly 100 percent of social groups in the 0.05–0 category of p values.[3]

Although females show similar reproductive rates *within* each pride, reproduction varies dramatically *across* prides, and pride size is an important determinant of individual reproductive success. Prides show a characteristic size-range in each habitat, with most females belonging to prides of intermediate size (figure 2.8). For example, most Crater females reside in prides of three to seven females (figure 2.8a) compared to three to ten females in the Serengeti woodlands (figure 2.8b). Although size ranges remained stable throughout the overall study period in the Crater and Serengeti woodlands, pride sizes on the Serengeti plains shifted after a sudden environmental change in 1997 (see chapter 10). Thus, whereas plains females typically belonged to prides of three to seven females from the 1960s to the late 1990s (figure 2.8c), pride sizes on the plains varied widely over the following decades with no clear size preference (figure 2.8d).

Reproductive rates within each habitat depend strongly on pride size, with intermediate prides enjoying the highest per capita reproductive success in the Crater, Serengeti woodlands, and the Serengeti plains from the 1960s to the 1990s (figures 2.8a, 2.8b, 2.8c). In each of these cases, the preferred pride sizes closely match the sizes with the highest per capita reproductive success (figures 2.9a, 2.9b, 2.9c). The lack of concordance in the plains prides from 1997 to 2014 (figure 2.9d) presumably resulted from an underlying unpredictability in the newly modified environment.

Recruitment of daughters is clearly sensitive to pride-size-specific reproductive rates. Although the dispersal of young females is largely triggered by the arrival of incoming male coalitions (see figure 3.3a), their apparent *willingness* to depart (or their mothers' supportiveness for them to remain) varies according to social group size. Brian Bertram (1975a) first showed that subadult females are more likely to disperse from larger prides, and we subsequently found that female dispersal depends on the daughters' cohort size as well as their mothers' pride size. Cohorts containing only one to two females almost always remain with their mothers rather than disperse to form an inviably small pride on their own, and, over all cohort sizes, young females are far more likely to remain when their mothers' pride would stay within the "optimal" size range after recruitment (figure 2.10a). The effect of maternal pride size is clearly illustrated by the contrasting dispersal patterns in the Serengeti plains and woodlands. Over the time period when the "optimal" pride size on the plains was three to six females versus three to eleven females in the woodlands, female cohorts in both habitats only rarely dispersed when their recruitment would have resulted in their mothers' pride totaling three to six females (a productive size range in both areas) (figure 2.10a). But whereas woodlands females mostly remained when their mothers' prides would thereby grow to seven to eleven females, over half of the plains-pride cohorts dispersed in the same circumstances: pride sizes of seven to eleven were

---

[3] The pressures that led to egalitarianism in female lions will be explored more fully in chapter 9. In brief, female lions are exceptionally well armed and can easily injure each other; these costs are amplified by their need to cooperate in aggressive encounters with outside males and neighboring prides. But regardless of cause, an essential consequence of egalitarianism is that any benefits from group living are shared equally by each female in the pride.

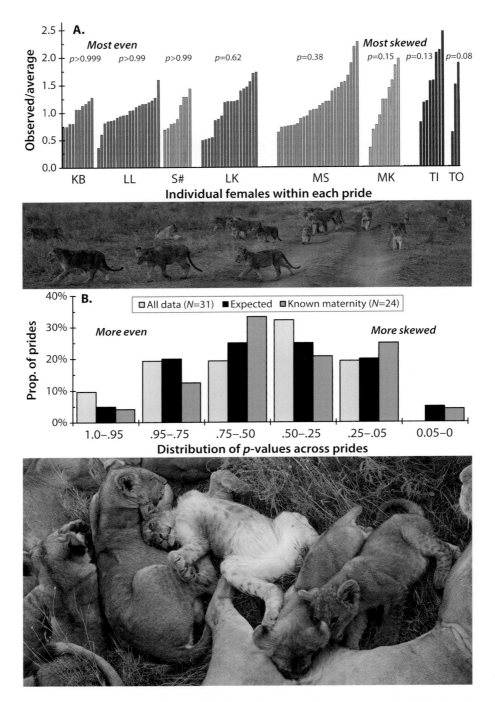

FIGURE 2.7. Reproductive skew in female prides. **A.** Relative reproductive performance of individual females in eight representative prides, including the three lowest and three highest observed levels of skew. Prides are ordered by increasing variation in reproductive rate; within each pride, females are presented in order of increasing reproductive success. Data are restricted to females that lived to at least five years of age and to prides that reared at least five surviving litters; plots control for pride-level differences by dividing each female's reproductive performance by the average for that pride. **B.** Overall distribution of prides showing respective degrees of skew. Probabilities are based on simulations that randomly assigned successful reproduction and controlled for individual differences in lifespan. Light bars include nursery groups where cubs from pooled litters were attributed equally to all candidate mothers; dark bars only include known maternities; neither distribution deviates significantly from the expected (black). Redrawn from Packer et al. (2001).

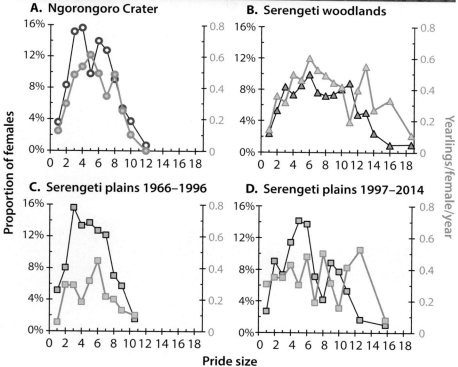

FIGURE 2.8. Proportion of females in prides of each size and pride-size specific reproductive rates. The distribution of pride sizes is the proportion of "female months," where a pride of ten females contributed 120 female months for each year of observation compared to 12 female months for each solitary. Yearling production is the number of surviving cubs born in prides of each size divided by the total number of female months for that size; pride size is averaged from birth to first birthday. The Serengeti plains prides have been separated (**C** and **D**) because of habitat changes in the late 1990s (see chapter 10). All datasets showed significant quadratic relationships with pride size except for 1997–2014 reproductive rates in the Serengeti plains.

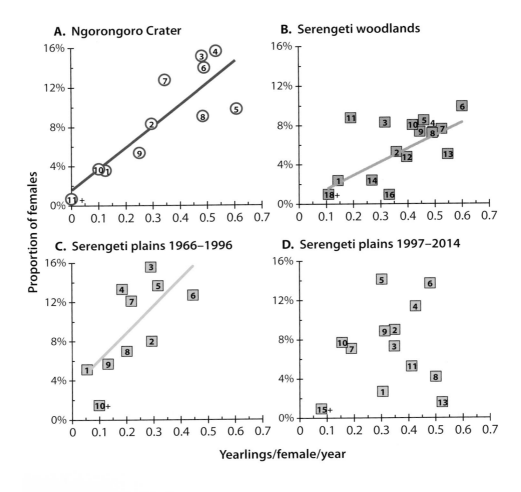

FIGURE 2.9. Pride size versus per capita reproductive rates. Females typically maintained their prides at sizes that maximized individual reproductive success ($p < 0.05$) except for the most recent years in the plains (**D**). Numbers indicate the pride size of each point.

advantageous in the woodlands but excessive on the plains. Cohorts in both habitats were equally likely to disperse when prides would have exceeded eleven females, a suboptimal size in both habitats.

Out of 1,011 adult females in the long-term study areas, 62 lived as singletons at some point in their lives, either because they were the last surviving females of once-larger prides or they had dispersed alone from their natal prides. Over 70 percent of these females remained alone for the rest of their lives, while only a small number managed to recruit a daughter and even fewer teamed up with neighboring solitaries or prides (figure 2.10b). The rarity of singleton females successfully "regrowing" their prides via recruitment would be expected from the poor reproductive performance of solitaries, but their reluctance to team up with neighboring solitaries or prides stands in striking contrast with the behavior of singleton males, who routinely form pairs and trios with unrelated coalition partners (see figures 2.13a and 9.10b).

Young females only become solitaries under exceptional circumstances. Figure 2.10c summarizes the origins and movements of the sixteen adult females that departed from their natal prides to live alone. Ten left when their prides had grown larger than the optimal pride size for that habitat, four were the only females in cohorts of subadults that were forced to disperse by incoming males (see chapter 3), and a fifteenth left when her young cubs risked exposure to incoming males. Thus, virtually every departing solitary left in the face of a significant short-term threat to her reproductive success.

Besides recruitment of maturing daughters, pride size occasionally ($n = 8$ cases) increased when singleton females moved to new prides (figure 2.10d). In almost every case, the change resulted in one or both prides moving closer to the optimum size for that habitat (e.g., a solitary joined a pair to form a trio, a Crater female moved from a pride of thirteen to form a pair with a solitary). Note that six of these eight solitary females joined recent offshoots of their natal prides, so their new companions were as closely related to them as typical pridemates and their ranges largely overlapped. The remaining cases involved two pairs of solitaries with adjacent home ranges, suggesting that females are only willing to team up with unrelated companions if they don't need to surrender their familiar home ranges.

Although solitaries were rare, they were extraordinarily valuable to our overall research program, as they essentially lived as leopards in a world filled with lions. Throughout this book, I will repeatedly explore *how* midsized prides gain such striking reproductive advantages, and, ultimately, *why* lions are the only social cat. Is it because of communal cub rearing? Cooperative hunting? Group territoriality? Each will be explored in turn, but, at this point, I just want to take the opportunity to highlight the obvious delight of females that have recently managed to escape the solitary life:

17-May-2008. Patrik Jigsved finds MS70 together with MK83 for the first time. MS70 is the last survivor of the MS Pride, seven years old, and solitary since her mother's death four years earlier. MK83, five yrs old, left the large (10-female) MK pride two years ago. The MK pride split from the MS pride in 1982 [figure 8.11], and the two prides have remained neighbors ever since. MK83 and MS70 are draped across the fallen trunk of a dead acacia tree, resting within 2 m of each other.

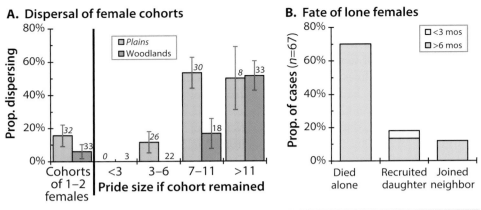
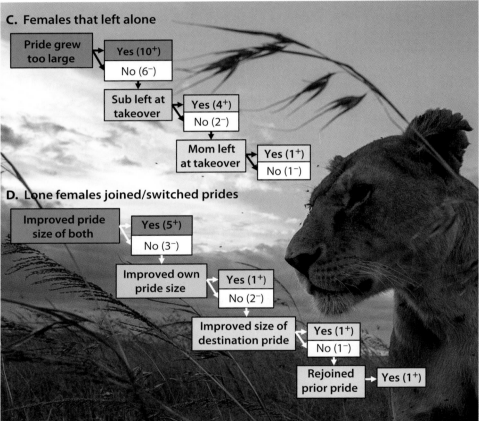

FIGURE 2.10. Pride size, female dispersal, and the fates of solitaries. **A.** Effect of cohort size and maternal pride size on female dispersal; sample size is number of cohorts. Most dispersal involved cases where recruiting daughters would have resulted in excessively large prides. Differences are significant between plains and woodlands for seven to eleven females, three to six and seven to eleven for the plains, and seven to eleven and greater than eleven for the woodlands. Redrawn from VanderWaal et al. (2009). **B.** After becoming solitary, singleton females typically spend the rest of their lives alone. Data are restricted to sixty-two adult females that were solitary for at least six consecutive months and include the last surviving members of their respective prides as well as females that dispersed alone from their natal prides. For three of twelve females that recruited a daughter, the younger female died within three months of reaching her second birthday. Five former singletons lost their new companion(s), thus becoming solitaries for a second time and eventually dying alone thus resulting in a total of 67 cases. **C.** Ten of sixteen solitaries left their natal prides when they exceeded the optimal size range; the remainder mostly left either when they were too young to mate with incoming males or when their cubs were at risk of infanticide. **D.** Most singletons joined or switched prides when the move improved the size of one or both prides.

1-Oct-2008. Patrik helps the veterinarian, Richard Hoar, replace MS70's collar. MS70 loses consciousness, and Patrik and Richard use their vehicles to shift MK83 away from the sleeping MS70, but MK83 is so reluctant to leave MS70 that it takes several minutes to shift her a safe distance away. Finishing with the collar, Patrik and Richard focus on collecting body measurements and blood samples then look up to see that MK83 has returned to within 20 m. Richard opens a large umbrella, the two men start shouting and making a fuss, but MK83 stays put. Patrik notes that MK83 is "obviously more worried about her best friend than about us! I have never seen this fearless behavior before."

After giving MS70 the antidote to the drug, they drive off ~100 m, and MK83 goes up to MS70. "The next day I checked them, they looked newly in love, lying together, rubbing their foreheads together, licking each other's necks. They must have really loved not being alone out there on the plains!"

The affection shown by MS70 and MK83 demonstrates the relief shown by females that have escaped a solitary existence. Nervous and vigilant, solitaries have little chance of maintaining exclusive territories in high-quality habitat, and they suffer far higher mortality if wounded by other lions (see chapter 8). Despite the fact that pride-living females typically spend about a quarter of their time alone (see below and chapters 4 and 7), only a handful of females ever lived as a solitary. Thus, group-living females may enjoy their time alone from time to time but only within the context of belonging to a larger social network.

Males

Male coalitions compete intensively with each other for pride residency, and Bygott et al. (1979) showed that larger coalitions gain considerable reproductive advantages. Updating their data, we found that larger coalitions of nomadic males are more likely to gain residence once they have reached three years of age (figure 2.11a); larger coalitions are also more likely to replace smaller resident coalitions than vice versa (figure 2.11b). Once resident, larger coalitions remain resident for significantly longer periods in each pride (figure 2.12a), gain more prides (figure 2.12b), and become co-resident in a greater number of prides (figure 2.12c). Consequently, larger coalitions father more cubs (figure 2.12d) to the extent that *per capita* reproduction increases with coalition size, though this advantage peaks at four males (figure 2.12e). Although moderate-to-large coalitions show a similar inverted U in per capita reproductive success as prides (on the assumption of equally shared reproductive success in each case), coalition partners compete with each other for access to receptive females, and reproduction appears to be highly skewed in the larger coalitions (see figure 5.1b); the implications of within-coalition competition will be explored in chapter 5.

Whereas only a small number of females ever teamed up with nonrelatives, solitary males readily formed partnerships with unrelated companions. This is so common, in fact, that nearly half of male pairs and trios contained nonrelatives (figure 2.13a). We only rarely observed the initial stages of these relationships, but they were among the true highlights of my time with the lions:

19-Apr-1981

14:08 Anne Pusey and I find two former resident males, Mwembe (aged 5.9 yrs) with Shabaha (8.7 yrs) about 15 m apart on the floor of the Ngorongoro Crater. Shabaha lost his coalition partner about 8 mos earlier, and Mwembe's two partners disappeared 6–7 mos ago. Both males have been nomadic after being ousted, carefully avoiding the coalitions that had killed their respective partners.

14:08 Shabaha moves to within 5 m of Mwembe; both flop on their sides.

14:25 Mwembe walks in a quarter circle around Shabaha then flops 16 m away.

14:50 Shabaha walks toward Mwembe, who raises his head and snarls slightly. Shabaha yawns and flops about 1 m away; Mwembe lowers his head.

16:38 Mwembe walks a few steps away to defecate then moves back again, flopping 7 m from Shabaha.

16:40 Mwembe moves to 9 m away then flops.

17:37 Shabaha walks up to 1 m from Mwembe, flops and rolls on his back.

17:53 Shabaha stretches and scratches his claws on the ground within 1 m of Mwembe; Mwembe snarls explosively. Shabaha lies down with his back to Mwembe.

21-Apr-1981

08:30 Mwembe and Shabaha are lying down 5 m apart.

08:48 Mwembe walks 50 m to drink at the Munge River, Shabaha sniffs where Mwembe had been resting, gives a Flehmen face[4] then drinks about 10 m from Mwembe.

09:00 Mwembe moves ~200 m to catch a wildebeest calf that had been chased by five hyenas then disappears into the reeds with the carcass. Shabaha watches but doesn't join.

22-Apr-1981

06:58 Shabaha soft roars twice, Mwembe once, looking at each other 15 m apart.

07:00 Shabaha starts walking while calling; Mwembe follows, calling, 15 m behind.

07:10 Shabaha lies down, Mwembe continues 25 m past him. Shabaha calls then walks parallel with Mwembe; Mwembe pauses, watching Shabaha, then walks on, before both stop 12 m apart.

09:10 After further calls and minor movements, Mwembe flops 2 m from Shabaha.

09:57 After drifting 3 m apart, Shabaha stands then flops 2 m from Mwembe.

10:00 Mwembe moves off 10 m to defecate then flops 2 m from Shabaha again.

26-Apr-1981

07:20 Mwembe abandons remnants of a kill and moves toward Shabaha who is ~60 m away.

07:29 Mwembe approaches Shabaha, who stands up and walks toward him. They stand about 2 m apart for a minute. Mwembe lies on his front. Shabaha head rubs Mwembe; Mwembe head rubs back; Shabaha walks 50 m away and lies down.

---

4 Curling the upper lip while sniffing helps transfer scents to the vomeronasal organ. Figure 3.2a shows an adult male with a Flehmen face.

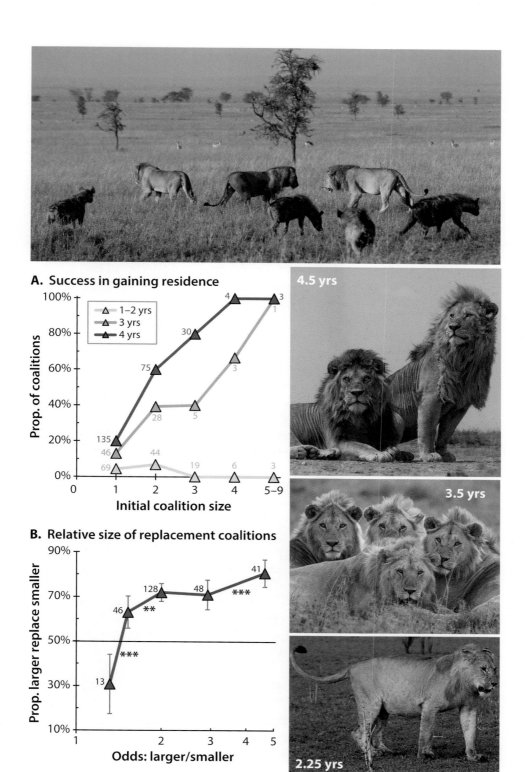

FIGURE 2.11. Effect of coalition size on gaining residence. **A.** The probability that a nomadic coalition gains residence depends on its size and age when first entering the study area. n = number of nomadic coalitions born in the study areas as well coalitions entering from elsewhere. Redrawn from Borrego et al. (2018). **B.** Larger coalitions are significantly more likely to replace smaller coalitions when they are at least fifty percent larger than their predecessors. n = number of replacements.

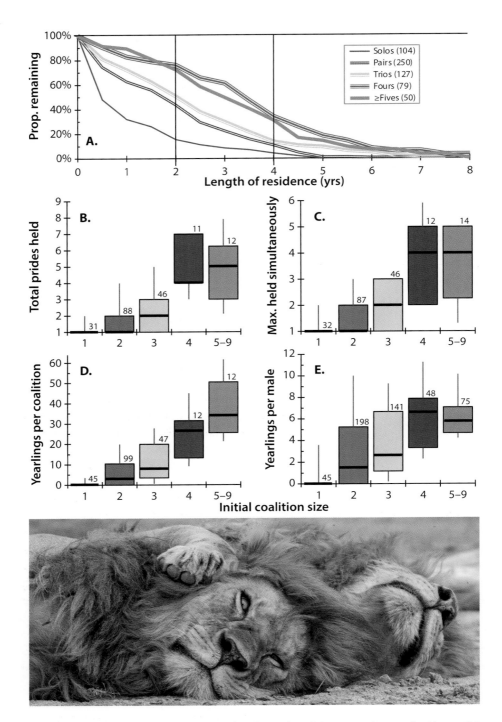

FIGURE 2.12. Lifetime reproductive performance of resident male coalitions. **A.** Persistence of resident coalitions within individual prides. Sample sizes refer to the number of residencies. Vertical lines indicate the average interbirth interval after surviving cubs (two years) and the residence time by which daughters reach reproductive maturity (four years). **B.** The total number of study prides in which each coalition held tenure. Sample size is the number of coalitions; data only include coalitions that sired at least one cub; coalitions with incomplete data were included if they scored higher than the median value for their respective coalition sizes. **C.** Maximum number of prides held simultaneously by each coalition during its overall tenure. **D.** Total number of surviving cubs per coalition. **E.** Per capita reproductive success versus coalition size; sample is the number of males. The overall increase is significant for all box plots in **B.–E.**

07:33 Mwembe walks toward Shabaha, who rises and moves toward Mwembe; they head rub each other and rub their sides together as they walk past.

07:38 Mwembe rolls over on his back, Shabaha rolls over beside him and starts head rubbing Mwembe, who head rubs Shabaha in return.

Unrelated partners are as attentive and affectionate toward each other as siblings, but these partnerships only form during the males' nomadic phase: once they become resident, they treat outsiders as rivals rather than as potential partners. This switch is well illustrated by a series of playback experiments performed by Jon Grinnell. As discussed more fully in chapter 8, the lion's roar often serves as a territorial display, so we could use playbacks to assess whether unrelated resident males would accept an additional male in their territory. When Jon broadcast the roars of an unfamiliar male to pairs of resident males, unrelated companions responded in the same way as two related males, approaching the speaker side-by-side and glancing at each other to a similar extent (figures 2.13b, 2.13c), and even attacking a taxidermically mounted male concealed behind the loudspeaker (Grinnell et al. 1995). Thus, whereas solitary males are receptive to forming new companionships when they are on their own, this openness comes to an abrupt end once they gain pride residency.

PRIDE PERSISTENCE VERSUS COALITION LONGEVITY

As seen in figure 2.14, the repeated pattern of female recruitment and dispersal enables prides to persist for generations as well to spawn daughter prides. One of the three prides that occupied the floor of Ngorongoro Crater in 1963 gave rise to a daughter pride that persisted for fifty years and, in turn, generated all of the prides in the current population.[5] Three of Schaller's nine study prides persisted until the end of the study in 2015, and twenty-one of the final twenty-nine prides descend from five of the nine original matrilines, while the remainder descend from prides that entered the study area from other parts of the Serengeti. Thus, like individuals, prides have lifespans and descendants, and pride-level survival and reproduction both depend on pride size: prides that contain at least five females can endure for decades, whereas most smaller prides typically disappear within a dozen years (figure 2.15a). Similarly, dispersing cohorts of five or more females generally succeed in establishing territories and eventually recruiting daughters, thereby allowing their new prides to survive for decades, whereas the lineages of smaller cohorts mostly die out in the first few years (figure 2.15b). As will be seen in chapter 8, this pride-level pattern of differential survival and reproduction largely results from continual intergroup competition across generations within a complex landscape that often prevents smaller daughter groups from either thriving in their mothers' shadows or successfully colonizing neighboring areas.

In contrast to female prides, male coalitions almost never persist for longer than a single generation (figure 2.16). Most coalitions either originate as members of the same

---

5 David Bygott and Jeannette Hanby initiated full-time monitoring in the Crater in 1974, but I later gathered enough notes and photographs from the first conservator of the Ngorongoro Conservation Area, Henry Fosbrooke, two early lion researchers, Pierre DesMeules and John Elliot, and tourists that visited the Crater between 1959 and 1972 to assemble a reasonably complete family tree of the Crater lions starting in 1963 (Packer, Pusey, et al. 1991).

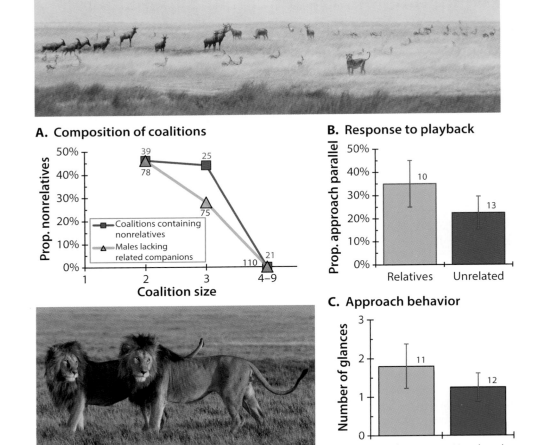

FIGURE 2.13. A. Percentage of coalitions of each size containing nonrelatives (squares) and individual males lacking related partners (triangles). Related males were born in the same natal pride; unrelated males originated from different prides. n = number of coalitions. Updated from Packer, Gilbert, et al. (1991). During approaches to the recorded roars of unfamiliar males, unrelated partners (**B.**) walked side-by-side and (**C.**) glanced at each other at the same rates as related partners. n = number of experiments. Redrawn from Grinnell and McComb (1996). *Middle photo:* Two unrelated coalition partners. *Bottom photo:* Four partners from the same natal pride.

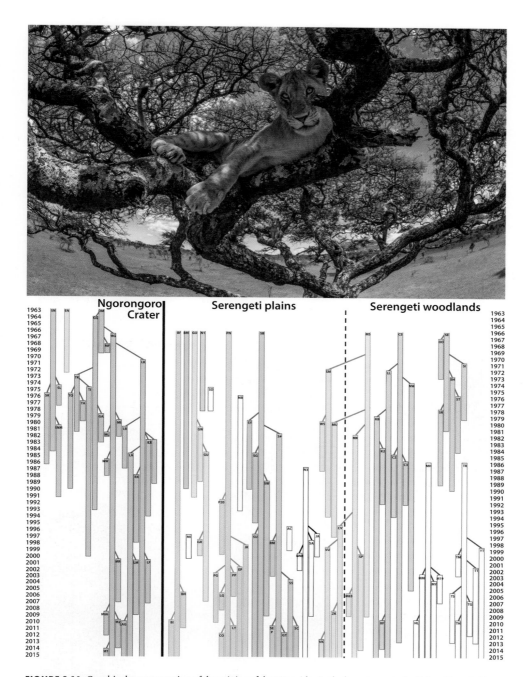

FIGURE 2.14. Graphical representation of the origins of the 104 prides in the long-term study. Colored bars indicate descent from prides that were present at the start of the respective studies (1963 in the Crater; 1966 in the Serengeti); unfilled bars represent groupings that descend from prides that originated outside of the study area. The width of each bar is independent of pride size, which ranged from one to twenty-one females. Each bar starts at the year of dispersal and ends at the death/disappearance of the final pride member; diagonal lines link the new pride with its maternal pride. The dotted line separates plains prides from woodlands prides, with the exception of the left-most blue bar (the KB pride), which shifted from the woodlands to the plains in 1983.

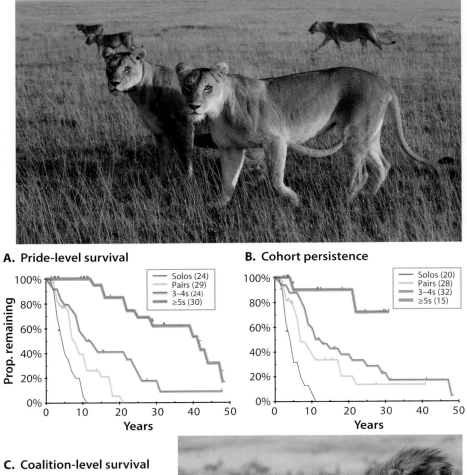

FIGURE 2.15. Group-level survival curves. **A.** Pride-level survival curves; prides classified according to average size from first observation to last. **B.** Post-dispersal survival of female cohorts; size based on average number of females over first twenty-five months after dispersal. **C.** Coalition-level survival of resident coalitions; coalition size based on number of males when first seen or at dispersal. Numbers in parentheses refer to number of groups. Tick marks indicate censored data for groups still under observation or groups that were known to have left the study area.

natal cohort or through the mergers of singletons and pairs (figure 2.13a), and sons only rarely join their fathers (see chapter 5), thus most male coalitions persist for the lifetime of their founding members, and the longest-lived coalition only maintained residence for twelve years (figure 2.15c). Besides the ephemeral nature of male lineages, figure 2.16 also illustrates two larger-scale patterns that will be discussed in detail in chapter 10. First, most of the Crater coalitions were born into this small population, and, in fact, descended from the same female lineage that eventually gave rise to every pride on the Crater floor, thus resulting in levels of inbreeding that have apparently heightened the Crater lions' vulnerability to infectious disease. Second, whereas the majority of resident coalitions in the Serengeti prides originated from outside the study area, the proportion declined through time, perhaps reflecting higher male mortality in other parts of the Serengeti ecosystem as anthropogenic impacts in these peripheral areas have intensified with the rapid growth in the human population surrounding the park.

KEY POINTS

1. Pride females raise their cubs together; the majority of daughters are recruited into the mothers' pride while the remainder disperse to form new prides, and sons always leave to become resident elsewhere.
2. The recurrent pattern of recruitment and dispersal results in female kin groups, but because resident males are typically unrelated to the females, kinship between maternal and daughter prides declines through time. Although the majority of coalition partners are closely related to each other, a sizable minority are unrelated.
3. Associations between mothers and offspring decline around the time that females give birth to their next litters; recruited daughters do not preferentially associate with their mothers or sisters, nor do resident males show any preference for their brothers.
4. Female lions do not show obvious signs of social dominance, thus pridemates equally share the benefits of group living, which are highest at intermediate pride sizes. Recruitment patterns of maturing daughters help maintain pride sizes within an optimal range. Once solitary, females typically remain alone for the rest of their lives, though a small proportion team up with neighboring solitaries and pairs.
5. Male lions similarly lack obvious signs of social dominance, though this varies across populations in Asia and Africa. Larger male coalitions enjoy greater reproductive success than smaller coalitions, singleton males frequently team up with unrelated partners, and unrelated partners are as cooperative as close relatives.
6. Female prides show differential survival and reproduction, with larger prides persisting for decades and giving rise to multiple daughter prides. Male coalitions typically only persist over the life spans of their founding members, as sons are rarely recruited into their fathers' coalitions.

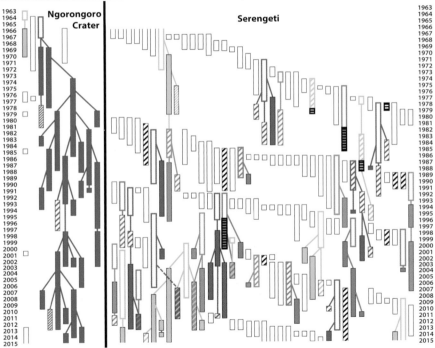

FIGURE 2.16. Graphical representation of the origins of the 243 resident coalitions in the long-term study. Unfilled bars represent coalitions that originated outside of the Crater or Serengeti study areas; those with thick colored edges indicate immigrant coalitions that fathered a descendant resident coalition; colored bars indicate coalitions that were born in the study area and are linked to their fathers' coalition by vertical/diagonal lines. The width of each bar is independent of coalition size, which ranged from one to nine males. Each bar starts at the year that coalition first became resident and ends at the death/disappearance of the final coalition member. Colored hatching indicates that the coalition partners had originated in different natal prides with the color matching their fathers' coalition; black-and-white hatching highlights unrelated males of unknown origins; fine hatching of two Crater coalitions indicates partners born in different prides and sired by the same resident coalition. Bars with horizontal black hatching indicate the five sets of sons that were recruited into their fathers' coalitions. The Serengeti is not subdivided into plains and woodlands habitats as males moved freely between habitats. Note that origins were unknown for the seven coalitions resident at the start of Schaller's study in 1966 as well as for most immigrant coalitions.

CHAPTER 3

INFANTICIDE AND EXPULSION

May we not suspect that the vague but very real fears of children, which are quite independent of experience, are the inherited effects of real dangers and abject superstitions during ancient savage times?
—CHARLES DARWIN

26-Jul-1987

10:30 a.m. Filmmakers Richard Matthews and Samantha Purdy find female BME with her three 4-mo-old cubs near the Boma Marsh on the Serengeti plains. BME is the lone mother in the BM pride, and the cubs' fathers have not been seen in the area since late April. Three females from a neighboring pride and a male named Smith appear over a rise 300 m away. BME runs off, crosses a gully and leaves her cubs in an open area.

10:33 The lead female arrives and sniffs the cubs, the other two females stop briefly, 20 m from the cubs, then all three females cross the gully in pursuit of BME, who is now 400 m away. Smith passes by about 50 m from the cubs. His attention is focused on the females, and he continues another 200 m without having noticed the cubs.

10:48 Smith is now 50 m from the cubs who are still in the same open spot. A patch of tall grass at the top of the gully blocks the view between the cubs and Smith. The three cubs are tightly bunched together, they are all thin, and one seems unsteady on its feet.

10:50 Downwind of the cubs, Smith gets up, and sniffs the air. He moves toward the cubs, spots them, and starts stalking. He recrosses the gully, bounds up to the cubs, and they split up. The weakest of the three cubs barely moves, and Smith bites its throat, holding it down with his forepaws, tilting his head back and forth, twisting its neck. After a few seconds, he looks up, briefly makes a Flehmen face, baring his teeth and sticking his tongue out, then bends down and crushes the cub's skull with his teeth. Dropping the first cub, he walks up to the second cub, bites its abdomen. He makes another Flehman face then bounds down into the gully, kills the third cub, carries it back up to the clearing and sees that the second cub is still alive. He bites the second cub on the head and neck, nuzzles underneath its limp body, finds its spine, breaks it, and makes another Flehmen face.

The death of BME's three cubs is not just an anecdotal case of aberrant male behavior. Infanticide is such a routine feature of lion biology that we had sent Richard and Samantha to the scene that morning because we were certain that the cubs were at risk. Infanticide inevitably follows from the basic reproductive conflict between male and female lions (Bertram 1975a). Whereas mothers of surviving offspring do not conceive again until their current cubs are about twenty months old, females conceive within a few months of the loss of their current litter (figure 3.1a). Males typically gain residence for only about two years before being ousted by another coalition; if they move voluntarily, they usually depart after an approximately three-year residence (figure 3.1b). Thus, males benefit from eliminating their predecessors' cubs as soon as they enter the pride and, if they are not ousted in turn, by remaining long enough to protect their own vulnerable offspring from a future takeover.

Figure 3.2 summarizes the impacts of male takeovers on the survival and dispersal of cubs and subadults. For infanticide to be advantageous to the perpetrators, the infanticidal males must sire the next set of cubs, and figure 3.2a shows the survival of cubs exposed to the seven male coalitions that were confirmed by genetic testing to have sired all the cubs born during their respective tenures in 1984–1987. None of the predecessors' cubs survived a month into the tenure of the new coalitions, and nearly a fifth of the vulnerable subadults (12–24 months of age) disappeared.[1] Over the same three-year period, the average monthly survival of the eighty-plus cubs that were not exposed to takeovers was 94 percent and the monthly survival of subadults was 98 percent.

Incoming males kill cubs of both sexes, but the fates of subadults ultimately depends on their sex, age, and dispersal status. Females that leave their mothers' prides between 2.0 and 3.0 and 3.0 and 4.0 years of age suffer lower survival than daughters that are able to remain with their mothers ($p < 0.05$) (figure 3.2b). The fates of subadult males are far more difficult to track, as an unknown proportion leave the study area, so it is almost impossible to distinguish death from dispersal. However, an extensive radio-tagging study by Elliot et al. (2014) found that the survival of dispersing males in Hwange National Park, Zimbabwe, increased from 0 to 100 percent between the ages of 2.5 and 3.0 years of age, thus the forced dispersal of males under 2.5 years of age amounted to a form of delayed infanticide. If each male disappearance is treated as a death in the Serengeti/Crater study, subadult males that are exposed to incoming males show substantially lower survival between the ages of 1.0 and 2.0 and 2.0 and 3.0 years than males whose fathers were still resident ($p < 0.0001$) (figure 3.2b). Males may also disperse even if their fathers are still resident, so the sex difference in figure 3.2b should not be taken as definitive; age-specific survival is best represented by the data in figure 1.20b.

Takeovers are the main driver of female dispersal (Hanby and Bygott 1987); females between the ages of 1.00 and 2.75 years frequently disperse within a few days or months after exposure to new males (figure 3.3a). But though the dispersal of young

---

1 Although these numbers may seem startlingly high, male tenure lengths are typically two years or longer in an undisturbed population (figure 2.12), thus relatively few cubs are vulnerable to incoming males, and infanticide has little impact on overall population growth. But excessive trophy hunting greatly increases the frequency of male takeovers, and infanticide can therefore drive down the population as a whole even where offtakes are restricted to adult males (chapter 12).

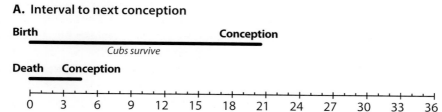

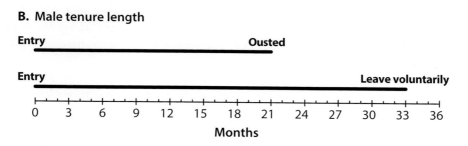

FIGURE 3.1. Typical time course of female reproduction and duration of male residency. **A.** Females typically do not conceive again until 21 months after the birth of surviving offspring but conceive within 4.5 months of their cubs' deaths. **B.** Coalitions that are replaced by larger rival coalitions maintain residency for approximately 21 months compared to about 33 months for males that departed voluntarily. Redrawn from Pusey and Packer (1994b).

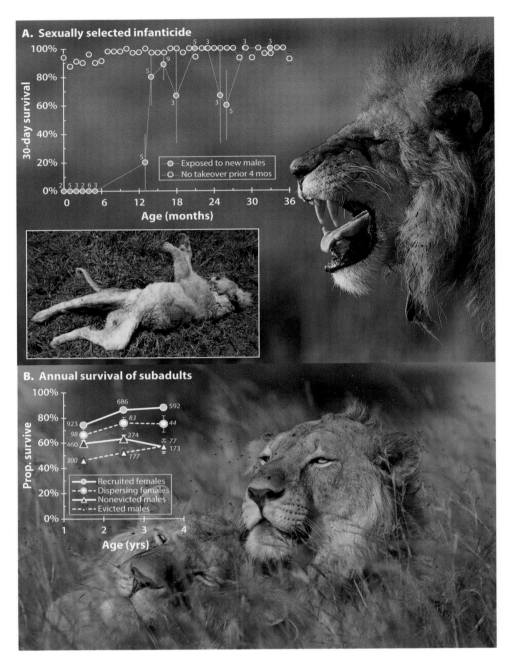

FIGURE 3.2. Impacts of male takeovers on cubs and subadults. **A.** Monthly cub survival following a male takeover; sample size is number of cubs. Analysis is restricted to coalitions whose post-takeover paternity was confirmed by genetic testing; all exposed cubs had been sired by prior coalitions. Redrawn from Packer (2001). **B.** Effects of takeovers on age-specific survival of subadult males and females. "Recruited females" and "non-evicted males" are individuals that had not been exposed to male takeovers by each age, and survival is measured from one birthday to the next. "Dispersing females" left to form new prides, and survival is to the first anniversary of their departure. "Evicted males" were exposed to new males, and survival is to the anniversary of the takeover. See text for statistics. Photos in **A.** show a cub that was killed by an incoming male and a male displaying a "Flehmen" face.

females generally coincides with the arrival of an incoming male coalition, their apparent *willingness* to depart (or their mothers' supportiveness for them to remain) varies according to social group size. Brian Bertram (1975a) first showed that subadult females are more likely to disperse from larger prides, and we subsequently found that female dispersal depends on the daughters' cohort size as well as their mothers' pride size (figure 2.10a). Cohorts of only one to two females almost always remain with their mothers rather than disperse to form an inviably small pride on their own, and females are far more likely to remain when their mothers' pride would stay within the "optimal" size range after recruitment, whereas they often disperse when their mothers' pride would have become excessively large (figure 2.10a).

Beyond the impacts on female mortality, departure from the maternal pride also affects their future reproductive success even if they survive to adulthood. Looking only at females that reached four years of age, those that had departed before their second birthdays had about a 10 percent lower chance of producing surviving offspring during their lifetime (figure 3.3b), and, of those that did manage to become successful breeders, females that dispersed before two years of age did not give birth to their first surviving offspring until they were nearly a year older than females that first bred in their natal pride or that departed at later ages (figure 3.3c).

AFTER THE STORM

While infanticide confers clear benefits to the incoming males, it inflicts such profound costs on the mothers that female lions have developed an array of counterstrategies both to protect their current offspring and to reduce their vulnerability to future takeovers. In introducing these adaptations, I want to first summarize the time course of post-takeover reproduction before discussing the details.

The loss of dependent offspring, whether by death or separation, causes mothers to return to sexual receptivity within a few days of a male takeover, as illustrated by the simulated reproductive patterns of a hypothetical pride in figure 3.4. Sexual receptivity lasts for a median of 4 days (1st and 3rd quartiles = 3 and 4 days) and the median interestrus interval is 16 days (quartiles = 13 and 19 days). Females cycle for about 100 days (quartiles = 80 and 110 days) before conceiving with the new set of males, followed by a 110-day gestation period. If the cubs survive, mothers experience about 1.5 years of postpartum amenorrhea. A proportion of these litters may be lost to malnutrition, disease, or predation, in which case the females resume cycling but subsequently conceive more quickly when mating with these now "familiar males" than during the first months following the entry of new males.

As will be described in more detail in chapter 5, an individual male forms a two-to-four-day "consortship" with a receptive female, preventing her from moving into close proximity to his coalition partners and effectively assuring that he sires all of the cubs in her litter (figure 2.3). However, a female will sometimes solicit matings from an additional male after her "primary" consort partner has lost interest and stopped guarding. These "secondary consortships" only last for a few hours, and the secondary male shows less possessiveness than his predecessor, so he is unlikely to father any of the cubs conceived during that cycle.

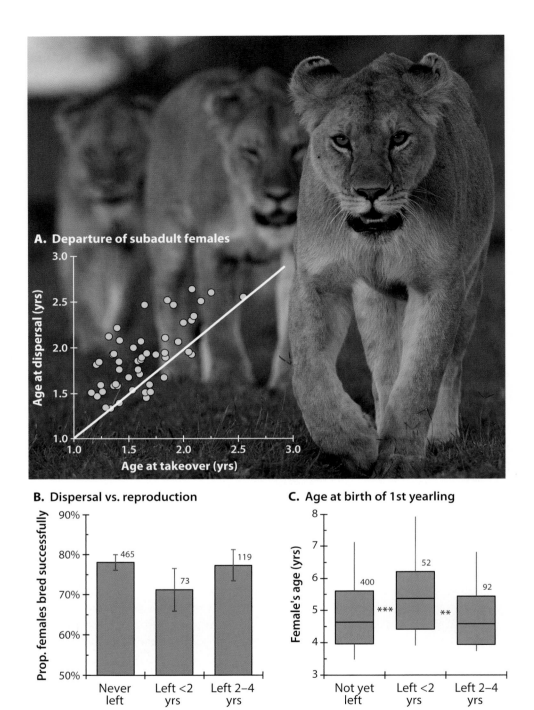

FIGURE 3.3. Context of female dispersal and consequences on reproductive performance. **A.** Age at takeover and age of departure for fifty-one dispersing female cohorts that were exposed to incoming males between the ages of 1.00–2.75 years. Diagonal line indicates departures on the same day as the takeover; all but eight of fifty-one cohorts left after the takeover ($p < 0.001$). **B.** Dispersal had only a minor effect on whether females that reached their fourth birthdays bred successfully during their lifetimes. **C.** Reproductively successful females that had left their natal prides prior to their second birthday gave birth to their first surviving offspring at significantly later ages than other females; females that dispersed between the ages of two to four years first bred successfully at the same age as those that bred for the first time in their natal pride. Sample sizes in **B.** and **C.** are individual females.

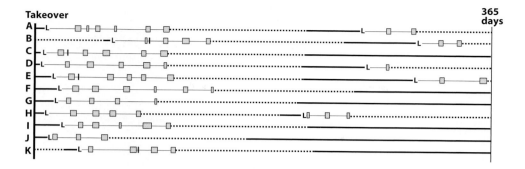

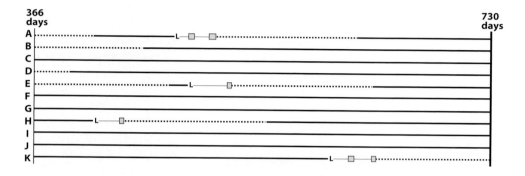

FIGURE 3.4. Simulated reproductive status in a pride of eleven females in the first two years following a takeover. Assuming that all eleven females lose their cubs at the takeover, the first few months are dominated by sexual cycling (pink boxes indicate estrus; thin lines show anestrus), followed by pregnancy (dotted lines) and eventual birth (heavy black lines). "L" indicates loss of cubs/yearlings. Redrawn from Packer and Pusey (1983a).

Consortships with recent immigrants show several striking characteristics: first, a female is more likely to initiate copulations by approaching her partner with a somewhat up-raised tail and a slightly lowered hind-quarters ("walking sinuously past" the male, as Schaller (1972) memorably phrased it); whereas consorting males that have been resident for at least a year typically initiate the majority of copulations by moving toward their partners (figure 3.5a). Females are also more likely to consort with multiple partners on the same day when consorting with unfamiliar males ($p < 0.001$) (figure 3.5b), and they are also more likely to engage in a secondary consortship with a different partner after a one-to-three-day gap, showing smaller peaks after one or two interestrus intervals (fifteen and thirty days, $p < 0.01$) (figure 3.5c).

When mating with long-term resident males, the female typically shows little direct interest in her mating partner before moving forward a few steps and lying down on her front while making a low rumbling sound. After the male ejaculates, the female then rolls on her back and sometimes swats the male's face in the process, but she otherwise shows little arousal. Besides frequently "walking sinuously past" her *new* mating partner during the first few months after a takeover, the female's heightened sexual motivation ("proceptivity" [Beach 1976]) is well demonstrated by the following two observations:

9-Mar-2005 13:54. Henry Brink finds three lions from the SU pride: SB66 (9.4 yrs), her adult daughter, SUE (4.7 yrs), and their resident male, BF16, who first became resident in mid-Dec. 2004. BF16 mates with SB66, while SUE rests a few meters away. BF16 tries to approach SUE, but she seems nervous and backs off. SB66 then mounts her daughter and makes thrusting motions (without contact between genitalia). After SB66 dismounts, the male quickly takes her place and mates with SUE. The entire sequence of SB66 mating with BF16, SB66 mounting SUE, and BF16 mating with SUE is repeated several times over the next few hours.

21-Sep-11 07:15. Photographer Nick Nichols finds newly resident male, C-Boy, mating with two adult females from the KB pride, K60 and K67. A third adult female, K81, solicits mating from C-Boy, but K60 mounts K81 instead (again without contact between genitalia). C-Boy and his companion Hildur entered the KB pride on 7-Sep-11.

These were our only observations of female-female mounting by mating females; females were far more sedate while mating with long-time resident males.

Despite initiating more copulations and consorting with more partners in the first few months after a takeover, females require several months longer to conceive with new males than with familiar males ($p < 0.001$) (figure 3.5d). A common explanation for mating with multiple males in infanticidal species is that females are attempting to "confuse" paternity, thereby reducing the risks that nonfathers will kill their offspring (Hrdy 1979; Palombit 2015). In Asian lions, two or more coalitions may share residence in the same pride, and members of each coalition are tolerant of the same cubs, possibly because of having mated with the same pride females, whereas "invading males" (that have not yet mated with the females) are infanticidal (Chakrabarti and Jhala 2019). However, in our study populations, such shared residencies are rare; cubs conceived

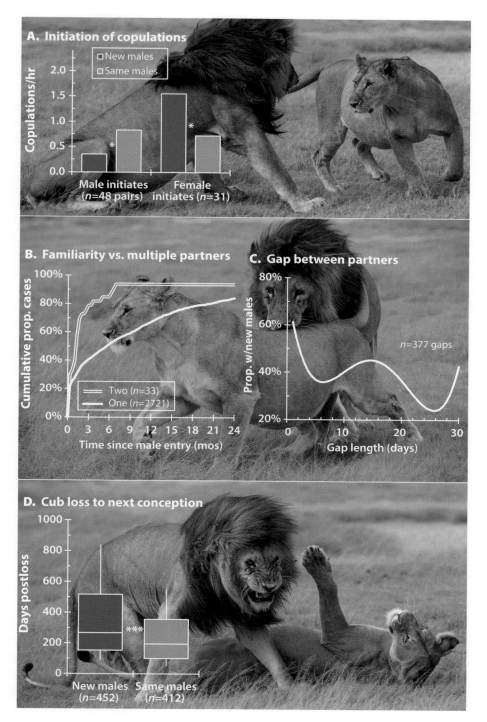

FIGURE 3.5. Female reproduction and mating behavior with new and "familiar" males. **A.** Familiar males initiate copulations at higher rates than do new resident males, whereas females initiate copulations at higher rates with the new males ($p < 0.03$). Data from Packer and Pusey (1983b). **B.** Females were far more likely to consort with multiple partners on the same day in the first few months after a takeover ($p < 0.001$). Sample size is the number of days females were seen consorting. **C.** The fourth-order regression line for all 377 cases where females were seen with different partners at each time span ($p < 0.01$, with all four components showing $p \leq 0.016$). The majority of one-to-three-day gaps involved new males; the peaks at fifteen and thirty days reflect the timing of the next two estrous cycles. **D.** Females take seventy-two days longer, on average, to conceive after losing cubs at a takeover ($p < 0.001$).

during these periods have significantly lower survival rates (chapter 5), and there is no obvious reason why female lions should largely restrict paternity confusion to the first few months following a male takeover. Instead, their heightened sexual behavior and reduced fertility may serve as a way to incite competition between rival coalitions and to "test" the staying power of the newcomers,[2] whereas the longer-term residents will have already been tested during the previous year (Packer and Pusey 1983b). As seen in chapter 6, females show a clear preference for males that can maintain residence long enough to successfully rear their cubs, and paternity uncertainty among coalition partners will be discussed in detail in chapter 5.

SYNCHRONIZING THE REPRODUCTIVE CLOCK

After synchronously losing their litters, pride females are simultaneously primed to breed again, and most cubs are born in the fifth and sixth month of the new coalition's tenure (figure 3.6a). Cubs born in the first 110 days post-takeover show significantly higher mortality than later-born cubs (figure 3.6b), presumably because a proportion had been sired by the preceding coalition and were soon killed by the new immigrants. The existence of these early born cubs provides evidence that lions do not show the equivalent of a "Bruce effect" (Bruce 1959) whereby female rodents spontaneously miscarry upon exposure to new males (otherwise, no births could have occurred until at least one full gestation length after the takeover). However, some of our estimated dates for male takeovers could well have been inaccurate, as we only collared a single female in each pride, and an incoming male could have interacted with her non-collared pridemates for several weeks before we first saw him, thus it is also possible that a small proportion of these "early born" cubs were sired by the new males rather than by the prior residents. If we only consider those cubs that were born at least 3.5 months after the presumed arrival of a new coalition, 60 percent of litters were born less than two weeks apart, compared to only 20 percent of litters born a few months later (figure 3.6c).

This pronounced birth synchrony predisposes females to pool their cubs into a stable and persistent crèche, a configuration that confers some of the strongest advantages of group living in female lions. Mothers do not simply resign themselves to the inevitable loss of their cubs when a male coalition attempts to take over their pride. As seen in this account, females sometimes die in their cubs' defense:

> **19-Mar-2004.** Kirsten Skinner finds an adult female, NN18, dead on the short grass plains and a pair of four-year-old nomadic males nearby. NN18's sole companion, NN22, and their six one-year-old cubs, SAA-SAF, are about 300 m away, escaping toward a kopje. A tour driver tells Kirsten how the two males attempted to take over the Snaabi pride but inadvertently killed NN18 when the two females protected their cubs.

---

2 Leopards are also highly infanticidal; female leopards show a similar delay in reproduction after losing their cubs to incoming males, and this may also allow females to test the staying power of potential fathers (Balme and Hunter 2013). Though male leopards are solitary, female leopards frequently mate with multiple partners, but this does not appear to inhibit infanticidal behavior of neighboring males.

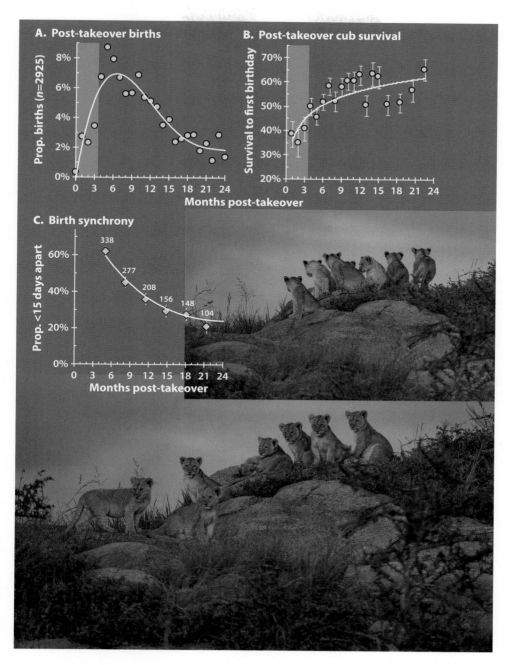

FIGURE 3.6. Post-takeover breeding synchrony. **A.** Frequency distribution of 2,925 cubs born each month after the most recent takeover. Gray shading indicates gestation length (110 days); cubs born within this period include cubs fathered by the prior residents plus inaccuracies in estimated dates of takeovers. **B.** Cub survival is significantly lower in the first 110 days following the takeover ($p < 0.001$); analysis includes 2,884 cubs that were not exposed to a subsequent takeover before their first birthday. **C.** Proportion of successive births less than fifteen days apart each three-month period following the most recent takeover. Birth synchrony declines through time ($p < 0.0001$), but approximately 25 percent of births occur less than fifteen days apart thereafter.

With only one female to tend the crèche, all but one yearling female, SAD, will die over the following months, but NN22 and SAD will persist as a pair for the next five years.

Fathers actively protect their offspring against outside males, but males and females spend considerable time apart from each other (see figure 1.18b). Figure 3.7a shows that males primarily associate with pride females during consortships and when feeding together at kills (see below). A substantial proportion of coalitions reside in two or more prides at the same time (figure 2.12c), which requires dividing their attentions between females in adjacent prides. Thus, in most circumstances, resident males are only found with females about 15 percent of the time in the Serengeti and 30 percent in the Crater, leaving mothers vulnerable to the kind of aggression by extrapride males that led to NN18's death. In the twenty-nine observed encounters between mothers and outside males, singletons almost always lost their entire litters whereas crèches of two or more mothers always succeeded in protecting most or all of their cubs (figure 3.7b). An example of group defense by a crèche is shown in figure 3.7c, where four mothers chased off an adult male.

The conflicts between mothers and nomadic males are well illustrated by a series of playback experiments performed by Karen McComb and Jon Grinnell. While mothers were largely indifferent to the recorded roars of their current resident males (the fathers of their cubs), roars of extrapride males elicited obvious signs of agitation and distress (e.g., frequent snarling, bunching together close to their cubs, and/or moving rapidly away from the speaker) (figure 3.8a). In fact, by the morning after each experiment, the mothers had almost always moved their cubs 4 to 5 km away from these dangerous situations. In contrast, and as discussed in detail in chapter 8, the mothers' responses to roars from outside females were focused and deliberate, as they invariably approached the speaker in a confrontational manner (figure 3.8a). When exposed to the recorded roars of three females, on the other hand, singleton males were either unwilling or slow to advance toward the speaker, whereas pairs of males rapidly approached (figure 3.8b). As also discussed in chapter 8, the corporate strength of three females is slightly greater than one male.

Though mutual defense provides clear advantages against infanticidal males, group living has complex consequences on cub survival. First, large prides are subject to more frequent takeovers than are smaller prides (figure 3.9a), either because larger prides are more likely to contain at least a few sexually active females that attract itinerant males to their pride or because larger prides are inherently more attractive to outside males. Takeover rates also vary with the size of the nursery group, being least frequent in prides with midsize crèches (figure 3.9b). The initial decline in takeovers with increasing crèche size presumably results from the greater corporate strength of larger female groupings, but, beyond a certain size, crèches may either be too attractive to outside males or such large groupings may be inherently less stable (see chapter 4). In the aftermath of a male takeover, survival of cubs to their first birthday is lower for singleton mothers than in crèches of two or more mothers (similar to the pattern in figure 3.7b), although crèche size does not strongly influence the survival of yearlings to their second birthdays (figure 3.9c). Yearlings are sufficiently mobile that one or more of the mothers in the crèche may accompany their retreating young

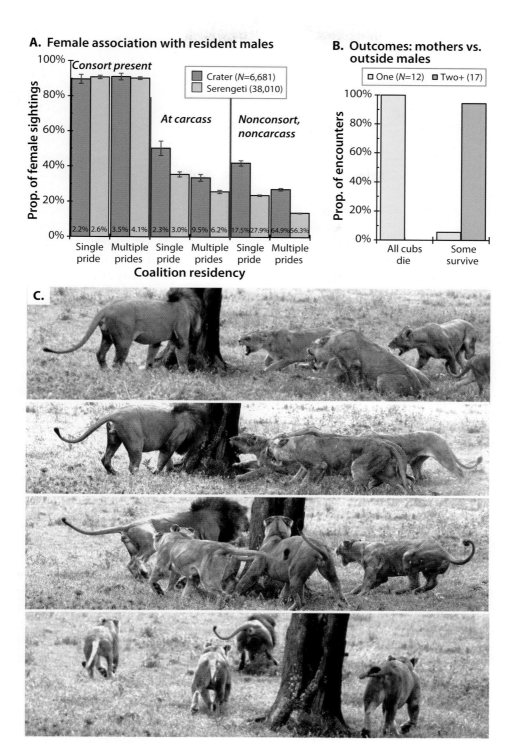

FIGURE 3.7. Female associations with resident males and confrontations with extrapride males. **A.** Associations with resident males were highest for consorting females, followed by females feeding at carcasses; associations were also higher in the Crater than in the Serengeti and when the coalition only resided in a single pride (also see chapters 4 and 9). Percentages at the base of each bar indicate proportion of all sightings in each habitat. **B.** Outcome of encounters between mothers of small cubs and extrapride males between 1966 and 2013. Singleton mothers were significantly more likely to lose their entire litters ($p < 0.0001$). Note that resident males were always absent in these observations. **C.** Four mothers chase off an adult male.

FIGURE 3.8. Responses to playbacks. **A.** Responses of mothers to the recorded roar of a resident male (potential father), unfamiliar male, or unfamiliar female. Whereas the mothers were indifferent to the roar of the resident male, they treated the outside female as a threat and showed obvious agitation at the sound of a potentially infanticidal male. Taken from McComb et al. (1993). **B.** Approaches by nomadic males toward the recorded roars of three females. Taken from Grinnell and McComb (1996).

to the edge of the pride range or join them in becoming nomadic until their second birthdays. Offspring survival is lower following the entrance of large male coalitions compared to singletons and pairs, but yearling survival is lowest after the entrance of quartets (figure 3.9d). Larger coalitions appear to be more efficient in killing cubs and separating mothers from yearlings, although coalitions of five to nine males are less successful than quartets in eliminating yearlings, suggesting that these very large coalitions may coordinate less effectively than members of smaller coalitions.[3]

THE ACADEMIC POLITICS OF INFANTICIDE

Brian Bertram wrote about infanticide in lions around the same time that Sarah Hrdy (1974) published her findings on Hanuman langurs (*Presbytis entellus*). Although male langur monkeys had previously been observed to kill infants, the behavior had largely been dismissed as a pathological response to overcrowding and human disturbance. Hrdy instead hypothesized that infanticide was an adaptive strategy whereby incoming males overcame the contraceptive effects of lactation amenorrhea. Hrdy's work was extensively cited by E. O. Wilson in his book, *Sociobiology* (1975), an exhaustive effort to meld the social sciences with evolutionary biology. In the 1950s and 1960s, primates had mostly been studied by psychologists and anthropologists with only a superficial awareness of evolutionary principles. "Biological determinism" was decried by many academics in the years following World War II, and opposition to sociobiological interpretations of human behavior raged through the late 1970s and early 1980s, with infanticide being one of the most contentious issues (Rees 2009).

But cats are not monkeys, so lions are not widely promoted as direct models for human evolution. Thus the same interpretation of the same behavior in the Serengeti lions scarcely registered with militant primatologists. Scarcely registered, that is, until a critique of Bertram's 1975 paper was published in the *American Anthropologist* in 1998 titled, "Infanticide by male lions hypothesis: A fallacy influencing research into human behavior" (Dagg 1998). We had published numerous papers on infanticide by then, and our work had not only confirmed Bertram's original observations but showed how consistently and pervasively cubs died at male takeovers. Infanticide wasn't something that only happened every now and then; it was a regular feature of the lions' reproductive cycle. But rather than take into account our more extensive dataset, Dagg primarily focused on Bertram's 1975 paper, which was restricted to the first few years of the Serengeti study. However, she was correct in stating that infanticidal males had not been directly confirmed to father the replacement cubs in their new prides. Hence, I looked specifically at cub mortality immediately following the takeovers by the males in our paternity study. These males were clearly infanticidal (figure 3.2a), and they fathered all the cubs born during their respective tenures (figure 2.3).

---

3 The data in figure 2.12 suggest that quartets in our study areas gain the highest per capita reproductive success, largely because the advantages of increasing coalition size reach an asymptote at four males. However, such large coalitions are rare in other parts of the lion's range, as described in chapter 9.

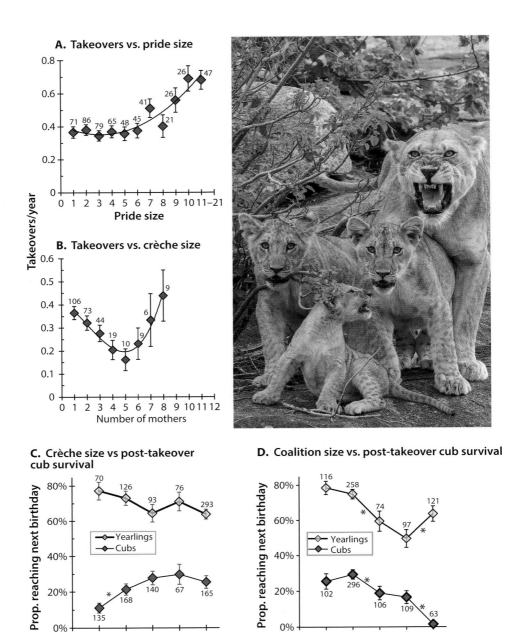

FIGURE 3.9. Risks and consequences of takeovers. **A.** Takeover rates show an overall increase with pride size. Sample size is the number of takeovers experienced by each pride size; error bars are based on the years of observation on prides of each size. **B.** Takeover rates are lowest for nursery groups ("crèches") comprising four to five females. Sample size is number of takeovers experienced by each crèche size; error bars are calculated as above. **C.** Survival to first or second birthday after a male takeover as a function of crèche size. Survival is significantly lower for cubs of singleton mothers than cubs of pairs ($p < 0.05$); sample size is the number of cubs or yearlings. **D.** Offspring survival varies with the size of the incoming coalition. Cub survival declines significantly as coalition size increases from pairs to trios and from quartets to five to nine males ($p < 0.05$). Yearling survival likewise declines between pairs and trios but increases between quartets and five to nine males ($p < 0.05$).

Lions have since been studied in many parts of Africa, and the threat of infanticide is ubiquitous. Females respond to unfamiliar males as a profound threat to their cubs' survival, and wildlife managers take great care to avoid introducing males into captive groups with small cubs. The evolutionary interpretation has had true predictive power: direct observations of infanticide were scarce in other carnivore species when we reviewed the topic in the early 1980s (Packer and Pusey 1984). But, to take just one example, detailed data on leopards in South Africa suggest that the proportion of cubs killed by replacement males may even be higher in leopards (>30%) than lions (~25%) (Balme and Hunter 2013). Most importantly, the constant threat of infanticide provides an essential cornerstone for our understanding of lion behavior—infanticide may not be the driving force behind the evolution of group living in lions—if it were, leopards should be just as social as lions, but infanticide clearly plays a fundamental role in crèche formation, which in turn provided the context for our detailed studies on cub rearing in the next chapter.

KEY POINTS

1. Long interbirth intervals render lion cubs vulnerable to infanticide as their continued survival prevents their mothers from breeding again for approximately eighteen months, whereas the typical reproductive life span of a male coalition is only twenty-four to thirty months.
2. Cubs less than six months of age suffer nearly 100 percent mortality in the first few months after a male takeover; older cubs are evicted by the new males and suffer substantial mortality post-dispersal.
3. By delaying reproduction and showing heightened sexual activity in the first months after a takeover, females appear to be inciting competition between rival coalitions so as to increase their chances of conceiving with a coalition that can maintain residence long enough to raise their next cubs to independence.
4. Simultaneous removal of dependent offspring aligns the females' reproduction, synchronizing births and allowing the formation of a crèche whereby mothers can collectively defend against infanticidal males.

CHAPTER 4

CUB REARING

*The lioness brings forth . . . generally two cubs at a time
and six at the very most but sometimes only one.*
—ARISTOTLE

What greater contrast exists in nature than between the strength and power of the female lion and the utter helplessness of her newborn cubs? We could easily recite a thousand clichés about the fierce protectiveness of the lioness and hardly do justice to the vision of a formidable killing machine capable of taking down a horse or cow unassisted before returning to her den to nurture a soft-boned ball of fluff the size of a beer can. But though we may be mightily impressed by the collective courage of a sisterhood that can chase away a potentially infanticidal male, the day-to-day duties of motherhood are no less important. The majority of cub fatalities stem from starvation and other mundane indignities. Thus, a mother must remain in a state of heightened alert for nearly two years, finding the next meal for her cubs, guarding against smaller predators, and remaining on constant standby in case an itinerant male tries to reset her reproductive clock. Her social life also swings from the most secretive and solitary times of her life to her most gregarious, thus we will follow females from pregnancy and parturition to the pooling of their cubs in a crèche, whereupon the egalitarian queens of beasts are faced with the intricacies of shared caregiving. We will also assess the impacts of grandmothers on their daughters' reproduction and measure the reproductive payoffs from litters of various sizes and sex ratios.

FROM THE DEN TO THE CRÈCHE

After fitting one to two females in each Serengeti study pride with VHF radio collars, we tracked each animal from a Land Rover, and the collared female and her companions were usually visible by the time we drove up. But mothers become highly secretive around the day of birth, often remaining hidden in thick vegetation or rocky outcrops where they are impossible to observe from a vehicle (figure 4.1a). This is well illustrated by the following account:

20-Feb-2013. Daniel Rosengren drives down the Girtasho River searching for radio-collared lions, when he picks up the mortality signal on the GT collar [indicating that the female hasn't moved in >12 hrs]. The collared female is still quite young, so he hopes to find her body to determine how she died. The radio signal leads him to a small but extremely dense bush. He takes out a handheld antenna and walks a complete circle around the bush. The signal always points toward the narrow bush, so he tries to peer inside, but it is too dense. He even tries opening the foliage with his hands, but all he can see is darkness.

28-Feb-2013. Daniel returns to the same general area a few days later. He detects the normal signal for the GT female rather than the mortality signal, and he is shocked to see that the collar is still attached to a fully living and healthy female. She is with two one-week-old cubs. Thus, she had previously remained hidden in that bush with her freshly born cubs when he had tried sticking his head inside. She hadn't moved or even let out a growl.

Collared females also separate themselves from their pridemates, remaining alone for much of the first few weeks postpartum (figure 4.1b). Ranging data from K82 (the GPS-collared female in figures 1.8 and 1.16a) illustrate her lack of movement for approximately five-day periods around the birth of her two litters in July 2013 and April 2015 (figure 4.2a).

VHF-collared females were only located a few times each week, but their movement patterns made it possible to estimate the location of 118 denning sites (figure 4.2b). These were almost all located along drainage lines or in rocky outcrops (wooded hills or "kopjes"). Given that we mostly collared a single female from each pride, we only rarely knew where pridemates denned with respect to each other. But of the five pairs of collared females that gave birth within four weeks of each other, their respective dens were located a median distance of 288 m apart (range 32–515 m). Thus, mothers den close to each other, but they keep their newborn litters separate.

Mothers are so careful to hide their cubs that we often only knew they had given birth from their swollen mammary glands and the nursing stains around their nipples. Over the course of the study, 387 litters died before ever being observed, and we were unable to infer the cause of death in these "lost litters" except for the 74 cases (19.1%) that occurred within one gestation length of a male takeover, which we attributed to infanticide.

After four to six weeks of isolation, mothers bring out their cubs to join the older cubs in the pride. The oldest cubs in each crèche are typically three to five months of age when they are first joined by younger cubs, and the later-born cubs generally join by three months of age (figure 4.3a). Over the course of the entire study period, 94.4 percent of 3,436 well-observed cubs belonged to a crèche, so we considered the behavior to be a fundamental feature of lion society. The mothers become each other's closest companions, associating far more with each other than with any other pride members. The crèche seldom remains in the same place for more than a few days, as the mothers search for fresh hunting grounds. When a kill is made, one or more mother will return to the cubs and lead them to the carcass, where they feed together. Despite moving every few days, the cubs' current location serves as the meeting point for the mothers, like a mobile den from which each mother may sometimes depart to hunt on her own (see figures 4.9 to 4.14).

Mirroring the advantages of midsized crèches at male takeovers (figure 3.9b), cub survival peaks in crèches containing four to six mothers, although the greatest

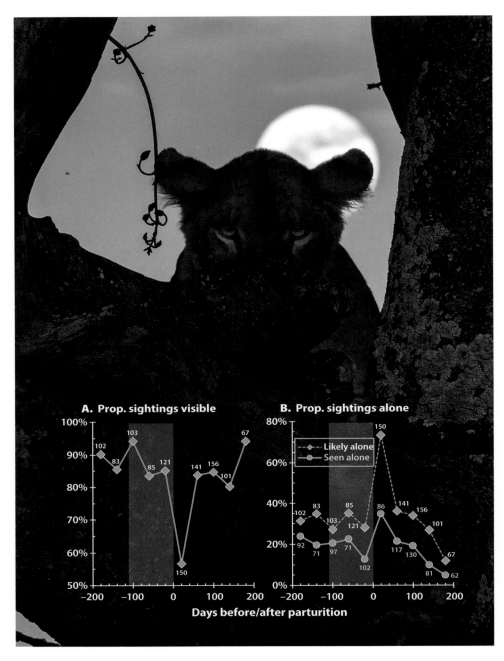

FIGURE 4.1. Changes in observability and sociality of radio-collared females six months before and after parturition. **A.** Collared females were less likely to be sighted when radio-tracked from a vehicle in the first week after giving birth. **B.** Collared females were more likely to be seen alone in the first weeks after giving birth; "likely alone" assumes that unseen females were always alone. Gray indicates pregnancy (110 days). Both redrawn from Packer et al. (2001).

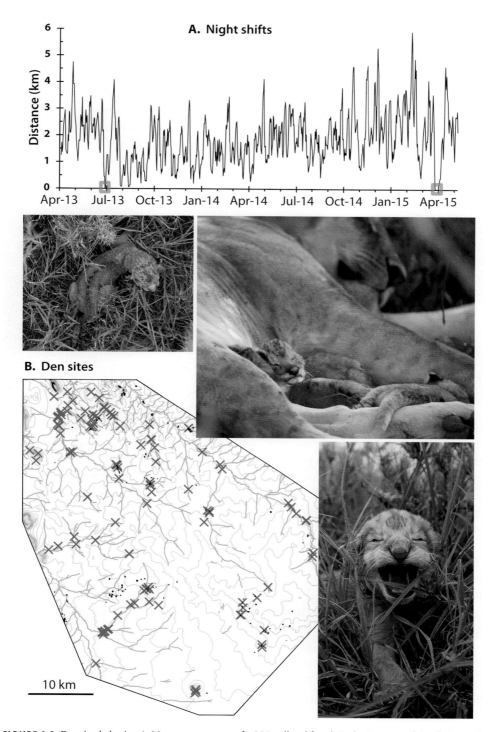

FIGURE 4.2. Denning behavior. **A.** Movement patterns of a GPS-collared female in the Serengeti plains. "Distance" is the four-day running average of straight-line displacements between locations at 18:00 p.m. each day. Red boxes indicate births of two successive litters: the first survived to two years of age; the second disappeared two to six weeks after birth. **B.** Den sites in the Serengeti study area. Estimates are based on a minimum of three sightings collected over a five-week span around the birth of a radio-collared female's litter; consecutive sightings must be less than 400 m apart with an overall standard deviation under 300 m. Blue lines are streams and drainage lines; faint green lines indicate elevation at 100 m intervals; black dots are kopjes (rocky outcrops). Note: the tiny cub in the green grass was abandoned when its mother transferred the rest of her litter to a new den.

improvements are achieved between one and three mothers and remains high even in crèches containing seven to ten mothers (figure 4.3b) despite the greater frequency of takeovers in the largest crèche sizes. The overall improvement in cub survival with crèche size cannot be explained by food intake: the mothers' average belly sizes are constant across crèche sizes, and, while cubs are significantly better fed when raised by pairs of females than by singletons, there is no further increase in larger crèches (figure 4.3c), nor is there any effect of crèche size on the variance in cub belly size, so cubs raised in crèches do not enjoy more consistent food-intake rates. Note that the findings in figures 4.3b and 4.3c are restricted to prides of three or more females, so these results are not due to pride size per se.

The crèche is the most stable social grouping in the pride, the heart of the lion's fission-fusion society until the cubs reach an age where they can fend for themselves, their mothers resume mating, and the whole cycle repeats again. Improved cub food-intake rates largely reach a maximum at two mothers (figure 4.3c) and the advantages of avoiding takeovers (figure 3.9b) and protecting cubs against incoming males (figure 3.9c) peak at four to six mothers, while cub survival remains high in even the largest crèches (figure 4.3b). The following sections explore two additional explanations why maternity groups might confer such strong benefits.

NURSING BEHAVIOR

Lion mothers are well known for their remarkable tolerance toward each other's cubs, even to the point of nursing any or all of the cubs in the crèche (Bertram 1976). Do cubs benefit from nursing from multiple females? One possible advantage would be the acquisition of "passive immunity" against a wider array of pathogens through exposure to immunoglobulins (Ig) produced by females of different ages. For example, older females may have been exposed to more disease outbreaks than have younger mothers and would therefore produce more diverse antibodies. However, the most relevant molecules, IgG and IgA, are transferred in the highest concentration in colostrum, which is only secreted during the first few days postpartum (Hurley and Theil 2011). Lion cubs are kept away from other mothers for the first few weeks of life; the cubs do not join the crèche until they are several months old (figure 4.3a) by which time the mothers' milk has been confirmed to be "mature" (Senda et al. 2010). Further, if "communal nursing" conferred strong immunological advantages, the behavior should be ubiquitous in nature, yet the lion is one of the few species where it has been documented (Packer et al. 1992).

To determine whether communal nursing plays an important role in crèche formation, Anne Pusey and I conducted detailed studies of over a dozen crèches—and we also observed the behavior of female spotted hyenas with their small (two-to-four-month) cubs. Like lions, hyenas keep their cubs together at the same location, but hyena cubs mostly remain below ground in the extensive tunnels of a communal den (Kruuk 1972), whereas lion cubs always remain above ground.

Female hyenas only nurse their own cubs, whereas lion mothers nurse their own offspring 70 percent of the time, despite the fact that non-offspring attempt to nurse at about the same rate in both species (figure 4.4a). Hyenas almost always resist the

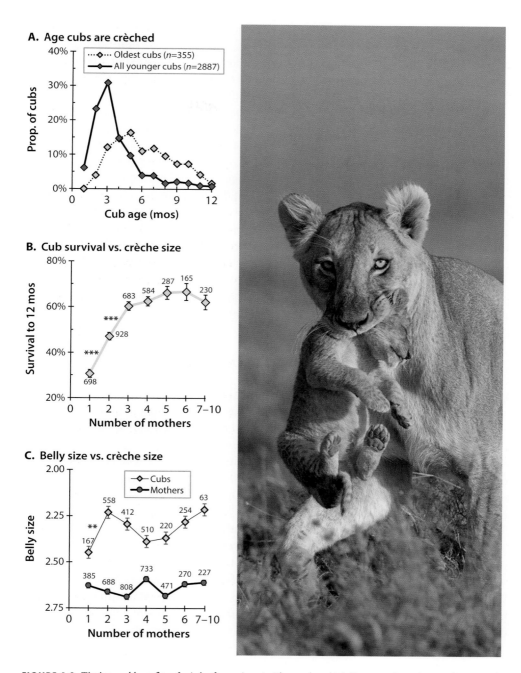

FIGURE 4.3. Timing and benefits of crèche formation. **A.** The age by which litters are brought together. **B.** Cub survival increases from one to three mothers in prides comprising at least three females. Crèche size is the median number of mothers each month when the cub was one to twelve months of age. n = number of cubs. **C.** Cubs reared alone are significantly thinner than cubs in crèches of two females in prides with at least three females ($p < 0.01$); mothers' belly sizes do not vary significantly with crèche size. Crèche size is the number of mothers in the month that belly size was measured. n = number of measurements taken more than five days apart.

nursing attempts by non-offspring, whereas they never resist their own cubs; in contrast, lions resist about 20 percent of attempts by offspring and non-offspring alike (figure 4.4b). Cubs in both species are more likely to initiate a nursing attempt when their mothers' heads are up—and presumably aware of their offspring's identity—than when attempting to nurse from other females (figure 4.4c), and females in both species are far more likely to groom (lick) their own cubs (figure 4.4d).

What can account for these striking differences between the two species? Hyena cubs spend most of their time untended in the safety of a well-excavated den that serves as their home base for the first year or two of life; hyena mothers only remain at the den long enough to nurse their cubs, and the Serengeti hyenas may remain apart from their cubs for days on end, traveling long distances in pursuit of prey (Kruuk 1972; Hofer and East 1993a) (see figure 12.4b). Hyena cubs are dependent entirely on milk until they are six to twelve months old (Holekamp et al. 1999) and may fast up to nine days at a time while their mothers are away (Hofer and East 1993b). In contrast, lion mothers lead their cubs to nearby kills, and lion cubs start eating meat when they are about three months of age (Schaller 1972) and thus rely far less on milk than hyena cubs. Whereas lactating hyenas return individually to the communal den for the sole purpose of nursing their cubs, female lions remain in close proximity to their cubs for most of the day. As a result, lion cubs are presented with the constant temptation to nurse from multiple sleeping females.

The parasitic nature of communal nursing is well illustrated by the behavior of mothers that have not yet brought their cubs to join the crèche. While taking a break from their newborn cubs, denning mothers will often socialize with the mothers in the crèche, but they are far more resistant to the nursing attempts of non-offspring, receive fewer attempts from non-offspring, and never nurse them (figures 4.5a, 4.5b, 4.5c). Thus, mothers whose young cubs are still entirely dependent on milk presumably follow a simple rule of thumb to remain highly attentive to the parasitic efforts of non-offspring while away from their own cubs and visiting the crèche.

A female lion has four teats, and cubs focus most of their nursing efforts on their own mothers (figure 4.5d), announce themselves less often to nonmothers (figure 4.5e), and more often wait until a nonmother is already nursing before attempting to join the other cubs on the nipple (figure 4.5f), apparently because mothers are less likely to terminate an overall nursing bout (either by rolling on her front or walking away) when being nursed by at least one of her own offspring (Pusey and Packer 1994a). The tolerance of the mothers for non-offspring varies according to two major factors. First, litter size: mothers with only one cub nurse non-offspring far more often than mothers with litters of four cubs (figure 4.6a). Second, kinship: mothers in crèches composed entirely of first-order kin (mother/daughter or sister/sister) most often nurse non-offspring (figure 4.6b). Milk production depends on recent food intake (figure 1.14a) but not litter size—the machinery of lactation is either on or off—so mothers with only a single cub have more milk to spare, and females are most generous when they are closely related to all of the cubs—thus minimizing the need to focus on the identity of each cub.

To summarize, lion cubs join the crèche at about the same time they start eating meat, thus milk makes up only a proportion of their diet. Cubs mostly nurse from their own mothers, but because lactating females spend so much time together with the

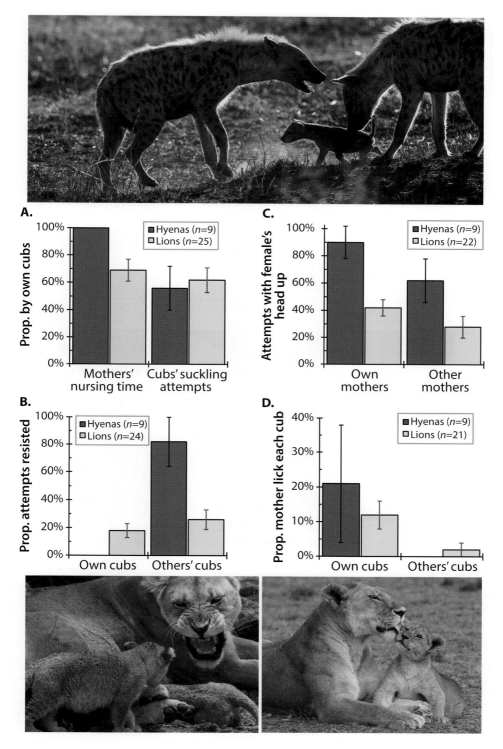

FIGURE 4.4. Nursing behavior of female lions and spotted hyenas. **A.** Proportion of total nursing time (left) and proportion suckling attempts (right) on the cubs' own mothers. **B.** Proportion of times that females resisted the nursing attempts of their own cubs versus the cubs of other females. **C.** Proportion of nursing attempts that occurred when the females' heads were up. **D.** Proportion of observation time that females licked their own cubs and the cubs of other females. n = number of females in all cases; see text for summary of significant differences between species. Redrawn from Pusey and Packer (1994a).

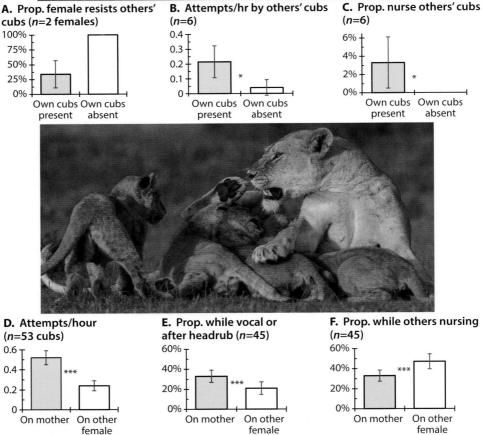

FIGURE 4.5. *Top:* Nursing behavior in the presence or absence of a female's own cubs. **A.** Female resistance to the nursing attempts by non-offspring. **B.** Frequency of attempts by non-offspring. **C.** Proportion of observation time that females nursed non-offspring. n = number of females; $p < 0.05$ for both **B.** and **C.** Redrawn from Pusey and Packer (1994a). *Bottom:* Nursing behavior of all cubs. **D.** Frequency of nursing attempts on their own mother versus the mothers of other cubs. **E.** Proportion of attempts where the cubs first announced themselves to the female. **F.** Proportion of attempts when another cub was already nursing from the same female. n = number of cubs; $p < 0.01$ for all three trends. Redrawn from Pusey and Packer (1994a).

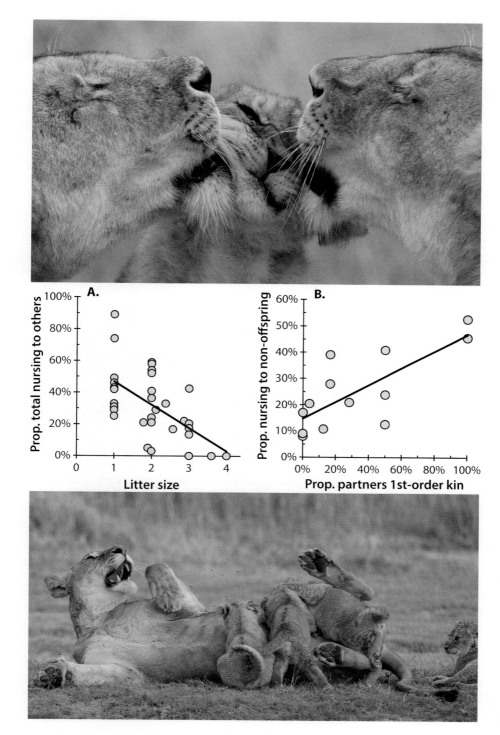

FIGURE 4.6. Factors affecting the extent of non-offspring nursing. **A.** The proportion of total nursing that goes to non-offspring plotted against the female's own litter size. **B.** The mean per capita proportion of nursing time by non-offspring plotted against the probability that all females were first-order kin. Dots represent different females; $p < 0.01$ for both trends. Redrawn from Pusey & Packer (1994a).

crèche, they are vulnerable to milk theft. The extent of non-offspring nursing partly depends on the cubs' ability to overcome the females' resistance, but females are more generous when they have milk to spare or when the non-offspring are their closest relatives. Thus, patterns of non-offspring nursing in lions reflects a combination of milk theft by cubs and differing levels of tolerance by the mothers. The young hyena cubs in our study, in contrast, relied entirely on their mothers' milk for nutrition, and they were unsuccessful in their attempts to nurse from anyone other than their own mother because of the females' greater attentiveness than shown by the sleepy mothers of a lion crèche.

NON-OFFSPRING NURSING ACROSS MAMMALS

Our field studies of the lions and hyenas inspired a broader investigation into non-offspring nursing, so we sent out questionnaires, searched the published literature and assembled data across one hundred different species. We found broad differences between "monotocous" species that typically give birth to a single offspring (e.g., primates, bats, marine mammals, ungulates) and "polytocous" species that routinely have litters (e.g., carnivores, rodents, and pigs). Non-offspring nursing is far more common in litter-bearing species than in species with a single offspring (figure 4.7a). Nursery groups occur in both types of species, but whereas monotocous females in large aggregations are *more* likely to be seen nursing non-offspring, litter-bearing females that live in larger groups are *less* likely to show the behavior (figure 4.7b). Observers were also more likely to consider non-offspring as being parasitic in monotocous species than in polytocous species (figure 4.7c).

Milk production in monotocous species is set by the nutritional requirements of a single offspring: a female only lactates when she is rearing that one offspring. In contrast, a polytocous female must produce enough milk to nurture an entire litter, but if she is currently rearing fewer than the maximum litter size, she can presumably "afford" to lose a fraction of her milk to non-offspring (as in figure 4.6a), and these losses will be least costly where the recipients are her closest kin (as in figure 4.6b). Coefficients of kinship inevitably decline with increasing group size, thus parasitic behavior is expected to become more common in species that typically form larger groups, whereas generosity should decline (figure 4.7b). Interestingly, spotted hyenas live in much larger social groups ("clans") than lions, occasionally reaching over eighty individuals, and hyenas also show a strong peak at two cubs per litter (Kruuk 1972), thus reducing the chance that a proportion of mothers will have milk to spare—and making them unlikely candidates for frequent non-offspring nursing.

Of the results in figures 4.4 through 4.7, any advantages from non-offspring nursing should be most obvious in large litters that have been reared together with small litters, as the largest litters in the crèche can gain supplemental milk from the mothers of singleton cubs. As discussed in greater detail below, cub survival depends on litter size, and litters of four cubs show nearly 50 percent higher survival when raised with singleton cubs than when mothers rear them on their own (see figure 4.15c versus 4.15b). However, mothers of four cubs are as likely to be found in crèches with mothers of three to four cubs as with one to two cubs, and cub survival is *reduced* when

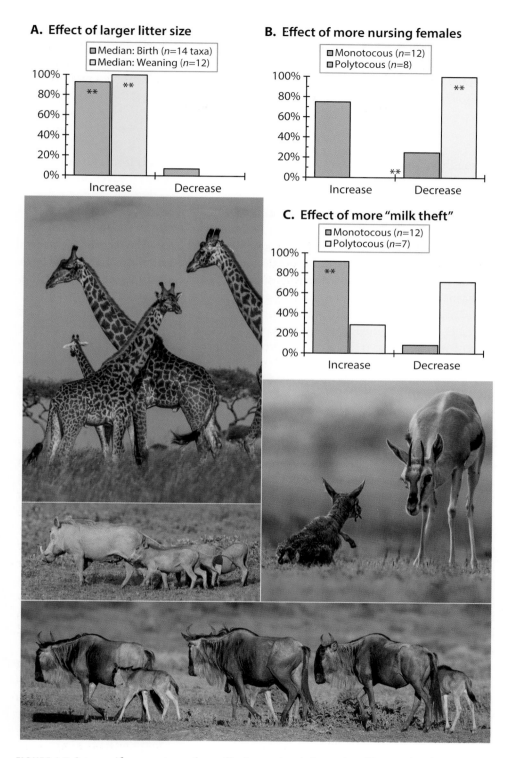

FIGURE 4.7. Interspecific comparisons of non-offspring nursing. **A.** Proportion of lineages that show more ("increase") or less ("decrease") non-offspring nursing in species with a larger litter size at birth and at weaning. **B.** Proportion with more or less non-offspring nursing in larger social groups. **C.** Proportion with more or less non-offspring nursing in species where the researcher classified the behavior as being primarily parasitic ("milk theft"). Asterisks within bars indicate significant deviation from 50:50; midgraph asterisks indicate differences between monotocous and polytocous species. All data from Packer et al. (1992).

litters of four are pooled with other large litters (figure 4.15c). Thus, non-offspring nursing seems more likely to be an inevitable byproduct of mothers merging their cubs together for some other reason rather than from a deliberate attempt to facilitate communal suckling.

BABYSITTING

Although we can use the belly scale to infer cub mortality rates from starvation (figure 1.14b) and peaks in cub mortality to infer infanticide at male takeovers (figure 3.2a), exposure to most other hazards are so rare—and so brief—that we cannot estimate their individual impacts on cub survival. However, lion cubs are clearly vulnerable to other carnivores. First, Schaller (1972) saw a female lion carry a small cub to a carcass that she had scavenged from a leopard in the Serengeti. She soon left to collect a second cub, but the leopard had remained nearby in the safety of an acacia tree. During the mother's absence, the leopard descended, captured the cub, carried it up the branches, and ate it. Second, during a disease outbreak in the Ngorongoro Crater in 2001 (see chapter 10), I watched as an orphaned six-month-old cub was ripped apart and eaten by two spotted hyenas. A lion project field assistant also found the paw of a lion cub in the Serengeti that had been brought to a hyena den; the cub had presumably been killed by an adult hyena. Third, although cheetahs have never been seen to kill lion cubs, we experimentally tested the willingness of an adult male and adult female to attack a stuffed-toy lion cub. The male stalked it from behind, bit it on the back of the head, carried it several kilometers, and then stopped and thrashed it around as if to make sure it was dead; in the second test, the female similarly bit the toy on the back of the head and carried to the top of a nearby termite mound (figure 4.8). Fourth, and perhaps most dangerous, are lions—not only infanticidal males but also females from adjacent prides. Pride boundaries shift with the season and with changes in relative pride size (chapter 9), and females have repeatedly been seen to kill cubs in neighboring territories. The following example illustrates the nature of these attacks:

> **20-Sep-1997 11:00 a.m.** Peyton West watches as six adult females from the GU pride enter the marsh near the Sametu kopjes. One of the GU females spots a two-week-old cub from the CV pride, runs at it "like a heat-seeking missile," kills it, and then climbs to the top of the biggest Sametu kopje and stands with the cub in her mouth. After a few minutes, she comes back down, drops it, and walks off. A second GU female immediately runs over, grabs the cub, and eats it.

Lions clearly recognize the threats posed by other predators, and their intense aggression toward spotted hyenas, leopards, and cheetahs (all of which they kill, but never eat, chapter 11), as well as their strict territorial defense against neighboring prides (chapter 8), presumably function, at least in part, to reduce predation risks on their cubs.

Crèche attendance, too, likely serves to protect vulnerable cubs by increasing the chances that at least one female will be in close proximity at any given time. Although mothers spend most of their time with their cubs, they may leave for several hours or

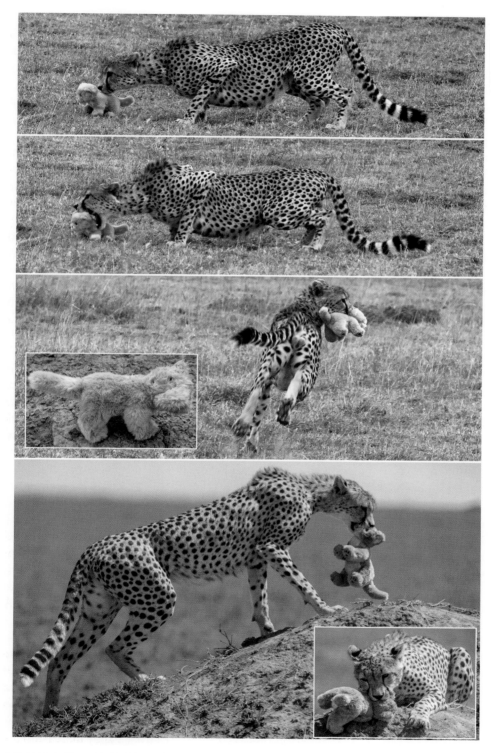

FIGURE 4.8. Serengeti cheetah attacks on a toy lion cub. Top three photos: attack by an adult male. Bottom photos: attack by an adult female. Inset shows puncture wounds on the toy's head.

days at a time, either departing individually or in varying combinations—sometimes leaving them completely unattended. The proportion of time cubs spend alone follows a U-shaped pattern with crèche size: cubs belonging to four to six mothers are the least likely to be on their own (figure 4.9a). A singleton mother is the sole provider for her cubs, so she must necessarily leave when she is actively searching for prey. Mothers in the largest crèches belong to the largest prides in each area and thus occupy the best-defended territories, so their cubs may have the least to fear from neighboring prides (chapter 8); alternatively, the mothers in the largest crèches may need to forage more extensively to compensate for greater within-group feeding competition. Females leave their cubs when searching for food, and foraging mothers are thinnest when they are in large groups (figures 4.9b and 4.9c). Pridemates typically have similar belly sizes (figures 2.6.a and 2.6b), perhaps predisposing hungry mothers to leave at the same time.

But once they have moved away from their cubs, each mother spends about 60 percent of her time alone (figure 4.9d). When a mother is with her cubs, she is significantly less likely to be apart from all the other mothers (figure 4.9d), although crèches of two females are more likely to leave a lone babysitter (approximately 40% of each mother's time with the cubs) compared to larger crèches (approximately 20%) (figure 4.9d). Thus, if mothers are all hungry at the same time, they may sometimes leave the crèche together, then split up and reaggregate upon their return to the cubs.

Figure 4.10a shows the grouping patterns of females when they are with their cubs: solid lines indicate crèches that deviate significantly from an expectation of equal group sizes; mothers in crèches of two to five females are generally all found together when with their cubs, but this pattern is weaker in crèches of six or more females. However, there is no real reason to expect each group size to be equally likely, especially if females move independently of each other. If each female typically spends x proportion of her time with the cubs, and if females typically join or leave the crèche on their own, then the probability that only one female will be present in a crèche of n females is $x(1-x)^{n-1}$, two will be present $x^2(1-x)^{n-2}$; and all n will be present x^n. Figure 4.10c shows the observed and expected distributions for groups of each size under these assumptions, and, except for crèches of two females, mothers are found in significantly *smaller* groups than expected. Each female spends most of her time at the crèche, so it is not particularly surprising that females are so often seen together when with the cubs; the surprise is to find just one or two mothers so often at the crèche. Figure 4.10b shows the difference between expected and observed for each group size for all crèche sizes. In every case, females are seen alone or in pairs with the cubs much more often than expected (trios are also more common than expected in crèches of six or more females)—and except for pairs, females are found in groups containing less than half of the mothers of the crèche more often than expected by chance. Thus, a proportion of females effectively serve as babysitters while the rest of the mothers are away foraging.

What factors predispose certain females to stay behind with the cubs? In several species, particular individuals take on the role of "sentinels" that watch for predators while their companions forage, and these sentinels are either better fed than their companions or have greater stakes in the survival of their vulnerable companions (Bednekoff 2015). In the lions, there is no obvious tendency for babysitters to have

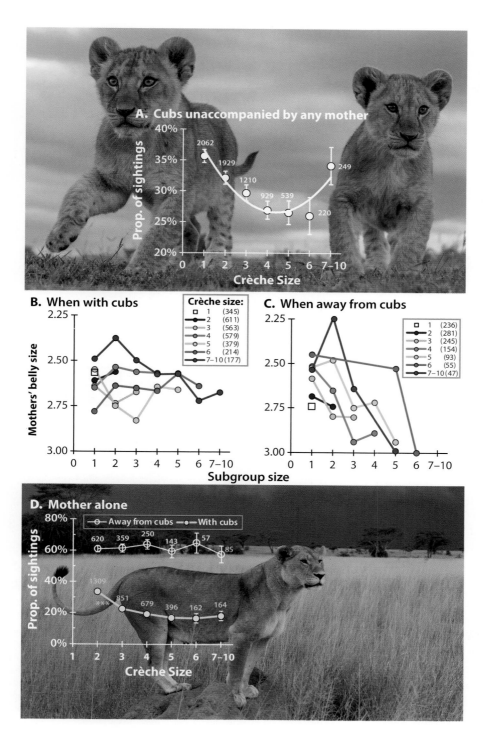

FIGURE 4.9. Grouping patterns of mothers and cubs. **A.** Proportion of cub sightings when the cubs were found apart from their mothers. Cubs raised by lone mothers and in crèches of seven to ten females were most likely to be seen unattended ($p < 0.005$). **B.** Belly sizes of mothers while with their cubs. The belly sizes of attending mothers did not vary consistently with subgroup size. **C.** Belly sizes of mothers when apart from their cubs. While away from the cubs, mothers were consistently thinner in larger subgroups than in smaller subgroups. **D.** Proportion of sightings where a mother was the only female in the group. Mothers spent more time alone with their cubs in crèches of two females (crèche size of two versus three females: $p < 0.001$); mothers spent far more time alone when away from the cubs than when with the cubs ($p < 0.0001$ for each crèche size). n = number of sightings.

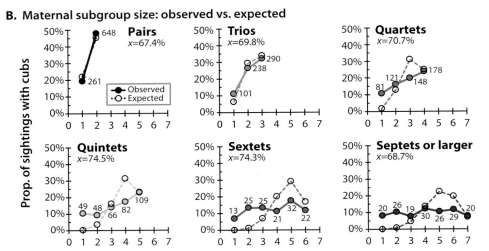

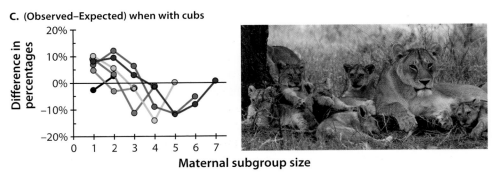

FIGURE 4.10. Female grouping patterns at the crèche. **A.** Relative proportion of sightings where females were found in subgroups of each size when with their cubs. "Relative proportion" is the observed number of sightings of each subgroup size multiplied by x/n, where x is the number of females in the crèche and n is the total number of sightings of a crèche of that size. Thus, if females were found the same number of times in each subgroup size, they would have a relative proportion of 1.0 for each. Females in crèches of two to five females were all found significantly more often in the maximum subgroup size ($p < 0.0001$ in each case), whereas females in larger crèches were found nonsignificantly more often in subgroups of four to six females. **B.** Observed and expected subgroup sizes based on the proportion of time that females typically spend with their cubs. The expected distribution is based on the proportion of time, x, that each female in a crèche of n females typically spends with cubs and assumes that females visit the cubs independently of each other. Thus, the probability that only one female will be present is $x(1-x)^{n-1}$, two will be present is $x^2(1-x)^{n-2}$, and so on. By this method, mothers in crèches of three to seven females formed significantly *smaller* groups than expected when they were with their cubs ($p < 0.01$); pairs were significantly more likely to be together ($p < 0.02$). n = number of independent sightings. **C.** Difference in percentage between the observed and expected subgroup sizes of mothers with cubs.

larger belly sizes (figure 4.9b), but instead the same individuals tend to remain behind while their companions are away from the crèche. In seven crèches where lone mothers were seen with the cubs on at least five separate occasions, one particular female was seen by herself more often than any of the other mothers, and this pattern was statistically significant in five cases. These persistent babysitters were typically the mothers of the youngest cubs or, otherwise, the mother with the most cubs (inset to figure 4.11). Not only are the youngest cubs most vulnerable to predation, but unweaned cubs and larger litters have the greatest need for milk. Thus, any persistent "division of labor" in babysitting is likely just a byproduct of the inevitable variations in birthdates and litter sizes within each crèche. Note that cubs in crèches with seven to ten mothers were more often found untended than were cubs in midsized crèches (figure 4.9a) despite having multiple babysitters (figure 4.11). Mothers presumably need to spend more time foraging in these outsized crèches in order to maintain nutritional levels that are comparable to mothers in smaller crèches (figure 4.3c), and none of the females would therefore have been able to serve repeatedly as babysitter. Intrapride feeding competition will be discussed in more detail in chapter 7.

To summarize, because mothers spend most of their time with their cubs, they are generally found together at the crèche in groups that are large enough to provide maximal protection against infanticidal males (figure 4.10a). However, one to three females sometimes remain with the cubs while the rest of the mothers are away foraging. This babysitting behavior reduces the time that the cubs are left untended and thereby protects against predation by other carnivore species and females in neighboring prides.

WHAT ABOUT GRANDMA?

Given the advantages of communal rearing, lions might seem like excellent candidates for an "adaptive menopause" whereby older females sacrifice their final years of fertility to enhance the reproductive efforts of their daughters and thereby produce more surviving grandcubs. While female lions do show a striking decline in fertility at advanced ages (litter size drops to a single cub by fourteen years of age [figure 1.21b]), and reproduction ceases altogether by seventeen years (figure 1.20c), older females rarely babysit unless they have cubs of their own, and they seldom contribute to the capture of large prey, so they seem to be more of a burden on the rest of the pride rather than a positive asset. Indeed, grandmother lions only enhance the survival of grandcubs when they are actively raising their own cubs (figure 4.12a). In addition, their current reproductive status does not measurably affect their daughters' litter sizes (figure 4.12b).

Orphaned cubs rarely survive to their first birthday, but neither a mother's death or her current reproductive status has much effect on the survival of her yearlings or older offspring, (figure 4.12c). Thus, even if older females were to die from engaging in late-age reproduction, "menopause" would not confer any obvious benefits on the survival or reproduction of their descendants. Instead, reproductive cessation is merely a consequence of so few females surviving to their mid-teens: selection becomes too weak to maintain the machinery of reproduction beyond a certain age (Packer et al. 1998). The timing of the drop in female fertility (approximately fourteen years) occurs

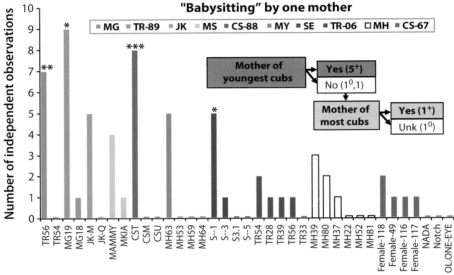

FIGURE 4.11. Crèches where a lone mother was seen with the cubs on at least five separate occasions. Observations collected more than five days apart are considered to be statistically independent. Asterisks indicate individuals that accounted for a significant proportion of the babysitting in that crèche. Initials refer to pride, numbers to year: thus, TR-89 and TR-06 indicate the Transect Pride in 1989 and 2006. **Inset:** Demographic characteristics of the persistent babysitter in all seven of the smaller crèches; none of the three largest crèches showed a similar pattern. Relevant data were unknown for the female with the superscript zero, as the litters had been merged in the crèche before we knew the maternity of each cub.

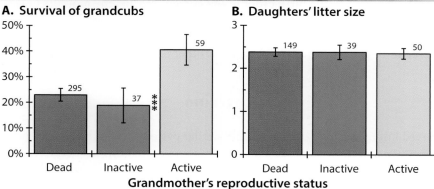

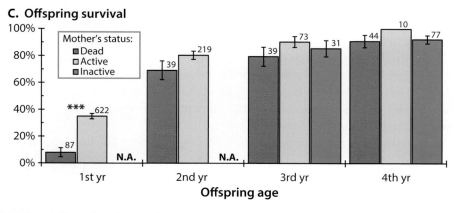

FIGURE 4.12. Effects of female survival and reproduction on descendants' productivity. Females are "inactive" if they have not reproduced during the previous twelve months and for six months thereafter; otherwise they are reproductively "active." **A.** Proportion of grandchildren in each litter that survive to their first birthday. Grandchildren of reproductively active grandmothers enjoy higher survival ($p = 0.0026$). **B.** Daughters' average litter size. **C.** Effects of female survival and subsequent reproduction on offspring survival. Survival is significantly lower for cubs orphaned during their first year ($p < 0.0001$), but the effect on orphaned yearlings is not significant. Mothers who give birth before an offspring's second birthday are considered to be "active" during its third year, otherwise they are "inactive." Male dispersal limits the analysis of older offspring to females, and female survival to three and four years of age is not influenced by their mothers' status. All redrawn from Packer et al. (1998).

at the age when a mother can raise one last set of cubs and therefore reflects the timing of her final contribution as a mother rather than the beginning of her life as a grandmother.

WHAT KEEPS THE CRÈCHE HONEST?

Orphaned *yearlings* show similar survival rates as their unorphaned peers (figure 4.12c), so if a mother's direct presence isn't necessary to ensure their survival, why doesn't she just "dump" her yearlings with the other mothers in the crèche and start again—in effect becoming a brood parasite and embarking on a second litter? Because orphaned *cubs* seldom survive, a dumping mother would have to wait until her prior litter's first birthday, thereby inflicting substantial mortality on her second litter: yearlings greatly depress the survival of small cubs (figure 4.13a)—presumably because of their considerable nutritional demands—and any female that dumped one litter to produce a second would receive very low returns from the younger.[1] Thus, the crèche is not only synergistic, but it is fundamentally honest, and mothers do not alter their interbirth intervals according to the birth order in the current crèche: the mothers of the first cubs in the crèche have the same interbirth interval as the mothers of the last-born cubs (figure 4.13b)—there is no jockeying to be the mother of the first-born cubs in the next crèche, or even to synchronize with the other females in the next crèche.

LITTERING

Observed litter sizes follow a remarkably similar pattern across the entire study area: litters of two and three cubs are consistently common in both the Serengeti and the Crater, in both good years and bad (figure 4.14a). Although cub *survival* varies with litter size during good years in the Serengeti woodlands and plains (figure 4.14b), litters of four or more cubs almost always produce the most surviving offspring, yet four-cub litters are never more common than singletons or pairs in any circumstance (figure 4.14c). Why are pairs so common? Are larger litters inherently more difficult for females to produce? If litter size depends on nutrition, litters should be larger in good years than in bad, but the distribution of litter sizes is unaffected by food availability within and between habitats (figure 4.14a), and there was no significant relationship between body size and litter size ($n = 465$ litters of 196 females that were measured between the ages of three and fourteen years).

Compared to lions in the Serengeti and Ngorongoro Crater, zoo lions show a much more even distribution of litter sizes with nearly twice as many singletons and quartets (figure 4.14d). But whereas litter size can be measured at birth in zoos, our litter-size

---

1 Mapema, the mother whose litter was adopted by her pride's crèche and raised a second litter after only an eleven-month gap (see footnote 2 in chapter 2) was a one-off. Rather than acting like a "brood parasite," Mapema was clearly distressed by the "loss" of her first litter. The only similar situation resulted from an interpride encounter between the TR and LL prides, where the two prides had cubs of roughly the same age and briefly fed together at a buffalo kill. When the two prides separated, three six-month-old LL cubs left together with the TR pride's crèche. The three "motherless" newcomers were tolerated by the TR mothers and cubs, but all died before their first birthday.

FIGURE 4.13. Birth order, cub survival, and interbirth intervals. **A.** Relative survival of three-month-old cubs when reared in crèches with older cubs. "Relative survival" is (survival in the presence of cubs of a particular age minus survival in their absence)/(survival in their absence). Analysis is based on 1,672 cubs in 46 prides; asterisks indicate significant effects ($p < 0.05$). The presence of same-aged (three-month-old) cubs raises cub survival by 28 percent; the presence of twenty-four-month subadults reduces cub survival by 42 percent. Redrawn from Packer et al. (2001). **B.** Birth order and interbirth interval. Interbirth intervals following synchronous births are the same as for mothers of the oldest and youngest cubs in a crèche. In all analyses of interbirth interval, data only include cases where the prior litters survive to two years of age, the successive litters are sired by the same resident coalition, and interbirth intervals are less than 920 days, as longer cases may involve undetected pregnancies. This analysis also excludes cubs born more than one year apart.

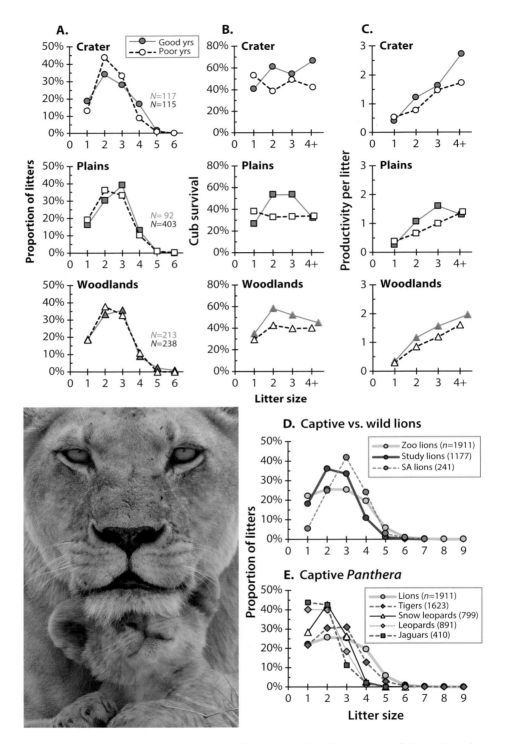

FIGURE 4.14. Distribution, survival and productivity of lion litter sizes. "Good" years are classified according to favorable population characteristics: years when the Crater population was relatively outbred and years when the Serengeti population grew in each habitat (see chapter 10). **A.** Litter sizes did not vary significantly within or between habitats. **B.** "Survival" is for each individual cub to its first birthday. **C.** "Productivity" is the number of surviving cubs per litter. n = number of litters. **D.** Comparison of litter sizes in zoo lions with wild populations. "Study lions" includes all data from the three study areas. "SA lions" are data from Miller and Funston (2014). **E.** Litter sizes from zoo animals. Lion litter sizes are significantly larger than tigers ($p \ll 0.001$) and the other three species. Data from Staerk et al. (2020).

measurements were only taken after the mothers brought their cubs out to the rest of the pride at about three to four weeks of age. In fact, the few times that we saw days-old litters were when mothers abandoned singleton cubs in open areas, which we interpreted as an adaptive behavior that allowed the mothers to replace lone cubs with larger litters in their next pregnancies (Packer and Pusey 1984). Given the 313 unseen "lost litters" that died outside of male takeovers, there is certainly scope for maternal abandonment to explain the rarity of singleton cubs, but what about the paucity of larger litters? During the first few weeks after giving birth, females shift their dens every few days, carrying one cub at a time to a new den, and we once observed the mother of four small cubs abandon one cub after safely transferring the rest of her brood (see photo in figure 4.2). Thus, assuming the zoo data more accurately represent litter size than our field data, the persistent peak of two to three cubs in the Serengeti and Ngorongoro may possibly result from a physiological predisposition to produce midsized litters that is further modified by the mothers' postpartum behavior.

Whatever the mechanism, the overall distribution of litter sizes is remarkably stable in our study populations, but why is the average size set at 2.4 cubs per litter? First, group living may confer advantages from producing larger litters as pride females are more successful than solitary females at rearing trios and quartets (figure 4.15a), and, in captivity, lions have larger litters than any other species in the genus *Panthera* (figure 4.14e). Among pride-living females, cub productivity depends on whether or not a female's litter is reared with other cubs of the same age: a quartet born more than ninety days before or after another litter produces the most surviving cubs (figure 4.15b), whereas cub productivity in closely spaced litters depends on the precise *combination* of litter sizes (figure 4.15c).

Figure 4.15c treats cub rearing as a game, where payoffs depend on the strategies (litter sizes) played by each participant. In these cases, the productivity of four-cub litters declines as their companions' litter size increases: four-cub litters do best when paired with singleton cubs but worst when paired with other quartets. The "unbeatable" litter size for synchronous births, therefore, turns out to be three cubs, as a trio attains a higher payoff when paired with another trio than any other litter size gains against a trio. This is the selfish or Nash equilibrium optimum (Maynard Smith 1982), as a population of females that only produced three-cub litters could not be invaded by mutants that produced an alternative litter size. However, pridemates are close relatives, so the litter-size game should tend toward a cooperative optimum, favoring the strategy that confers the greatest benefits to the population as a whole. But trios also gain a higher payoff against other trios than pairs against pairs or quartets against quartets, so three-cub litters would be expected to prevail, as trios cannot be invaded by either larger or smaller litter sizes in either a selfish or cooperative population and could thus be described as the evolutionarily stable strategy (ESS). Most cubs are born synchronously (figures 3.6a and 3.6c), thus trios should be the most common litter size in these populations—yet pairs are as common as trios. In an earlier study (Packer and Pusey 1995), we suggested that litters of two cubs could invade a population of trios and vice versa, but our original analysis was based on only 239 litters, whereas figure 4.15c includes 355 litters.

A female that gives birth out of synch with the rest of her pride would do best to have a litter of four cubs; synchronized females should have trios. Solitary females

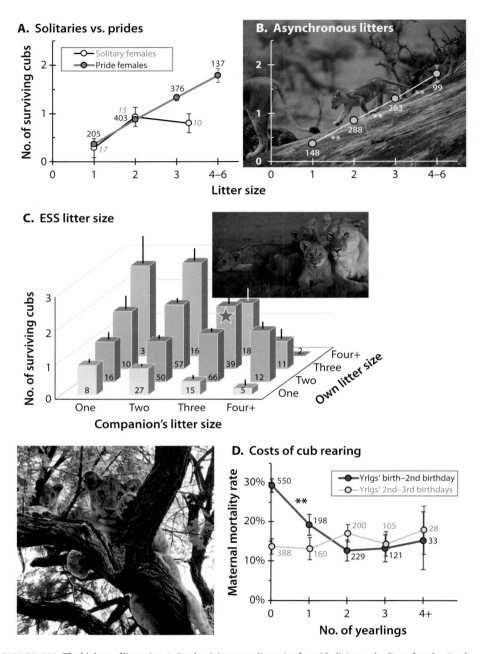

FIGURE 4.15. The biology of litter size. **A.** Productivity versus litter size for pride-living and solitary females. Productivity peaks at four cubs for pride-living females; productivity peaks at two cubs in solitaries. **B.** Larger asynchronous litters are more productive than smaller litters. Each additional cub significantly increases the number of surviving offspring in litters born more than ninety days before/after the next litter in the crèche ($p < 0.01$). **C.** Communal cub rearing transforms litter size into an evolutionary game. Bars indicate the number of surviving offspring produced by each female when paired with companions that produced litters of each size. Data are restricted to cubs born within ninety days of the prior/next litter in the crèche. Three-cub litters receive a significantly higher payoff when played against other trios than any other litter size receives against trios (e.g., the payoff of trio versus trio exceeds that of pair versus trio: $p = 0.0151$). No other litter size fares best against itself compared to the payoffs of any alternative litter size (e.g., the payoff of pair versus pair is significantly lower than trio versus pair: $p < 0.01$). Three-cub litters also gain the highest payoff of any litter size played against itself (e.g., the payoff of trio versus trio exceeds pair versus pair: $p < 0.01$), thus trios are also the cooperative optimum. **D.** Raising larger litters is not more costly to mothers. Mothers that fail to raise any surviving offspring suffer higher mortality during the second year postpartum than successful mothers ($p < 0.01$), but there is no effect of offspring survival in the third year.

reach an asymptote at two cubs, but the vast majority of females live in prides, so nothing predicts the preponderance of two-cub litters. What's going on? "Productivity" only considers the number of cubs surviving to their first birthday, so it is possible that rearing larger litters inflicts significant long-term costs on their mothers. First, larger litters might be expected to require longer periods of maternal investment, but interbirth intervals do not lengthen significantly with the number of surviving offspring ($n = 119$ intervals of 110 females). Second, successful reproduction reduces subsequent maternal survival in a number of mammalian species (Hamel et al. 2010), but female lions show the opposite trend: mothers that rear the *fewest* offspring suffer the highest mortality during the second-year postpartum (figure 4.15d), perhaps because unhealthy mothers are unable to rear their offspring or the social/environmental factors (e.g., male takeovers, small pride size, poor prey availability, etc.) that impair cub survival also harm the mothers. Direct maternal investment ends after about two years, so if females exhaust themselves by rearing larger batches of cubs, they might suffer greater risks of mortality the following year, but these death rates do not vary with the number of surviving offspring (figure 4.15d).

Litter sizes in southern Africa are far larger than in our three study areas (figure 4.14d). Frustratingly, the southern African lions show the precise litter size of three cubs that is predicted by the game-theoretical analysis in figure 4.15c. However, comparable demographic data are not available from these populations, nor is it known if these females abandon their smaller litters. Determining the factors influencing litter size isn't just an academic exercise. In the future, many lion populations will be confined within small fenced reserves, where lions show such rapid population growth (Miller and Funston 2014) that reserve managers are forced to use contraceptives or surgeries to *limit* reproductive rates (see chapter 12). If we could better understand why the Serengeti lions have such small litters, we could better inform management practices in the fenced reserves.

IT'S A GIRL! IT'S A BOY!! IT'S THREE BOYS AND A GIRL . . .

Because sexual reproduction inevitably bestows equal reproductive success on males and females, mothers are expected to divide their investment equally between sons and daughters (Fisher 1930). Thus, if sons were inherently more costly to raise to maturity, sex ratios should be female biased and vice versa. I will first examine this pattern during the primary period of maternal care then consider the full life spans of mothers and offspring. Figures 4.a and 4.b show that cubs and yearlings maintain larger belly sizes than their mothers at virtually every age up to two years. Sons show higher food intakes than daughters, and the patterns are broadly similar in the Crater and Serengeti, though "peak food intake" is reached at the earliest age in the Crater, and the peak is least pronounced in the dry season in the Serengeti. Thus, sons would seem to be more expensive than daughters to rear until independence.

On the other hand, if males suffer higher intrinsic mortality, sons may actually be "cheaper" as a greater proportion will die before receiving their full allotment of maternal care (Fisher 1930). Excluding deaths associated with male takeovers, cub mortality varies strikingly from year to year, with almost every cub surviving during good years and almost none surviving in harsh years (see figure 10.5c). Figure 4.16c shows that males indeed suffer slightly higher mortality under most circumstances—except during extremely harsh years when overall cub survival falls below 20 percent, and males are nearly three times as likely to survive as females. We observed little difference in nursing behavior between males and females, but a few obviously malnourished females appeared unable to compete against the slightly larger—and more aggressive—males, especially when feeding from carcasses. Meat becomes the sole source of nutrition after cubs are weaned at six to eight months of age, so the reversal in cub mortality in figure 4.16c likely results from males outcompeting females during periods of severe food shortage. But, overall, lions do show the mammalian norm of higher mortality in males than females.

Relative investment in sons and daughters is complicated by the fact that lions produce litters rather than single cubs, thus mothers often rear sons and daughters together, and, if one cub dies prematurely, the mother will continue to look after the survivors. But, as would be expected from the males' greater food consumption and their longer periods to reach full size (figure 1.20a), interbirth intervals are longer following litters with a higher proportion of surviving sons: average interbirth intervals are 600 days following all-female litters, 690 days for mixed-sex litters and 760 days for all-male litters (figure 4.16d). Compared to females (figure 4.16c), males suffer 18.4 percent higher mortality before their second birthday, but surviving males in all-male litters receive 28.7 percent more care than females in all-female litters. Maternal mortality does not vary significantly with the sex of their surviving offspring ($n = 491$ surviving litters), so a reasonably comprehensive estimate of the mothers' total investment in each sex can be obtained by simply summing the number of days devoted to the care of sons versus daughters. Each surviving female in an all-female litter is assumed to receive 600 days of care, males and females in mixed-sex litters receive 690 days of care, and males in all-male litters receive 760 days of care. Any cub that dies before these respective ceilings only receives care up to its age of death. After summing the total number of days of maternal care for all 3,850 cubs in the long-term study, the 1,927 females received a total of 871,848 days of care, whereas the 1,923 males received 860,287 days of care. Thus, the birth sex ratio was 49.95 percent male, and males received 49.66 percent of the total care, which seems pretty fair to both sexes, overall.

Though the basic fundamentals of meiosis and chromosomal sex determination suggest that mammals inevitably produce equal numbers of X- and Y-bearing sperm, dozens of studies have shown evidence of birth sex ratios that deviate significantly from 50–50 (e.g., Clutton-Brock and Iason 1986). Although, many of these findings subsequently disappeared with more extensive data (e.g., Silk et al. 2005), evidence has accumulated that biased birth sex ratios are physiologically plausible, either through hormonal mechanisms operating at conception (Navara 2013), effects of inbreeding on the proportion of Y-bearing sperm (Malo et al. 2017) or differential mortality during pregnancy (Geffroy and Douhard 2019). In many mammalian species, sex ratios are predicted to be sensitive to "local resource competition" whereby mothers produce fewer of the nondispersing sex (usually females) as they will compete more with their mothers for food and shelter *as adults*, and hence eventually "cost" more than dispersing sons in the sense that Fisher originally had in mind (Clark 1978, Silk and Brown 2008). As we saw in chapter 2, lions show the typical mammalian pattern of

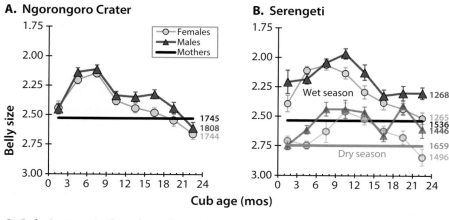

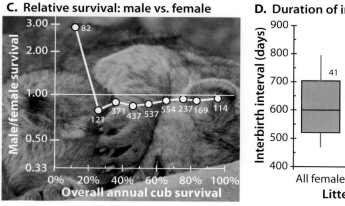

FIGURE 4.16. Costs of sons vs. daughters. Average belly size of males and females at each age in **A.** the Crater and **B.** Serengeti. Overall averages are presented for mothers in each habitat/season as their belly sizes did not vary with the age of their cubs. Sons show consistently larger belly sizes than both daughters and mothers ($p \ll 0.001$). Serengeti cubs are thinner during the dry season (July–December) than the wet season (December–April). n = number of belly measurements. **C.** Sons suffer higher mortality than daughters except during periods of extreme prey scarcity. "Overall annual cub survival" is the proportion of all cubs (sexed and unsexed) born in that habitat in a calendar year that lived to twelve months of age; the analysis excludes years with fewer than twenty at-risk cubs and prides that experienced male takeovers during that year. "Male/female survival" is male survival divided by female survival during the same year; the y-axis is plotted on a logarithmic scale so that, for example, 3.00 and 0.33 are equidistant from 1.0. Females are consistently more resilient than males, except in years with the lowest overall survival. n = number of cubs. **D.** Mothers have significantly longer interbirth intervals following the birth of male-biased litters ($p = 0.001$, $n = 119$ surviving litters). Intervals following all-female litters are significantly shorter than mixed-sex litters ($p < 0.01$) and any litter containing a male ($p < 0.001$). Results remain significant after controlling for the nonsignificant effects of litter size. n = interbirth intervals.

male-biased dispersal, with all sons eventually leaving and about three-quarters of daughters remaining (figure 2.1). But daughters mostly stay in their mothers' pride when the pride thereby remains within an optimal size range (figure 3.3d), thus adult daughters are seldom "costly" companions, and the overall birth sex ratio in the study lions was almost exactly 50–50.

However, instead of local resource competition, we previously suggested that lions might alter their sex ratios in circumstances where sons would benefit from local resource *enhancement* (Packer and Pusey 1987): whereas per capita reproductive success in females generally reaches an asymptote by about three individuals (figure 2.8), and cohorts of one to two females are generally able to join their mothers' prides (figure 2.10a), males gain substantially more benefits from belonging to coalitions of four or more males, but these only originate from cohorts that disperse together from the same natal pride (figure 2.13). Thus, litters of three to four cubs could potentially generate greater numbers of grandchildren if they mostly consisted of males. Inbred males are known to produce an excess of X-bearing sperm (Malo et al. 2017) and the largest male coalitions originated from exceptionally large cohorts of surviving young (figure 5.5d), so it is noteworthy that larger litters are indeed male biased, but only when the Crater population was thriving compared to periods of chronic inbreeding (figure 4.17a, also see figure 7.15). A similar male bias was also seen during bursts of rapid population growth as opposed to prolonged periods of population stability in the Serengeti (figure 4.17b). Infanticide synchronizes births during the first months after a takeover (figure 3.6a), and young males in these pride-level crèches would also benefit from greater numbers of like-sexed companions, for example, cousins and half-siblings, to form coalitions of four or more males (figure 5.5c and 5.5d). The sex ratio of cubs born in the first 120 to 300 days after a takeover is significantly male biased (53–47), whereas later births are slightly female biased (49–51) (figure 4.17c). Thus, regardless of the timing or the mechanism, female lions do seem to raise more sons in circumstances when they or their prides can generate larger cohorts of sons.

KEY POINTS

1. After hiding their newborn cubs in dense vegetation for four to six weeks, mothers bring them out to join a crèche of two to ten females where cubs enjoy higher survival and better food intake than cubs reared alone.
2. Cubs nurse from their own mothers to varying degrees, depending on a combination of milk theft and differing levels of tolerance by the other mothers. The extent of this tolerance is highest when females have milk to spare, and the nursing cubs are their close genetic relatives. Similar patterns are seen across mammalian species.
3. Mothers generally remain at the crèche in groups large enough to protect against infanticidal males, and often leave one to three females with the cubs while the rest of the mothers are away foraging, thereby reducing the time that cubs are left vulnerable to predation by other carnivore species and neighboring females.
4. Females experience an abrupt decline in reproduction at approximately fourteen years of age, but elderly females do not suffer increased mortality costs of

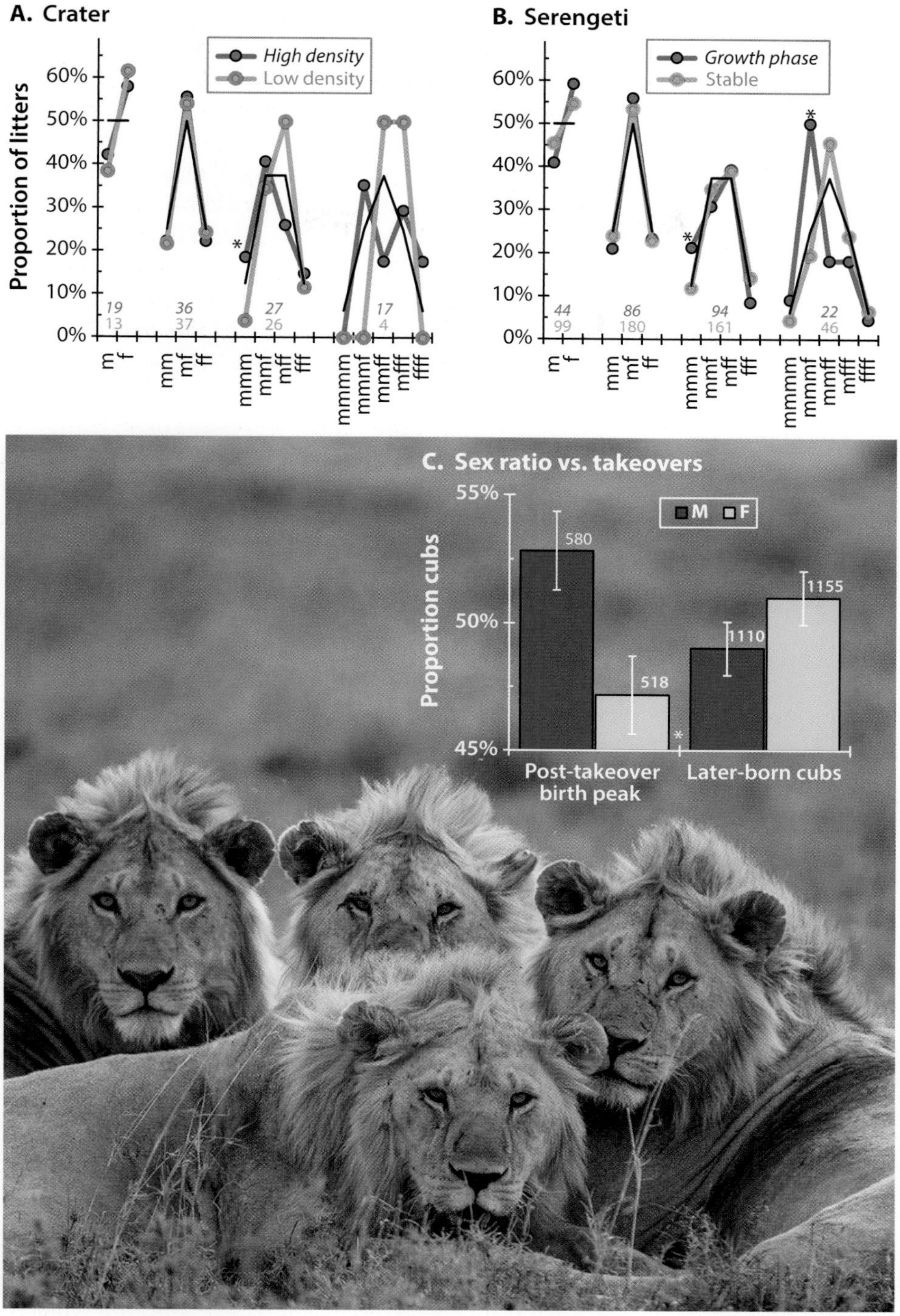

FIGURE 4.17. Male-biased sex ratios in large litters and post-takeover cohorts. Observed and expected litter composition in **A.** the Crater and **B.** Serengeti. A significant excess of three-male litters in trios and quartets was observed when the Crater population was large and relatively outbred and when the Serengeti population was undergoing rapid growth. n = number of litters of each size under each circumstance. **C.** Sex ratios of cohorts born 120 to 300 days after the takeover and later-born cubs. Cohorts born during the post-takeover peak are significantly male biased. n = number of cubs.

reproduction, nor do postreproductive females enhance the fitness of grandchildren or older children. Instead, reproductive cessation appears to result from senescence.
5. Litter size averages 2.4 cubs across all three study areas despite differences in food intake. Whether born synchronously or asynchronously, litters of three to four cubs produce greater numbers of surviving offspring. Though lion litter size is higher than in other *Panthera* species, it is unclear why litters in the Serengeti/Crater are smaller than predicted.
6. Overall birth sex ratios are very close to 50–50, as is the average investment in sons and daughters. However, sex ratios are significantly male biased in circumstances where cohorts would be likely to give rise to large male coalitions.

CHAPTER 5

MATING COMPETITION AND PATERNAL CARE

> It is the *green-eyed monster* which doth mock
> The meat it feeds on; that cuckold lives in bliss
> Who, certain of his fate, loves not his wronger
> —OTHELLO, ACT 3

31-Mar-1981

14:30 Two nomadic males, Weirdo and Macho,[1] rub past each other, urine-mark by scraping the ground with their back claws, and Macho flops on top of Weirdo as they lay down about 25 m from three members of the Naabi pride. The resident male, Barney, guards Ndub, as her pridemate Ndua rests nearby.

14:39 The nomads move to within 15 m of the consorts. Barney charges at Weirdo who veers away toward Ndua. Ndub avoids Macho's approach, and Barney stays close behind her but turns twice to charge the intruder.

14:40 Weirdo falls in behind Macho, and Macho leads the pursuit until Barney turns to fight. Macho rolls over on his back, and the two males bite each other's faces. Weirdo catches up; Barney retreats toward the females. Weirdo and Macho keep after Barney, and he veers away from the females. The two males chase him about 1.5 km with Weirdo now leading the pursuit, 15–20 m behind Barney, and Macho 25–45 m further back.

14:47 Barney escapes and heads for a wooded hillside, bleeding at the mouth. Weirdo and Macho turn back toward the two females.

14:51 Weirdo and Macho arrive within 200 m of Ndua and Ndub, who run away briefly but then stop. The two males come within 80 m; the females avoid again with Ndua falling behind Ndub. The males follow with Weirdo in the lead. Weirdo turns and briefly chases Macho, forcing him to veer off. Weirdo continues to within 20 m of Ndua and stops. Macho is now 40 m from the females.

15:11 Ndua gets up and walks away; Weirdo follows then sniffs where she had been sitting and Flehmens. She moves again, and Weirdo follows, stops, sniffs, and Flehmens at the second spot.

15:15 Macho comes to the first spot where Ndua had sat, sniffs and Flehmens.

---

1 Weirdo got his name for carrying around a large stick like an outsized tawny dog the day we first saw him; *Macho* happens to be the Swahili word for "eyes."

15:26 Weirdo again approaches Ndua; she withdraws but this time rolls over on her back; Weirdo again follows, sniffs the ground and Flehmens.
15:29 Weirdo chases Ndua then marks the ground.
15:31 Ndub approaches Weirdo; he starts toward her, but she runs away in a low slinky manner.
15:33 Ndub approaches Weirdo and solicits mating but then runs off; Macho sniffs another spot where Ndua had sat.
15:39–15:59 Ndua and Ndub both come up to Weirdo and solicit matings multiple times but always run off before making contact.
16:11 Weirdo is 3 m from the two females; Macho is 10 m away; Ndub starts to approach Macho, but Weirdo gets up, she veers off, and he chases Macho away from her.

Once resident, a male lion has two overriding goals: father as many offspring as possible then assure that they survive. The first goal creates tension within his coalition and considerable maneuvering for mating access, as seen in the example above, while the latter goal raises a number of questions. A mother clearly knows the identity of her own cubs but does a father distinguish between his own cubs and his companions', or does he help raise all the cubs in his pride? After cubs reach maturity, mothers seldom face the prospect of mating with their sons and typically welcome their daughters as pridemates, but do fathers ever mate with daughters and team up with their sons?

WITHIN-COALITION COMPETITION

At first glance, membership in a large male coalition would appear to be the best of all possible worlds, sitting at the top of the food pyramid, easily gaining residence in multiple prides then remaining resident for years and years as his coalition fathers more and more cubs per capita. But a zoological Voltaire might doubt that life was quite so rosy for every member of these successful coalitions as there are good reasons to wonder whether "per capita" accurately captures the reality of the situation. After all, coalition partners don't just politely take turns mating with each female in their prides; they instead compete among themselves for exclusive multiday access—one possessive male per receptive female for two to four days at a time—and some males may therefore end up the big winners while others are left out. Although we saw little evidence of unequal feeding access between coalition partners, any given meal isn't particularly valuable compared to a receptive female—the ultimate limit on a male's reproduction.

A consorting male remains in a heightened state for days at a time, maintaining constant vigilance of "his" female, preventing her from approaching any of his coalition partners, snarling at anything that might approach too closely. Once he has established "ownership" of the female, his companions defer to his possessiveness in the same way that rivals respect owners at feeding sites at a carcass, again likely reflecting the high costs of fighting (Packer and Pusey 1982). The consorting male's coalition partners do often position themselves nearby, especially when few cycling females are available elsewhere (figure 5.1a), but as long as one male is clearly the owner, his partners remain at a distance, overt aggression is rare, and the roles are reversible: the same male may be the owner in one consortship and the rival the next. Fights mostly occur when ownership is unclear or if two consort pairs come into close proximity (and thus both males are "owners") (see photograph in figure 5.3), and this respect of ownership does not differ regardless of whether the partners are related or unrelated to each other (Packer and Pusey 1982). Littermates are almost always full siblings (figure 2.3), so the zealous mate guarding of the consorting male is usually successful in securing all the paternity available from any given reproductive opportunity—but what happens over a larger time scale? Are some individuals more successful than others? Or do coalition partners contrive some sort of overall equity?

What We Learned from the DNA

The paternity data suggest a higher level of within-group reproductive skew among male coalition partners than seen in female pridemates, but this inequality is only obvious in larger coalitions (figure 5.1b). Thus, while pairs of unrelated males gain synergistic benefits from group formation, the direct reproductive advantages from living in larger coalitions are restricted to the most successful individuals, which likely explains why solitary males are more careful to team up with males of similar age as themselves when forming trios (where there is a risk of being a nonbreeder) than when forming pairs (where both males are able to breed) (figure 5.1c). The even greater skew in the only quartet in the paternity study (figure 5.1b) provides one possible explanation for why unrelated males never team up to form coalitions of four or more: the risks of being a nonbreeder are simply too high, whereas related males in such large coalitions still gain inclusive fitness effects by enhancing the reproductive success of their brothers or cousins (Reeve et al. 1998).

What about Those Larger Coalitions? Hints from the Behavioral Data.

Given the relatively small sample in the genetic survey (seven coalitions) and the absence of paternity data from any of the "super coalitions" of five to nine males, I examined consorting activity across every coalition that was observed an average of at least five consortships per male (figure 5.2). Note, though, that these data have several important limitations. First, although every female is impregnated by a consorting male, only a proportion of consortships results in a pregnancy (Bertram 1976), thus many of the observed consortships were nonreproductive. Second, as described in chapter 3, females sometimes solicit matings from an additional male after their "primary" consort partners have lost interest and stopped guarding. These "secondary consortships" are brief, and a secondary male shows less possessiveness than his predecessor, so he is unlikely to father any of the cubs conceived during that cycle. Our routine monitoring only rarely involved return visits to the same pride on successive days, so we mostly saw females during a single day of receptivity in any given cycle, thus we seldom knew if a particular partner was a primary or secondary consort. Thus, it isn't particularly

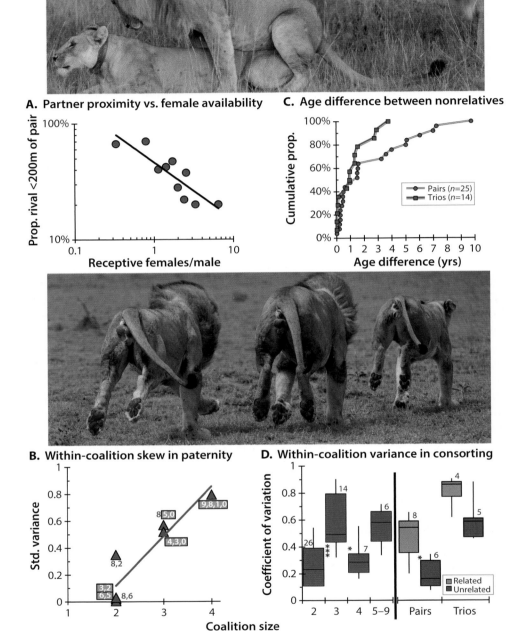

FIGURE 5.1. A. The proportion of sightings where nonconsorting males ("rivals") were seen less than 200 m of consort pairs declines with the number of potentially cycling females available to the rivals elsewhere. n = ten coalitions. Redrawn from Packer and Pusey (1982). **B.** Standardized variance in individual reproductive success increases with coalition size. Numbers by each point indicate total cubs fathered by each male; boxes enclose related partners. Redrawn from Packer, Gilbert, et al. (1991). **C.** Age differences between unrelated partners in pairs and trios. Across all coalitions, unrelated pairs show significantly higher variability in age difference than do trios (F = 5.42; p = 0.001). **D.** Coefficient of individual variation (standard deviation/mean) in consorting activity varies with coalition size and relatedness. Analysis restricted to coalitions with at least five separate consortships per male, excluding secondary consortships. n = number of coalitions.

surprising that only four of the seven coalitions in figure 5.1b showed a positive correlation between *actual* paternity and *observed* consorting activity. However, the consort data do provide potential insights on a dozen different coalitions of four to nine males—and the overall analysis suggests several intriguing possibilities.

As with the paternity data, within-coalition variation of consorting activity is higher in trios than in pairs—and the data suggest higher skew in related pairs than in unrelated pairs (figure 5.1d), as would be expected if unrelated males are only willing to team up if they can thereby gain direct reproductive benefits. However, within-coalition variation in the consorting activity of quartets is lower than for trios (figure 5.1d), which directly contradicts the paternity data for the quartet in figure 5.1b. Although it is not possible to determine whether the quartets in figure 5.1d truly experience lower reproductive skew than trios, such a pattern could plausibly result from the greater number of prides typically controlled by quartets than by trios (figure 2.12c): if multiple females from different prides are receptive on the same day, multiple males would inevitably be able to consort simultaneously, thereby reducing the level of within-coalition skew (figure 5.3a), and, after controlling for this effect, the residual skew in consortships is similar between trios and quartets (figure 5.3b).[2] In addition, the estimated proportion of secondary consortships (where a second male was seen consorting with the same female less than four days after a prior male) also increases with coalition size (figure 5.3c), and although I excluded all *known* secondaries from these analyses, our consorting data for the largest coalitions inevitably includes a greater number of *unrecognized* secondary consortships, thereby potentially overrepresenting the true reproductive activity of any subordinate males.

But assuming that consorting activity does actually provide a rough indication of the males' reproductive success, the data suggest two further ways that larger coalitions may reduce the effects of greater within-group competition. First is the possibility of reproductive "queuing" (Kokko and Johnstone 1999) whereby subordinate males eventually rise in rank and achieve higher mating success, as seen in a quartet called "The Killers" that remained in the study area for seven years: the oldest male in the coalition, S#93, consorted more than any of his partners in 2008, S#99 consorted the most in 2010, followed by S#94 in 2012 and S#98 in 2013 (figure 5.4a). Overall, these males had the second-lowest overall reproductive skew of any quartet (figure 2.15), despite the fact that one male outperformed the rest in nearly every year. Males rarely feed while consorting (figure 5.3d), so highly active individuals may inevitably "burn out" after too much continuous sexual activity;[3] quartets maintain residence for a median of six years compared to only four years for trios (figure 2.2), thus quartets would be more likely to survive long enough to benefit from reproductive queuing.

Secondly, males in the largest coalitions also show the most divergent mating preferences, with individual males selectively consorting with different females. For example, in the coalition of nine "MS" males, MS10 frequently consorted with K06, MS12 with L12, and MS18 with SMB (figure 5.4b). Such divergence was typical of the

---

2 Asian lions live in much smaller prides than in the Serengeti and Ngorongoro (see chapter 9), and male coalitions show higher levels of skew in mating activity than seen in figure 5.2, presumably because fewer receptive females are available at any one time (Chakrabarti and Jhala 2017).

3 Anderson et al. (2021) found that top-ranking male baboons show high levels of DNA methylation, indicating accelerated rates of aging. Similar effects would be expected in male lions that engaged in frequent consortships.

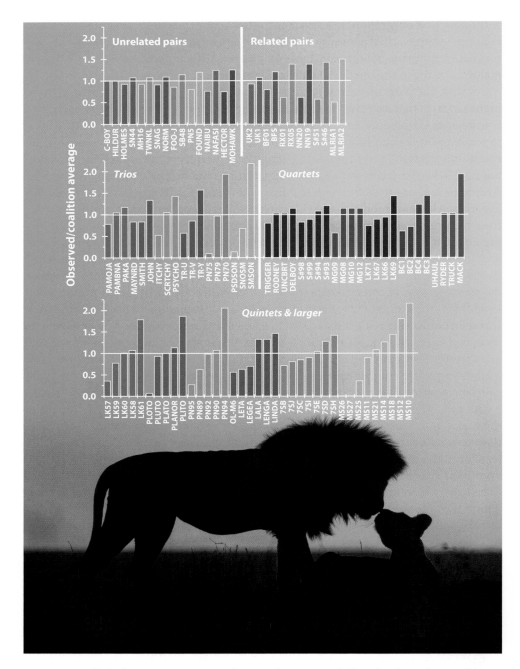

FIGURE 5.2. Individual consorting activity in coalitions with an average of at least five observed consortships per male. *Top:* All pairs of known kinship. *Middle and bottom:* Larger coalitions; trios and quartets only include the two lowest, two highest, and two midmost levels of skew. For each coalition size, groups are ordered by increasing variation in consorting activity; within each coalition, males are presented in order of increasing consorting success.

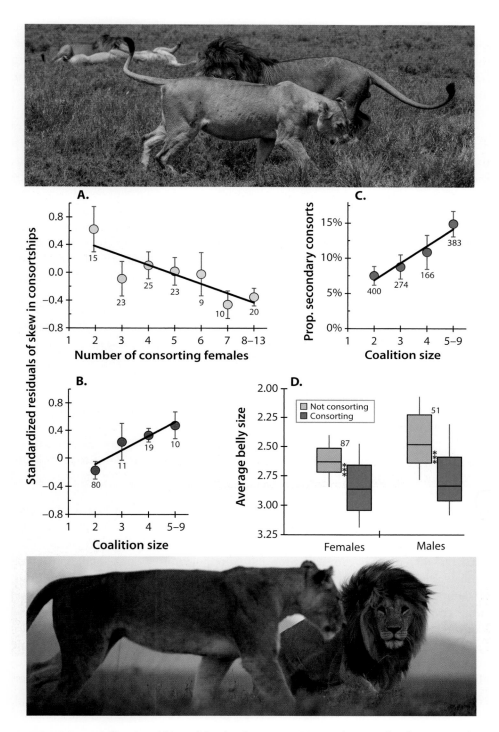

FIGURE 5.3. Factors influencing within-coalition skew in consort activity. **A.** After controlling for current coalition size, an increase in the number of receptive females significantly decreases skew. n = number of coalitions where partners were observed to consort at least twice per male in a particular year, excluding secondary consortships. **B.** After controlling for number of females, skew increases significantly with current coalition size. n = number of years as before. **C.** Proportion of secondary consortships increases significantly with coalition size. n = number of consortships. **D.** Males and females are significantly thinner during consortships. n = numbers of males and nonpregnant females whose belly sizes were measured at least three different days when consorting and three days when not consorting. Photographs: Consorting males attempting to alter the movements of their partners.

largest coalitions: concordance in consorting preferences declined significantly with increasing coalition size (figure 5.4c), and this decline is again associated with access to a greater number of females (figure 5.4d). Almost all of the coalitions of four or more males maintained simultaneous residency in multiple prides (figure 2.12c), and the divergent preferences for K06, L12, and SMB in the nine-male coalition in fact resulted from their subdividing themselves between the KB, LL, and SM prides.

But if the typical skew in quartets really is lower than indicated by the paternity data, we are left with the question of why males only recruit unrelated companions up to a maximum of three. If we return to a per capita accounting of reproductive success, the individual advantages of adding a fourth partner would be considerable. Perhaps recently formed partnerships cannot achieve the necessary level of coordination for forming a truly effective quartet—maintaining any sort of group-level cohesion while holding multiple prides and responding collectively against outside threats may only be possible for four-male groups that have spent their entire lives together. Or perhaps an unrelated trio is so successful in gaining residence that the males opt for the immediate payoffs from gaining residency and switching into territorial mode before finding a like-aged fourth partner. If I had to choose, I'd say that the answer probably lies somewhere in the middle: an inherently greater level of reproductive skew in larger coalitions combined the greater difficulties in group-level coordination plus a certain amount of impulsiveness when presented with the opportunity to gain residence.

ONE FOR ALL AND ALL FOR ONE?

Regardless of the behavioral mechanisms that allow larger lion coalitions to work together, it is truly remarkable that coalition partners have never been known to seriously injure or kill each other. In contrast, male chimpanzees, which also form male-bonded social groups that compete against groups of neighboring males (Goodall 1986), have repeatedly been seen to maim or even kill their erstwhile partners (Wilson et al. 2014). In several such cases, the killers teamed up to attack a high-ranking opponent thereby increasing the survivors' subsequent access to mates (Massaro et al., in review). Although we have never seen male lions seriously injure their coalition partners when directly competing for mating opportunities, such encounters could conceivably take place in other contexts or otherwise remain unobserved. Thus, I examined the circumstances whenever resident coalitions lost partners and tested whether the losses occurred during time periods when intracoalition competition was likely to have been particularly intense.

Once resident, a male never leaves his coalition to become a singleton, so if a single coalition member disappears, we can be certain that the "missing" male has actually died. As shown in figures 5.3 and 5.4, intracoalition competition increases when partners have access to fewer receptive females. Generalizing this pattern to the total number of females (regardless of reproductive state) available to each coalition, I tested whether partners were more likely to die/disappear when their coalitions had access to smaller numbers of females. I first calculated the total number of prides where each coalition was resident each month over a two-year period, centering around the month of the companion's disappearance. I then compared the coalition's residence

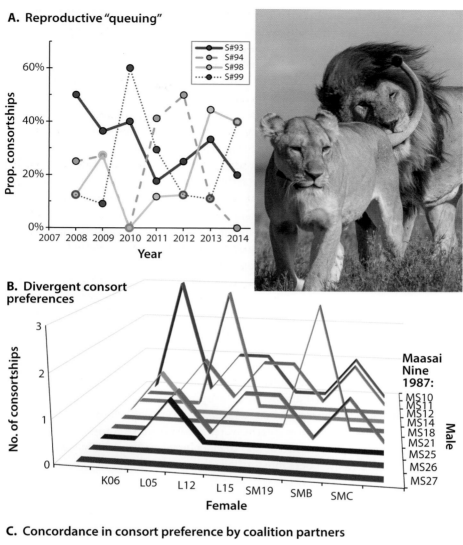

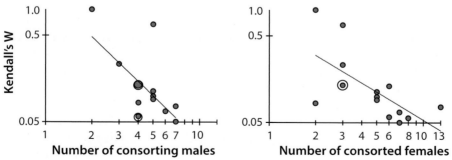

FIGURE 5.4. Factors reducing reproductive skew. **A.** Proportion of consorting activity by each male each year in the Killers. **B.** Total number of consortships by each male with each female in the coalition of nine males in 1987. **C.** Concordance in annual consort preferences declines with the number of (*left*) males and (*right*) females. Kendall concordance values were calculated for all coalitions with at least [2 · (no. females) + 2 consortships] in the same year. Note that W < 0.5 indicates discordance rather than concordance.

patterns in three time periods: first, the 1.5 months prior to and following the day of the partner's disappearance (the three-month "disappearance timespan"); second, the ten months prior to the disappearance timespan ("before"); third, the ten months following the disappearance time span ("after").

Comparing the "before" and "after" time spans, it's clear that the loss of a partner from a pair has far stronger impacts on the surviving singleton than do losses on the survivors of larger coalitions: over half of the singletons completely lost residence in the following ten months, whereas very few survivors from the larger coalitions were ousted from all their prides (figure 5.5a). However, a substantial number of these surviving pairs, trios, and so forth, lost residence in at least one pride—and about a third actually increased their holdings, either gaining residence for the first time or annexing additional prides around the time of their partners' deaths. Taking over or surrendering a new pride almost always involves competition between rival coalitions, and, altogether, about two-thirds of partner losses in coalitions of three to seven males were associated with either losing or gaining new prides, thus these deaths almost certainly involved intercoalition competition.

But what about within-coalition competition? Here we would expect deaths to relate to the number of *females* rather than to the number of *prides*: partners compete with each other for consortships with individual females, and, even though some males may spend more time with one pride than another, each coalition partner is free to associate with every female in the coalition's domain. Thus, if within-coalition competition were ever to lead to lethal aggression, I would expect it to be most common in the largest coalitions (where multiple males might sometimes compete for a single mating opportunity) and that a substantial proportion of partner losses would occur when these males recently experienced a *reduction* in available females. However, the trend is in the opposite direction: deaths were relatively more common in large coalitions in months when they gained access to *more* females (figure 5.5b)—which, again, suggests that coalition members mostly died while acquiring more females from rival coalitions rather than competing with their partners.

Thus, whereas male chimpanzees are capable of truly dastardly behavior toward their comrades, lion coalitions seem to be truly lion-hearted when it comes to their relationships with their companions. Though partners-in-arms may angrily spar with each other when ownership of a receptive female is unclear, such confrontations largely result in facial scratches and superficial wounds (see figure 6.3), while their most damaging attacks are solely inflicted on outsiders (see figure 8.2). Beyond the risks of personal injury, coalition partners are vital companions in the larger battle of "us" versus "them," so any male lion that fought too viciously with his companions would soon be vanquished by a more cohesive coalition (see chapters 8 and 9).

WHY ARE LARGE COALITIONS SO RARE?

Given the unity of purpose among coalition partners and the striking advantages from forming coalitions of four or more males, it may seem surprising that only 24 percent of 507 resident males belonged to quartets or larger, whereas 48 percent were singletons or pairs. However, the comparative rarity of these larger groupings stems from

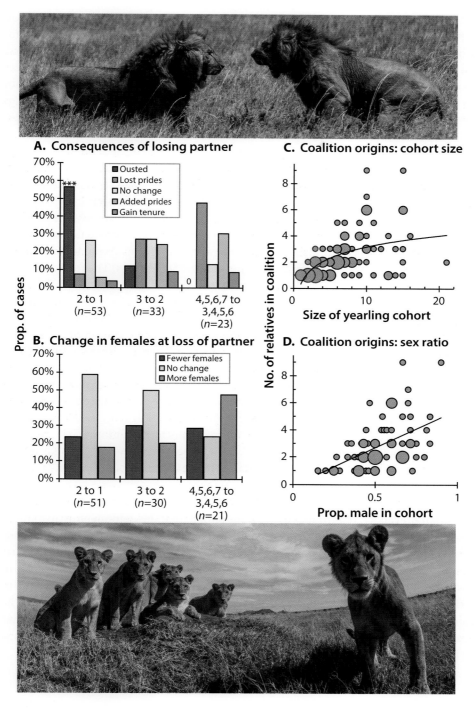

FIGURE 5.5. *Left:* Loss of a coalition partner versus the total number of females in their pride(s). **A.** After losing a partner, pairs were significantly more likely than larger coalitions to completely lose residence in the following ten months. For larger coalitions, the loss of a partner coincided with either losing or gaining a proportion of their prides. n = number of coalitions that lost a single partner over a six-month period. **B.** Compared to pairs and trios, coalitions of four or more males were slightly more likely to lose a partner during times when they gained rather than lost access to additional females. *Right:* Factors determining the number of related companions in resident male coalitions. **C.** Coalitions of six to nine males only arose from cohorts of ten to fifteen yearlings. **D.** Large coalitions mostly arose from male-biased (i.e., the proportion of males exceeded 0.50) yearling cohorts; sample restricted to cohorts of at least four yearlings. Circles are scaled in proportion to sample size.

the limitations of pride demography. First, large coalitions can only emerge from large, male-biased cohorts (figures 5.5c, 5.5d). Second, large cohorts are inherently unstable: eleven of thirteen cohorts containing eight or more males split up (figure 5.6a), and males born more than twenty months apart always separate (figure 5.6b). Males in small, same-aged cohorts, on the other hand, almost never part, hence post-split cohorts show the same distribution of size and age disparity as intact cohorts (figures 5.6c and 5.6d). Thus, coalitions of four or more males are rare because prides seldom generate large enough cohorts of like-aged sons.

Once resident, the effect of even a slight difference in age continues to affect the relationships between coalition partners. Coalitions also show fission-fusion grouping patterns, and subgroups in the largest coalitions only average 1.11 males (figure 5.7a). Across all coalitions, age difference is the most important factor determining partners' associations with each other: the variance in spatial associations across all pairwise combinations of partners is significantly higher in coalitions where males vary in age by more than a month (figure 5.7b). The sociograms in figure 5.7 summarize the associations of coalition partners in eight representative coalitions. The four coalitions on the left all show significant variation in associations (figures 5.7c, 5.7d, 5.7e, and 5.7f), whereas relationships on the right do not differ significantly from each other (figures 5.7g, 5.7h, 5.7i, and 5.7j). In most cases, subdivisions directly follow the age difference (e.g., agemates LKJ and LKK associated more with each other than with LKS, LKV, LKX, and LKY, which in turn associated more with each other, figure 5.7e), and subgrouping in the coalition of nine males (figure 5.7f) predicted the eventual splitting of the five younger males from the older four to form two separate coalitions. But some relationships transcend age (figure 5.7d), and age difference does not always result in asymmetric relationships (figure 5.7i).

FATHER(S) OF THE PRIDE

Paternal care is widespread in fish and birds, but mammals are *defined* by their extraordinary levels of maternal care, with mammalian males seldom contributing to childcare except in monogamous species, such as marmosets and canids (Clutton-Brock 1991). At first glance, lions seem like a typical polygynous mammal with intensive maternal care and disengaged fathers, but male lions play an essential role in protecting their cubs against infanticidal rivals and in capturing prey large enough to feed their entire pride (see chapter 7). However, paternity is divided between multiple males, and certain individuals may have sired more offspring than others. Do males bias their care-giving behavior toward specific cubs, or do they treat all the cubs born during their residency the same?

Who's Your Daddy?

We have never seen resident males lick, groom, or carry food to individual cubs, but fathers could conceivably confer greater protection to individual offspring by preferentially spending more time with them. Figure 5.8 shows the extent to which

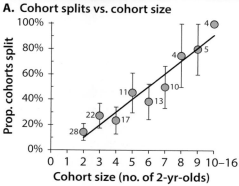

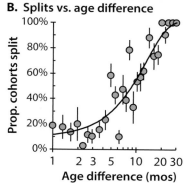

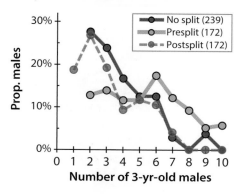

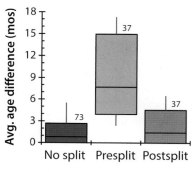

FIGURE 5.6. Factors affecting whether male cohorts split up. **A.** Larger male cohorts are more likely to split. n = number of cohorts of two-year-old males. **B.** Males born less than five months apart almost never separate, whereas males born more than twenty months apart never remain together (Total sample size = 114 cohorts). **C.** The distribution of post-fragmentation group sizes is almost identical to the distribution of cohort sizes that did not split. n = number of three-year-old males. "Proportion of males" is weighted by the number of males in each cohort. Note that 81.3 percent of the thirty-two post-split solitary males were at least five months younger/older than the rest of their cohort. **D.** Post split, the average age difference was identical to cohorts that never split. n = number of cohorts of three-year-old males.

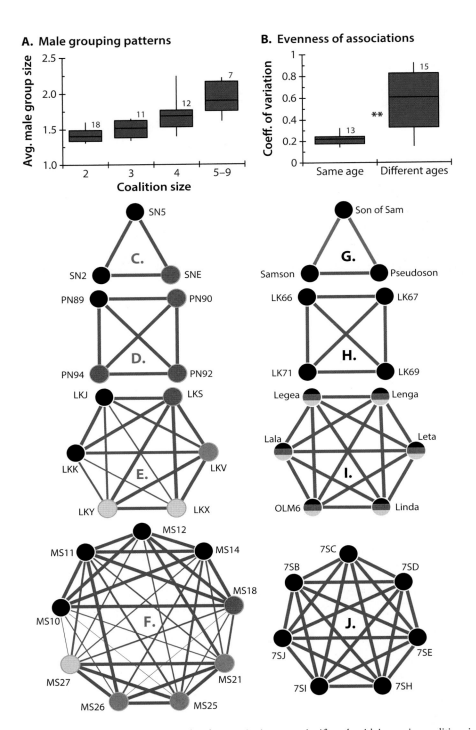

FIGURE 5.7. Male-male associations. **A.** Male subgroup size increases significantly with increasing coalition size to an average of about two males. *n* = number of coalitions observed on at least thirty occasions. **B.** Same-aged males show more similar spatial associations with their coalition partners than do males born more than a month apart. *n* = number of trios and larger that were observed on at least fifty occasions. **C.–F.** Representative coalitions of each size with significant variation in spatial associations. **G.–J.** Representative coalitions with relatively homogeneous relationships. Red lines: associations between unrelated partners. Blue lines: associations between related partners. Black circles: oldest male(s) in each coalition. Gray circles: younger males; the youngest cohort is indicated by the lightest shade of gray. In **I.**, the six males were known to have been born over a six-month period, but they were not individually recognized until fully mature.

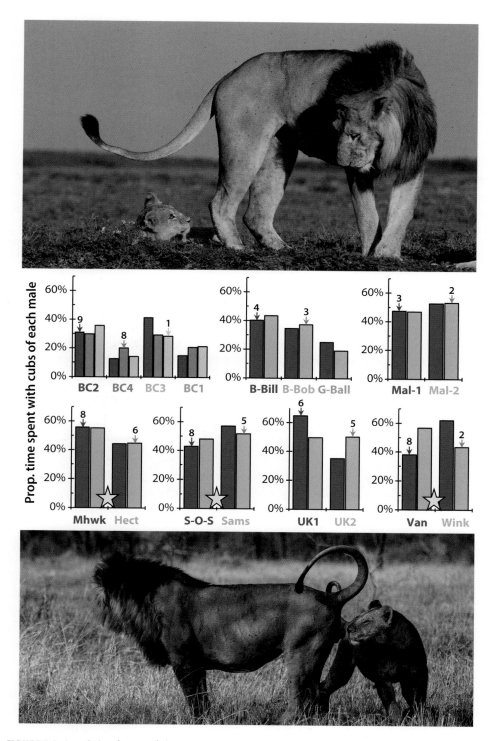

FIGURE 5.8. Associations between fathers and cubs. Fathers do not preferentially associate with their own offspring. Y-axes show average proportion of time each coalition partner was seen with each cub out of all observations that males were seen with that cub; bars separate each male's average associations with each of his own cubs versus each of his respective partners' cubs. Arrows indicate each male's own offspring; n = number of cubs fathered by that male. Stars highlight unrelated companions. Note: if coalition partners spent equal time with each cub, the maximum values for the quartet would have been 25 percent, for the trio 33 percent and for the pairs 50 percent. One coalition male disappeared before his coalition was observed with any cubs.

each male in the paternity study associated with his cubs. Only eight out of these seventeen males spent more time with their own cubs than with their partners' cubs, and the pattern was similar for related and unrelated companions (six of eleven related partners versus two of six unrelated partners). Note that since crèche-mates typically remain together, there is little opportunity for males to associate with one cub without associating with the entire crèche, thus any differences in time spent with offspring versus non-offspring are inevitably small. Note, too, that only three of seven of the most reproductively successful males in each coalition spent the most time with cubs. The paternity data, therefore, provides no evidence that males associate with individual cubs—or an entire crèche—according to their actual or probable degree of kinship. Instead, resident males seem to show a general tolerance toward all the cubs sired during their tenure regardless of their genetic contribution to the brood.

In terms of genetic relationships, the offspring of competing coalitions are no different than the cubs of unrelated coalition partners, yet males are infanticidal toward the former and protective toward the latter. What causes males to switch from one mode of behavior to the other? In studies of captive rodents, normally infanticidal males stop behaving aggressively toward newborn pups approximately one gestation length after mating—whether or not a particular newborn is his own offspring (vom Saal 1985; Perrigo et al. 1990). As seen in figure 3.5, female lions show heightened sexual activity during the first few months after a takeover, and this behavior may assure that every member of an incoming coalition has mated before the conception of the first batch of cubs. However, females seldom mate with multiple males after the males' first year in residence (figure 3.5), so copulation may not be necessary to inhibit infanticidal behavior toward later-born litters, and inhibition may instead be perpetuated by continued cohabitation (see Elwood and Kennedy 1994).

Although we never continuously followed a female day-and-night throughout a complete estrous period, there is one clear example where we can confirm that an incoming male changed his behavior from intolerant to protective without ever copulating with females of the new pride:

26-Apr-1982 07:00. I find five new males (Nafasi & Naibu, and their 3.75-yr old sons, SMD, SME & SMF) with three adult females and a subadult male of the BF pride. Nafasi guards Barabara about 300 m from the main group. Naibu guards Barbie, and Barbara rests beside SME and SMF. SMD chases away the subadult male then starts back toward the main group. Barbara crouches down and growls as if SMD were a trespassing female. SMD looks like a female from a distance, but SMD is a genetic male—cell culture has revealed an absence of Barr bodies, which are diagnostic of XX females. Even though SMD lacks external genitalia and a mane, SMD has normal levels of testosterone and shows typical male behavior: with raised neck and tucked chin, SMD walks to within 20 m of Barbara. She responds to SMD's approach and solicits mating. SMD straddles her briefly, but intromission is impossible. SMD stands beside her for a few moments then walks to a nearby bush and urinates in the typical posture of a male marking his territory, but SMD's urine streams in a low arc beneath the branches.

22-Nov-1984 10:00 a.m. SMD is with Naibu, SMF, two BF females, and a 2-mo old cub, BFG. SMD's behavior toward BFG is indistinguishable from the other two males.

Coalitions often reside in multiple prides simultaneously (see below), and, when annexing a second or third pride, males remain tolerant of their progeny in the first pride, even during periods when they kill their predecessors' cubs in the new prides. Thus, male lions do not have a simple on/off switch where they are either infanticidal toward every cub they encounter or universally tolerant. Instead, their infanticidal behavior depends on their personal experience with the mothers of the cubs, as seen in the following anecdote involving a coalition of four Serengeti males during an almost absurdly complicated period in their reproductive career:

16-Jun-1980 06:10. Anne Pusey finds Sura, a female from the SE pride, vigorously defending her three 6-week-old cubs against a male named Plato. Less than 50 m away, one of Plato's coalition partners, Planor, is consorting with a female from the SR pride. Plato and Planor belong to a coalition of four males that are simultaneously resident in six different prides and have resided in the SE and SR prides for 6 and 7 mos respectively. As Planor continues to guard the SR female, Plato overcomes Sura's defenses and kills two of her cubs. The third cub escapes into the brush. Sura eats both of the victims. Planor has made no attempt to intervene, despite the cubs having been fathered either by himself or one of his other two coalition partners. Plato had only been seen with the SE pride on a single occasion two weeks earlier, and he is the only member of the coalition that has not previously been seen with Sura. Until today, Plato has mostly been seen with the MS pride (where his coalition has been resident for 3 years).

3-Oct-1980 17:50. Plato mates with Sura, as his companions, Plito and Ploto, feed from a buffalo kill alongside a female from the MS pride.

A similar case was reported in a coalition of six males called the Mapogos in South Africa's Sabi Sands Conservancy (Huerta 2015). The Mapogos simultaneously resided in a large number of prides, and all but one of the males, named Mr. T, moved freely between each pride. Mr. T had only infrequently associated with his coalition partners over the preceding year, and he killed several of their cubs when he first entered a recently annexed pride. Like Plato, Mr. T maintained membership in his coalition with no further infanticide in any of the resident prides.

The Responsibilities of Maintaining Residency

To provide effective protection against infanticidal adversaries, males must remain in the general vicinity of their cubs (figure 3.7b). But their propensity to reside simultaneously in multiple prides not only dilutes their availability to the females in each pride (figure 3.7a) but also to their cubs (figure 5.9a). As discussed in chapter 8, adjacent prides remain closer to each other when they "share" the same coalition than when they are controlled by separate coalitions (figure 8.9e). Indeed, the average spacing between shared prides, approximately eight km, is within the audible range of the lion's roar, so mothers should generally be able to alert their wayward husbands when they need reinforcements against outside threats, which may explain why the cubs of co-resident coalitions suffer only a slight (albeit significant) reduction in survival compared to the cubs of single-family males (figure 5.9b).

Offspring spend less time with their fathers after they are fourteen to fifteen months of age (figure 5.9a), by which point they are mobile and far less vulnerable to infanticide (figure 3.2a). Before their first birthday, cubs spend more time with males from larger paternal coalitions (figure 5.9c), presumably because there are more males to monitor their pride(s). Time spent with fathers also increases with the size of the mothers' pride (figure 5.9d), thus, at least some of the reproductive advantages from forming larger coalitions and larger prides may result from the greater availability of paternal protection in encounters with outside males.

Resident coalitions are typically only ousted by larger rivals (figure 2.11b), but co-resident coalitions like the Mapogos and Plato-Plito-Pluto-Planor may reduce their holdings voluntarily, as evidenced by replacements in abandoned prides often being smaller coalitions ($n = 76$ cases). The choice of which pride to abandon is largely based on the need for continued paternal care: downsizing males typically surrender cub-less prides while retaining prides with an average of three cubs (figure 5.10a). In the subset of cases where males do abandon a pride with cubs, the coalition typically surrenders a pride containing fewer females, and, in almost all of the remaining cases, the males abandon prides where they had been resident for over two years (i.e., prides that would soon contain sexually mature daughters) (figure 5.11a). Thus, downsizing coalitions generally recognize the need to provide protection to vulnerable cubs (figure 5.10a), and otherwise base their departure decisions on future reproductive opportunities both by remaining with the largest prides and avoiding father-daughter matings (figure 5.11a, also see below).

Resident coalitions almost always follow a simple sequence of entry, exclusive residency, and permanent departure ($n = 510$ coalition tenures). But in twenty-two cases, a resident coalition only departed temporarily, being succeeded by a second coalition that maintained residency for a median of forty-seven days (range: 2–165 days) before the return of the first coalition, and, occasionally, a second departure and a further back-and-forth ("yo-yo"). In almost every case, one or both of the yo-yo-ing coalitions were co-resident in multiple prides (figure 5.11b), and, on average, these periods of uncertainty persisted for 1.31 years (range: three weeks to three-plus years). The eventual outcomes are mostly determined by coalition size (with the larger coalition winning sixteen of twenty-one cases where coalitions differed in size), but the five exceptions are revealing: the losing larger coalitions had typically been resident for at least four years (figure 5.10b)—and had fathered mature daughters (figure 5.11c), again suggesting that males avoid father-daughter matings by surrendering (albeit reluctantly in these particular examples) prides to competitors in smaller coalitions.

These back-and-forth residencies sometimes persist long enough for the losing coalition to sire offspring that are exposed to the winning rivals. Though the losers' cubs suffer higher mortality than vulnerable cubs born before or after the resolution of the yo-yo (figure 5.10c), the fact that so many of these cubs survive the back-and-forth

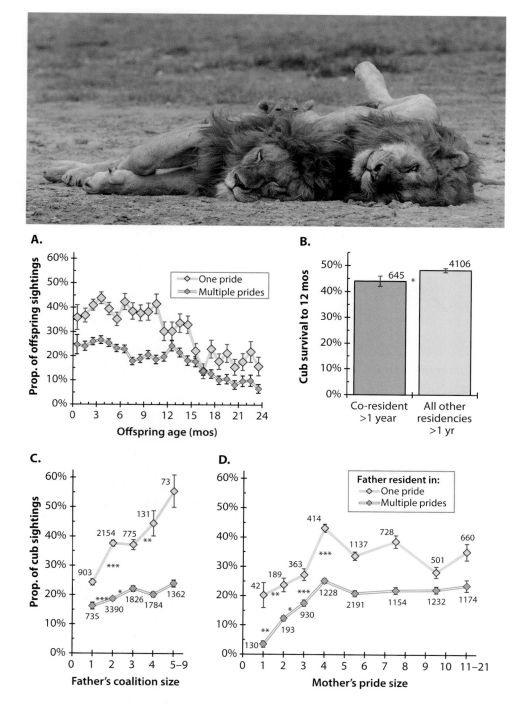

FIGURE 5.9. Fathers and offspring. Effects of co-residency on the proportion of observations with fathers are highly significant ($p < 0.0001$) in all tests. **A.** Resident males spend less time with their offspring as the cubs approach independence ($p < 0.0001$). **B.** Cubs suffer lower survival if their fathers remain co-resident for over a year. **C.** Males in larger coalitions spend more time with cubs, but the increase is stronger in coalitions that are resident in only one pride. **D.** Fathers spend increasing amounts of time with cubs as pride size increases to four females. In **A.**, **C.**, and **D.**: n = number of sightings. In **B.**, n = number of cubs.

A. Voluntary departures

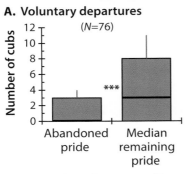

B. Tenure length at start of "yo-yo"

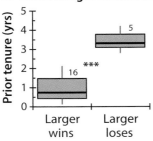

C. Cub survival vs. unstable residency

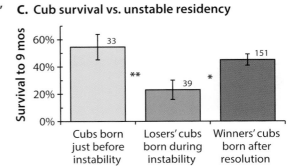

FIGURE 5.10. Uncertain residencies. **A.** Down-sizing co-resident coalitions abandon prides containing significantly fewer cubs than the prides they retain. n = number of coalitions. **B.** Following an unstable residency ("yo-yo"), larger coalitions that prevail against smaller coalitions have typically resided in that pride less than 1.5 years, whereas coalitions that lose to smaller rivals have usually resided greater than 3 years. n = number of coalitions. **C.** Cub survival before, during, and after yo-yos. To be considered, cubs born before the instability must be less than nine months of age when the usurpers first enter; losers' cubs must be conceived during the losers' tenure and be under nine months of age at the end of the yo-yo; winners' cubs are born after the final resolution. Cubs born before the yo-yo have higher survival than losers' cubs; winners' cubs have higher survival than the losers'. n = number of cubs. Top photo shows a resident male with cubs born during a period of instability that were not fathered by his coalition.

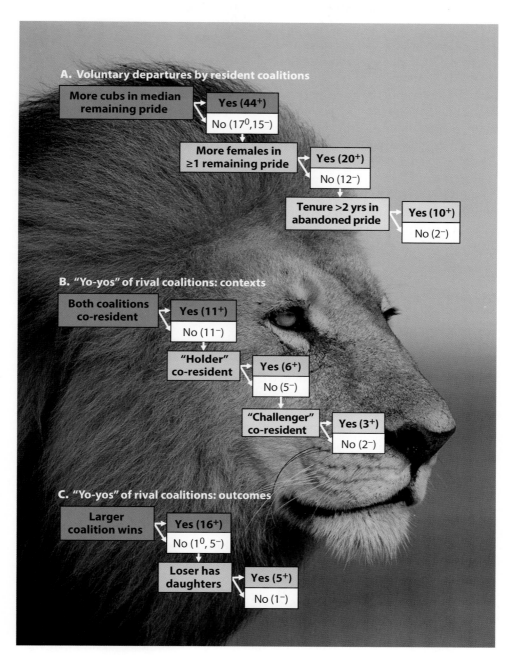

FIGURE 5.11. Flow charts of uncertain residencies. **A.** There were seventy-six cases where a larger coalition was co-resident in multiple prides when it was replaced by a smaller coalition. Thus, the males presumably emigrated voluntarily and chose between preexisting residencies. In forty-four cases, the coalition surrendered prides with fewer cubs than resided in the rest of their prides. In the remaining thirty-two cases, the males lacked cubs in any pride in seventeen cases and surrendered *more* cubs in fifteen. In twenty of these thirty-two cases, the males retained at least one larger pride, but they retained smaller prides in the other twelve. In ten of these twelve cases, the males had resided in the abandoned pride for at least two years. **B.** In half the twenty-two yo-yos, both coalitions were co-resident in multiple prides; in nine of the remaining eleven cases, one of the two coalitions was co-resident. **C.** The larger coalition won sixteen of twenty-two yo-yos, and in five of the other six cases, the larger coalition abandoned prides containing mature daughters to smaller coalitions.

127

between the two coalitions (including cubs born shortly before the onset of the yo-yo) suggests that the competing males are less infanticidal than usual. Toleration toward each other's cubs likely results from several factors: first, the relatively high survival of the initial residents' cubs suggests that the challengers are less confrontational during their first entry into the pride—perhaps recognizing that the mothers might be able to solicit help from the initial coalition. Second, cuckolded winners may tolerate the interlopers' cubs because they have retained the familiarity with the pride females that they had developed during their prior residency. Given that yo-yos are so rare (approximately 4% of all residencies) selection may be too weak to result in an innate safeguard against this sort of cuckoldry—on the other hand, the costs of making mistakes may be sufficiently high to favor a blanket strategy of tolerance toward any cubs born to familiar females, similar to the generalized provisioning behavior by reed warblers in cases of brood parasitism by cuckoos (Davies et al. 1996).

I LOVE YOU, DAD, BUT...

Nearly 40 percent of paternal coalitions of four or more males successfully maintain tenure in the same pride long enough (approximately four years) to potentially mate with sexually mature daughters (figure 2.12a). However, as will be discussed in chapter 10, father-daughter inbreeding is virtually nonexistent in the Serengeti and relatively uncommon even in the isolated and inbred Crater population. Father-daughter matings are largely avoided in three ways. First, maturing daughters are far more likely than their mothers to engage in consortships with extrapride males (figure 5.12a). Older females are also somewhat more likely to mate with extrapride males after their resident coalition has remained with their pride for over 2.90 years; the coincident increase in outside matings by these "nondaughters" may result from the daughters attracting more males to the area and/or from the aging resident males being less able to keep the area clear of rivals. Either way, daughters show strikingly greater sexual interest in unfamiliar males than do their mothers. Second, when their fathers are still resident, dispersing females leave around the age of sexual maturity, whereas females that leave at male takeovers are significantly younger (figure 5.12b). Coupled with their greater interest in outside males, these maturing females often seem to be directly motivated to disperse *in order to* find unrelated mates. Third, as described above, co-resident coalitions frequently abandon prides containing maturing daughters (figures 5.10b and 5.11a, 5.11b, and 5.11c), and, when males leave larger prides to take up residence in smaller prides, they typically depart after a residence of at least two years (figure 5.12c). This latter pattern is particularly pronounced in coalitions of three or four males, which otherwise only rarely swap larger prides for smaller ones (figure 5.11a). Thus, long-time residents often clear the way for their maturing daughters to mate with unrelated males.

In contrast to daughters, sons sometimes retain life-long relationships with their fathers. For example, the intersex lion, SMD, along with SME and SMF, took over the BF pride in partnership with their fathers, Naibu and Nafasi, and the five lions remained together until the older males' disappearance. Although we only observed five cases of young males joining their fathers, these mergers were quite common if

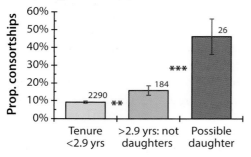

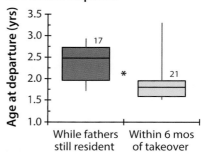

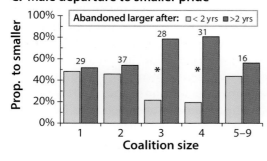

FIGURE 5.12. Fathers and daughters. **A.** Consortships with extrapride males. The proportion of consortships where females are guarded by extrapride males increases significantly after a resident coalition has resided in a pride for more than 2.9 years (the shortest possible tenure before daughters reach sexual maturity) ($p < 0.01$). Daughters are far more likely to consort with extrapride males than nondaughters ($p < 0.001$). n = number of consortships. **B.** Age of female dispersal. Females that emigrate while their fathers are still resident leave at sexual maturity, whereas females that disperse at male takeovers depart at significantly younger ages ($p = 0.02$). n = number of dispersing cohorts. **C.** Male abandonment of larger prides for smaller prides. Males that move from larger prides to smaller prides typically switch after a two-year residency (91 of 141 cases, $p < 0.0001$), but this is mostly due to the behavior of trios and quartets ($p \leq 0.05$). n = number of departures.

the paternal coalition had been forced to disperse from its only pride when the sons were between 3 and 4 years of age and the fathers were between 8.5 and 10 years of age (figure 5.13a). Over the entire study period, there were only thirteen cases matching this age distribution at the time of the fathers' departure, and sons almost always joined fathers that had only been resident in a single pride, but almost never joined if the fathers were co-resident in multiple prides (figure 5.13b). Thus, the fathers had to be young enough to take over yet another pride and the sons had to be old enough to contribute to their fathers' corporate strength. Once males have been ousted from their one and only pride, they become nomads that no longer show any form of territorial behavior, and they will readily team up with unrelated companions (see chapters 2 and 9), but an even better addition to a fading middle-aged coalition would be their own sons.

KEY POINTS

1. DNA analysis suggests that reproductive skew increases with coalition size. Although paternity was evenly divided within pairs, only two males per coalition fathered offspring in the two trios and one quartet, perhaps explaining why solitary males readily form pairs with unrelated companions but are more careful to find same-aged partners when forming trios and never form groups of four or more.
2. Consorting behavior is more highly skewed in trios than pairs and is also higher within pairs of related males than of nonrelatives. However, mating is less skewed in quartets than in trios, perhaps because quartets hold simultaneous tenure in multiple prides for prolonged periods, allowing greater opportunities for reproductive queuing and divergent mating preferences by coalition partners.
3. Despite pervasive competition for mating opportunities, there is no evidence of increased male mortality during periods of higher within-coalition competition. Instead, resident males mostly die during periods of intense between-group competition.
4. Demographic processes limit the number of like-aged males in a cohort and hence restrict the number of large coalitions in the population. Males born at least five months apart are most likely to split apart after dispersing from their natal pride and also associate less often with each other after gaining residence.
5. Rather than recognizing particular cubs as their own offspring or having experienced mating with particular females, male lions appear to rely on a rule of thumb based on familiarity with pride females to switch from infanticidal behavior to tolerance for cubs sired during their residency.
6. Males trade-off between protecting their current cubs and siring additional cubs in adjacent prides, though the costs of becoming co-resident in multiple prides appears to be small. When aging males surrender one or more of these prides, they give priority to protecting their cubs, followed by maintaining access to greater numbers of females.

FIGURE 5.13. Fathers and sons. **A.** Cumulative age plots of father-son coalitions. In all five cases, at least one son was between 3.0 and 40 years of age and at least one father was 8.5 to 9.8 years old. Ages are from the dates that the father-son coalitions entered their first prides. n = total number of sons and fathers. **B.** Proportion of fathers and sons with the same age structure as in (A) that teamed up, classified according to the fathers' residency. Sons were significantly more likely to join fathers that had only been resident in a single pride ($p < 0.05$).

7. Father-daughter matings are rare as daughters show heightened interest in extra-pride males when their fathers are still resident in their pride, and resident males are more likely to abandon a pride when it contains their adult daughters.
8. Although recruitment into the paternal coalition is rare under most circumstances, three-to-four-year-old sons routinely join their fathers' coalitions when the fathers have been ousted from their only resident pride at 8.5 to 10 years of age.

CHAPTER 6

THE LION'S MANE

The sparrow is sorry for the peacock at the burden of its tail.
—RABINDRANATH TAGORE

The lion's mane is a trait so iconic, so totemic, why else would the lion have been portrayed as the universal *King* of Beasts? Western heraldry, sports mascots, corporate logos—these images all show the male of the species, and they all accentuate the male's mane. The tiger may be larger, but the tiger is a creature of the forest—its stripes are meant to conceal. The lion's mane is a *proclamation*—a look-at-me-and-be-damned statement of male privilege. Surrounded by that crown of hair, the male's face appears foreshortened, imparting an aura of anthropomorphic power and royalty.

But what is the mane *for*? What purpose does it serve?

Male lions live a life of struggle; they face the constant threat of annihilation; they battle with enemies, squabble with companions. Do their manes protect their throats and necks from their opponent's teeth and claws? In 1871, Charles Darwin wrote, "The mane of the lion forms a good defence against the one danger to which he is liable, namely the attacks of rival lions." Or could the lion's mane simply be an *ornament* like the peacock's tail, a device to impress the ladies? Although Darwin was the first to recognize the importance of female preferences for exaggerated traits, it wasn't until 1972 when George Schaller wrote *The Serengeti Lion* that anyone seriously considered the possibility that a male's mane might signal his *quality* as a prospective mate: "The mane makes the male look impressive; it enhances his appearance, especially during a strutting display in front of another lion which may either be impressed or intimidated depending on its sex." However, Schaller then went on to write: "An important secondary function of the mane is protection during fights."

How can we distinguish between these alternatives?

First, we characterized the mane of each male in the three study populations. Individual mane hairs can reach 23 cm in length, some are black, and some are nearly white. The relatively short hair closest to the face tends to be orange, the longer chest and neck hair is usually dark, though some males are blond all over. Some manes look like mohawks, some flow evenly around the neck and shoulders. To quantify these differences, Peyton West organized panels of undergraduates at the University of Minnesota to score the color and length of each male's mane from ID photographs taken between 1964 and 2000. She then averaged and normalized the scores of all parts of the mane so that the mean across all males was 1.0. Second, we needed to estimate the risks incurred from fighting. We only rarely observed serious fights, but whenever we found a lion, we noted every visible wound—a bite wound on the back, punctures

TABLE 6.1. *Variables influencing mane length and coloration of males above the age of four years*

Independent variable	Effect		Interpretation
Length n = 126 males			
Born in woodlands	Negative	**	Shorter in warmer/humid habitat
Mane darkness#	Positive	**	Darker hair less prone to breakage
Annual temp 3–4 yrs of age	Negative	*	Shorter in warmer years of development
Male injured in photo	Negative	*	Loss of mane when wounded
Darkness n = 114 males			
Born in Crater	Positive	***	Darker in colder/food-rich habitat
Monthly temp when photographed	Negative	***	Lighter in warmer months
Born in woodlands	Negative	***	Lighter in warmer/humid habitat
Log (Age)	Positive	*	Darker with age (mostly from 4 to 6 yrs)
Darkness n = 68 males†			
Resident in plains	Negative	**	Lighter in food-poor habitat
Log serum testosterone	Positive	*	Darker with higher testosterone

Source: Reformatted from West and Packer (2002).
Note: Models are multiple regression; all variables are also significant in univariate analyses.
Mane color was included because dark mane hairs are thicker and presumably less vulnerable to breakage.
† Results reported separately because of the limited number of hormonal assays.
*** $p < 0.001$; ** $p < 0.01$; * $p < 0.05$

on the legs, gashes on the face and flank. The damage persists for weeks and leaves obvious scars on the survivors; dying lions often carry the fatal markings of combat. Third, we needed to determine whether the lions themselves were impressed by the length or coloration of a male's mane, so we contacted a Dutch toy maker, International BonTon Toys, to make a set of life-sized plush-toy lions that we could use to test the preferences of female and male lions when given a choice between large manes and small or dark manes and light.

It is convenient to think of the lion's mane as being analogous to a human beard, as both reflect similar physiological changes during puberty. Appearing at about the same age that testosterone levels start to rise, manes continue to grow and darken until about four years of age (figures 3.1a, 3.1b, and 3.1c). But unlike a beard, a mane can fall out if the male is badly wounded (see bottom inset to figure 6.1b) and manes are significantly shorter after wounding (table 6.1).[1] Mane color is not inherited from father to son (West and Packer 2002) but instead varies with diet, habitat, and testosterone (table 6.1), sometimes darkening or even becoming lighter well after the male has reached adulthood (figure 6.1d). Some coalition partners have manes that are virtually indistinguishable from each other, whereas at the other extreme a short-maned male

    1 David Bygott and Jeannette Hanby first noticed this phenomenon in the Crater when a wounded male completely lost his mane; they named him "Frog." We observed several additional instances of complete mane loss (including the male in figure 6.1b) but more commonly saw large patches of hair falling out from the neck or shoulders. In figure 6.1b, virtually every fully adult male with a mane length rating of less than or equal to 0.75 had suffered from wounding in the previous few months. Mane hair grows about 1.25 cm per month, so some of the longer-maned older males may have recovered from an earlier bout of mane loss (West and Packer 2002).

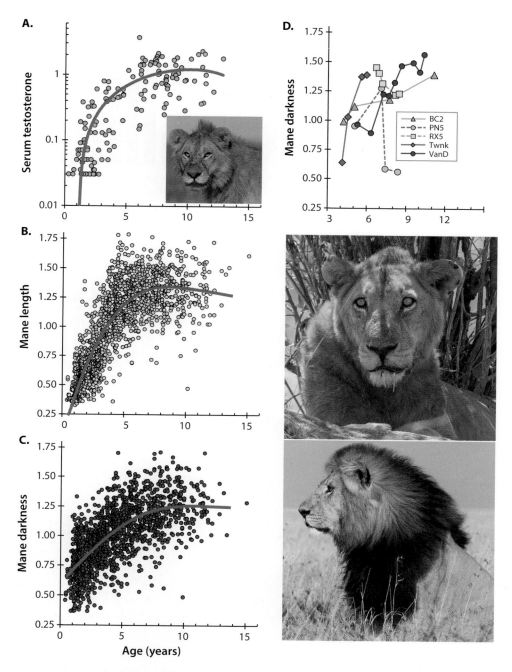

FIGURE 6.1. Mane development. **A.** Serum testosterone increases from one to four years of age. **B.** Mane hair begins to grow at one year and reaches full size around four years. **C.** Manes darken with age, but mane color varies from the earliest stages of growth. **D.** Age changes in adult mane color. Of forty-nine males photographed at least four times after four years of age, thirty-five grew darker with age. The typical pattern of darkening is illustrated by BC2, but some males became lighter (PN5 and RX05), and some grew darker over a short period of time (Twinkle and Van Dyke). All graphs contain multiple observations from the same individuals. **A.–C.** redrawn from West and Packer (2002); **D.** modified from West (2005). Note: the male in the middle photo completely lost his mane several weeks earlier.

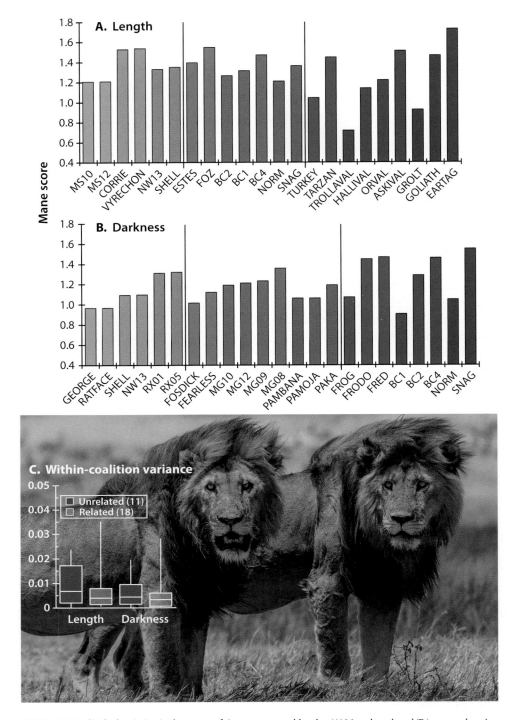

FIGURE 6.2. Individual variation in the manes of six-to-ten-year-old males. (**A**) Mane length and (**B.**) mane coloration vary within and between coalitions. Partners in the three left-most coalitions in each panel are most similar to each other, the middle three coalitions show intermediate levels of variation, and the right-most are the most variable. **C.** Differences in mane length and darkness between unrelated partners are similar to the variation between related partners.

may be paired with a long-maned companion (figure 6.2a) or a light-maned male with a black-maned partner (figure 6.2b), and, on average, mane length and color vary as much within coalitions of unrelated companions as within coalitions of close relatives (figure 6.2c).

MANES AND WOUNDING

Does the lion's mane primarily function as a shield? Males are certainly wounded at higher rates than females (see figure 8.6a), but lions target their opponents' backs and hindquarters during gang attacks on outsiders (figures 8.1 and 8.2), and coalition partners inevitably bite and scratch each other's faces during face-to-face, one-on-one fights over access to receptive females (figure 6.3). Lion-inflicted wounding (a puncture, cut, gash, or bite) is disproportionately common on their legs and faces—even in females and subadults, both of which lack manes—and subadults are also bitten more often than expected on their backs or hips (figure 6.4a). Males appear to only rarely be wounded on their shoulders, chest, neck, and forehead, though mane hair likely obscures all but the most conspicuous wounds in these areas. But across all age-sex classes, the mane area does not appear to be a particular target for attack, nor are wounds to the neck and shoulders especially dangerous: a wound to the mane area is no more likely to be fatal than a wound to the rest of the body (figures 6.4b, 6.4c).[2] Indeed, if the mane were so important as a shield, it would be unlikely to fall out when a male has been seriously wounded—thus providing no protection at a time of serious danger. Further, male leopards, tigers, and so on also fight with each other to gain and maintain access to breeding territories, yet none of these species have manes. Compared to these other species, lion males certainly face unique competitive interactions with coalition partners over mating access with individual females, but (a) this competition rarely escalates to overt aggression, as rivals respect their consorting partner's temporary "ownership" of a receptive female (chapter 2, also see figure 5.3); (b) even when such fights do occur, contact is largely restricted to the face (figure 6.3) rather than the neck, chest, and shoulders; and (c) the advantages of coalition formation are so strong (figure 2.12) that seriously injuring a coalition partner would reduce the long-term reproductive success of the winning partner, thus male lions have little incentive to engage their coalition partners in potentially fatal aggressive competition (see chapter 9).

Although we cannot rule out the possibility that the mane may have guarded against wounding at some stage in the lion's evolutionary history, we rejected the hypothesis that the lion's mane *primarily* serves a protective function and suggested instead that any minor reduction in wounding to the neck area is merely a byproduct of some other function (West and Packer 2002; West et al. 2006). But if the mane isn't a shield, what purpose does it serve? Could it instead be a way of showing off the male's prowess, his *quality*? After all, mane hair grows at about a centimeter a month (the same as human hair), and, given that the mane can fall out so easily, the mere presence of a large mane at least indicates that the male hasn't been badly wounded in several years. But what else can be deduced about a male from his mane?

---

2 The most obvious impacts of wounding are on solitary females and, to a lesser extent, solitary males (figure 8.6).

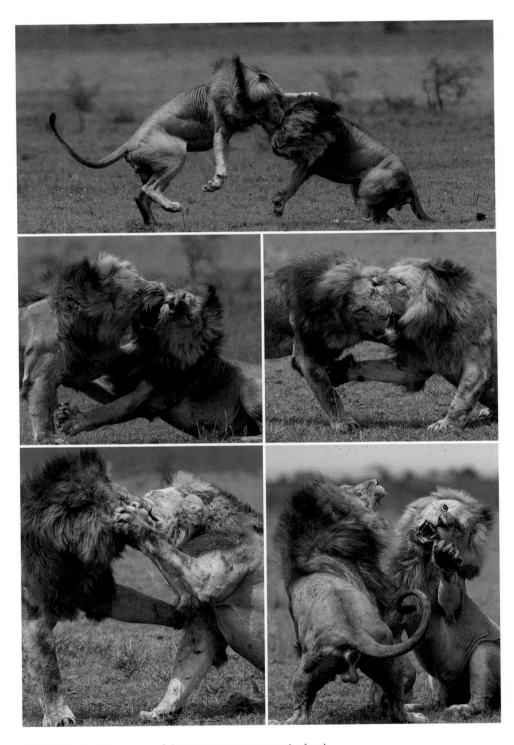

FIGURE 6.3. Coalition partners fighting over access to a receptive female.

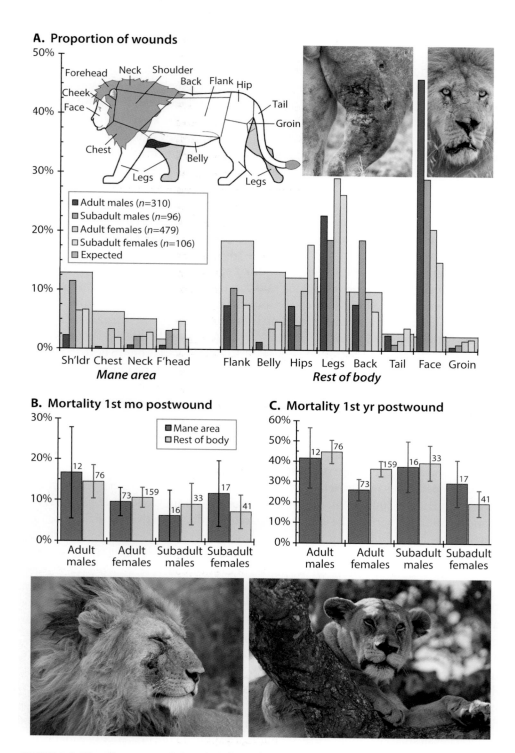

FIGURE 6.4. Wounding patterns. **A.** Location of wounds on the bodies of males, females, adults, and subadults. Expected proportion of wounds based on the surface area of each body part (gray bars). n = number of wounding events. Data from West et al. (2006). Mortality risks from wounding on different parts of the body: lions wounded on the mane area were no more likely to die in (**B.**) the following month or (**C.**) the following year (as purulent wounds may fester for months) when compared to lions wounded on other parts of the body. n = number of wounded animals. Data from West et al. (2006).

MANES VERSUS MALE QUALITY

Lions are easily overheated. Find any pride of lions in the middle of the day. If they're awake, they will almost certainly be panting. Watch a female lion chase a wildebeest or zebra, and the hunt often ends once she starts to overheat. Watch an adult male run any distance at full speed, and you'll soon see him foam at the mouth. Winter coats are well known for their excellent insulative properties that allow mammals to survive the arctic winter (Zimova et al. 2018), so a male lion sporting a full mane in the African sun is like wearing a fur vest in the middle of summer with his mane advertising that *he can take the heat.* It has long been recognized that sexual selection can favor the evolution of condition-dependent signals that honestly demonstrate how well each individual male can withstand environmental stress (Maynard Smith and Harper 2004), and the mane is an excellent indicator of the male lion's ability to cope with a meaningful physiological challenge.

The relationship between mane size and temperature has long been recognized: the British explorer, Frederick Selous (1908), noted that lions in the hotter parts of Africa have smaller manes. Now-extinct populations of Cape and Atlas lions lived at the colder limits of the species' global distribution and possessed truly extravagant manes, while lions in coastal equatorial African climates barely have manes at all (Caputo 2002). Indeed, climate and temperature underlie variations in both the length and coloration of manes even in the Serengeti and Ngorongoro. Our three adjacent study areas differ from each other in temperature and humidity: the Serengeti woodlands are at about 1,500 m elevation *vs.* 2,000 m in both the Serengeti plains and the Crater floor. The woodlands are closest to the humidity of Lake Victoria; the Crater is surrounded by highlands, where temperatures can approach freezing. Temperatures also vary with the season, being coolest in the first months of the dry season (June and July) and warmest just before the onset of the rains (October), and, like the rest of the globe, temperatures in northern Tanzania have increased over the past half century (West and Packer 2002). Males born in the hotter, humid woodlands have shorter manes than those born in the cooler study areas, as do males that were three to four years of age during hotter years, and males have darker manes in the coolest habitats and during the coolest months of the year (table 6.1). Patterson et al. (2006) also found similar effects of temperature on the manes of captive lions.

Peyton West used a thermal camera to directly measure the surface temperature of forty-three different lions in the Serengeti and Ngorongoro. She waited until dusk and pointed the camera at each lion's flank when it first stood up in the evening (inset to figure 6.5a). Controlling for ambient temperature, males have a significantly higher surface temperature than females (figure 6.5a). Thermal cameras only measure surface temperatures, so this difference presumably reflects the greater radiant heat being dissipated by the parts of the body that are not covered by the mane. Males in Kenya's hot and humid Tsavo National Park are often maneless, and we found that their surface temperatures are the same as in the Tsavo females (West and Packer 2002). Sex differences in surface temperature, therefore, result from the presence of a mane, not from any metabolic difference between males and females.

Peyton's thermal-camera study also revealed that surface temperatures are significantly higher in dark-maned males and in lions with larger belly sizes (West and

Packer 2002). Dark mane-hairs are significantly thicker than blond mane-hairs (figure 6.5b), which increases the insulative properties of darker manes. To test whether males avoid the heat-related costs of digestion by reducing their food intake in the hottest months of the year, we focused on the lions in Ngorongoro Crater, where prey abundance and average belly size are essentially constant throughout the year (figures 1.5 and 1.13a). As would be expected from the digestive thermogenesis generated by a high-protein diet (Halton and Hu 2004), Crater males feed significantly less in hotter weather, whereas Crater females feed at a constant rate throughout the year (figure 6.5c), and dark-maned males alter their food-intake rates to a much greater extent than do light-maned males (figure 6.5d). Across all three study areas, males have darker manes in areas of greater food richness (darkest in the Crater, lightest in the plains) (table 6.1), so it is possible that the males' seasonal variations in food intake (figure 6.5d) may help explain why manes tend to be lighter during the hottest time of the year (table 6.1).

In Zimbabwe's Bubye Valley Conservancy, Trethowan et al. (2017) implanted miniature temperature-sensitive bio-loggers to measure the lions' core-body temperatures. They found that the two sexes maintained similar average core temperatures over the course of their entire study, but they also showed that males drink considerably more water than females (drinking for longer periods and making more visits to water holes each day), which presumably allows the males to compensate for their greater thermoregulatory demands via higher levels of evaporative cooling. Trethowan et al. (2017) also confirmed that medium- and dark-maned males experience higher maximum core-body temperatures than light-maned males, that feeding at a kill raises the lions' core-body temperatures, and that the effect of feeding on body temperature is greater in males than females.

Finally, elevated body temperatures are known to cause abnormalities in mammalian sperm. Surface temperature of a male's flank is strongly correlated with the surface temperatures of his scrotum, and the darkest-maned males have significantly higher levels of sperm abnormality than light-maned males (West and Packer 2002). However, the magnitude of this effect is so small that it is unlikely to cause any measurable effect on fertility. As will be seen in chapter 10, a history of persistent inbreeding has had a substantial effect on sperm abnormality in the Crater lions,[3] yet litter size is almost exactly the same in the Crater as in the Serengeti (figure 4.13a). Thus, though darker manes confer clear physiological costs, these appear to be too small to impair reproductive performance in the subset of males that are able to grow them.

A male's mane tells a story. If it's long and flowing, he probably hasn't been seriously wounded in years. If it's very dark, he has high levels of testosterone, a steady supply of food, and an ability to withstand heat stress (table 6.1). But does any of this translate into greater numbers of descendants? Although we couldn't detect any relationship with mane length, dark-maned individuals are significantly more likely to survive the first year after being wounded, dark-maned coalitions maintain pride residence for significantly longer periods of time, and their *yearling* offspring are significantly more likely to reach their second birthdays (table 6.2). Cubs of dark-maned

---

[3] Samples from nine Crater males contained 50.5 percent abnormal sperm compared to 24.8 percent from eight Serengeti males, mostly because of a bent midpiece or a tightly coiled flagellum (Wildt et al. 1987).

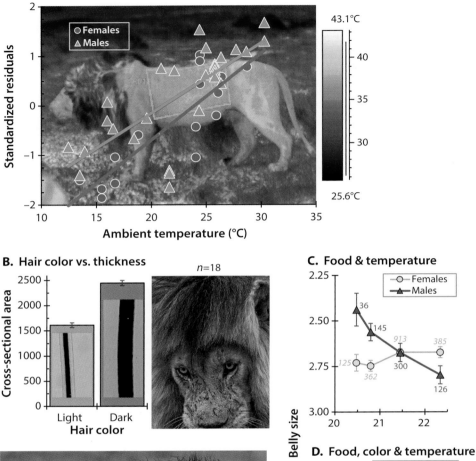

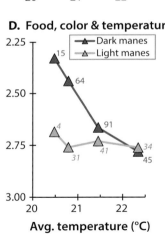

FIGURE 6.5. Manes and heat stress. **A.** Residuals of surface temperature after controlling for distance of the lion from the thermal camera. Each point is a separate individual. *Background photo:* Infrared image of Serengeti male. Image color denotes surface temperature; green box outlines the sample area. **B.** Dark mane hairs are significantly thicker than light mane hair in eighteen males. **C.** Females do not alter their food intake at different temperatures, whereas males significantly decrease their intake in the hottest months. **D.** Dark-maned males feed significantly more than light-maned males in the coldest months but not in the warmest months. *n* = number of sightings. All data from West and Packer (2002). Thermal photo and microscopic images by Peyton West.

TABLE 6.2. *Fitness parameters associated with mane darkness*

Independent variable	Effect		Interpretation
Annual survival *n* = 110 wounded males			
Age when wounded	Negative	***	Younger more likely to survive
Coalition size	Positive	*	Larger coalitions better able to survive
Mane darkness	Positive	*	Dark-maned males better able to survive
Lifetime tenure *n* = 71 coalitions			
Coalition size	Positive	***	Larger better competitors
Average darkness	Positive	*	Darker better competitors
Yearling survival *n* = 443 yearlings			
Exposure to male takeover	Negative	***	Mortality at takeover
Mane darkness of paternal coalition	Positive	*	Darker better protectors
Maternal mortality	Positive	*	Orphans less able to survive
Cub survival *n* = 155 litters			
Ngorongoro Crater:			
Exposure to male takeover	Negative	***	Mortality at takeover
Mane darkness of paternal coalition	Positive	**	Darker better protectors
Serengeti Plains:			
Exposure to male takeover	Negative	***	Mortality at takeover
Mane darkness of paternal coalition	Negative	**	Darker more possessive at kills

Source: Reformatted from West and Packer (2002).
Note: All variables are also significant in univariate analyses.

coalitions are also more likely to reach their first birthday in the Crater, but the cubs of dark-maned coalitions are less likely to survive in the Serengeti plains (table 6.2). The effect of mane color on cub survival likely reflects male behavior at kills. Recall that males dominate all other age-sex classes at kills (figure 2.5a), and that carcass size is typically largest in the Crater and smallest during the dry season on the plains (figure 1.6). Although dark-maned males exclude adult females from larger carcasses, they allow their cubs to feed for significantly longer periods than do light-maned males, but dark-maned males exclude everyone, even their own cubs, from small carcasses (West and Packer 2002).

The lack of any significant relationship between mane *length* and reproductive success may seem surprising, but a male that has never lost a fight might actually have avoided getting into any fights in the first place. Thus, a long mane could potentially reflect cowardice rather than fighting prowess. More important than hair length, coloration along the shafts of the mane hair provides a continuous chronology of his physiological state over the past few months, and his current physical condition can be assessed from the color of his hair, even if it's just a few centimeters long (think of the three-day stubble of a sexy male film star). Although mane hair sometimes lightens after wounding (e.g., Snaggle-puss in figure 6.1d), dark-maned males are generally able to recover from their injuries (table 6.2).

Across all three study areas, dark-maned coalitions are more successful at maintaining tenure and protecting their yearling offspring, but dark-maned coalitions only significantly enhance first-year cub survival in the Crater, whereas the cubs of dark-maned coalitions suffer lower survival on the Serengeti plains. While the Serengeti woodlands and Ngorongoro Crater both produce excess subadults that disperse to other areas, the harsh plains habitat is a population "sink" (figure 9.2f) that is continuously replenished by lions entering from elsewhere—thus any offspring that reach maturity on the plains face an uncertain future. From the point of view of the entire population, the advantages of higher survival in a source population will far outweigh the costs of lower survival in a sink (Dias 1996). Further, 18.5 percent of 130 coalitions that gained residence in a plains pride also resided in one or more of our woodlands study prides at some point—and a proportion of the remainder almost certainly gained residence in a woodlands pride located outside our long-term study area sometime during their reproductive career—so any reduction in cub survival suffered during their residency on the plains would have been balanced by the advantages from improved yearling survival while residing in a woodlands pride.

WHAT DO THE LIONS THINK?

Mates and rivals can be confronted with coalition members displaying conspicuously different manes (figure 6.2a, 6.2b). Do females prefer longer manes over shorter manes? Dark over light? Are males more intimidated by one type of mane rather than another? To find out, Peyton West set out pairs of life-sized toy lions with manes that varied in either length or coloration (photos in figure 6.6) and broadcast recordings of feeding hyenas to attract nearby lions.[4] Peyton was careful to test the behavior of single-sexed groups, as we expected the females' responses to contrast sharply with those of the resident males. And, indeed, females advanced straight toward the dummies, sometimes waving their tails as they do when approaching a potential mate, and sometimes even sniffing under a dummy's tail. Males, on the other hand, treated the dummies as opponents, being much more cautious and sometimes circling around so as to approach from behind, even "pussy-footing" by threading their paws very carefully through the tall grass so as not to make a sound. Females showed a significant preference for the darker-maned dummies, whereas males consistently approached the lighter manes (figure 6.6a). Females showed a slight preference for long-maned males, while males significantly preferred shorter-maned dummies (figure 6.6b). The contrast between the two sexes was significant for both coloration and length.

Can we accept these patterns as a legitimate reflection of lion behavior? Consorting males only guard one female at a time, so on the few occasions when there are

---

4 These experiments followed a long tradition of presenting inanimate objects to animals. In the earliest studies, ethologists were able to trigger responses in fish, lizards, and birds even with crude wooden models (reviewed by Tinbergen 1948). Although it might seem doubtful that any mammal could be so easily manipulated, wild chimpanzees had been shown to respond to a stuffed leopard by beating it with sticks (Kortlandt 1972), and we had used a taxidermically mounted lion to elicit aggressive responses by resident males (Grinnell et al. 1995, chapter 8). Our stuffed lion was quickly damaged, and we lacked a source of replacements for Peyton's choice experiments, so we used a set of life-sized models custom-built by International BonTon Toys. However, each lion could only be fooled once: in contrast to a playback experiment where an unseen stranger could have conceivably departed by the time the lions arrived, the dummy was available for close-up inspection and was obviously not real. Thus, we took the dummies to Tsavo in order to obtain an adequate sample size (see Packer 2015).

more receptive females than resident males in a pride, an "excess" female is freely able to move to any male she chooses (bottom photo in figure 6.6). In thirteen of fourteen such occasions, the excess female mated with the darkest male in the coalition (but showed no preference for longer manes) (figure 6.6c), thus mirroring the results of the dummy tests. We never saw a fight between rival coalitions that was analogous to the male dummy tests, but when we broadcast the roars of unfamiliar females to coalitions of resident males, the darkest male led the way significantly more often (figure 6.6c) and dark-maned males had significantly higher levels of testosterone (table 6.1), suggesting that by preferring the lighter-maned dummy, males were opting to approach the less threatening opponent.

WHY LIONS? WHY MANES?

The clear preference of female lions for dark-maned males, over an evolutionary timescale, would provide just the sort of selective pressure that could result in the King of Beasts becoming a mere Boy Toy burdened with a costly trait that signals his current physiological "quality" to the ladies. A female that encourages a dark-maned male to remain resident in her pride will improve her chances of successfully rearing her next set of cubs and passes on the "honest indicator" as well as her mating preference to the next generation. As much as possible, we staged the dummy tests with one subject at a time, as the lions were only fooled once by a life-sized stuffed toy. So as far as the males knew, they were approaching a potentially fatal situation of only one of "us" versus two of "them," and they seemed especially sensitive to any difference between their fake opponents as they responded to both coloration and length.

But why lions? All cats compete intensely for mates, which is why male leopards, tigers, and jaguars are presumed to be so much larger than females (Gittleman and Van Valkenburg 1997). But the lion is the only felid where females are routinely exposed to multiple males and, hence, able to make any sort of active mate choice. In nonhuman primates, the most extreme examples of conspicuous decoration are the bright coloration of male mandrills and the face-pads of male orangs, and both species have social systems where females are exposed to multiple males (Dixson et al. 2005; Grueter, et al. 2015). Ancient European cave paintings portray the extinct cave lion as a group-hunting predator, so cave lions were presumably as social as modern lions, but the paintings also show that male cave lions were maneless (figure 6.7) (Packer and Clottes 2000). There is no way of knowing whether cave lions also formed male coalitions, but, *if not*, females would have been unable to choose between companions in the same way as modern lions and thus could not have fueled the selective pressure favoring the evolution of male ornamentation. However, the lack of heat stress in Ice Age Europe also meant that a mane would not have been as useful as an indicator of male quality. Modern lions in cool climates can grow manes that extend much further back on their torsos and even cover parts of their bellies. But cave lions show clear thermal adaptations in terms of a thick undercoat of hollow-coiled fur (Boeskorov et al. 2021), thus a mane might not have imposed sufficient thermal costs to have imparted any useful information about the male's quality. On the other hand, thick, cold-adapted body fur might not have adequately radiated excess heat from the un-maned

FIGURE 6.6. Manes and behavior. **A.** Female lions significantly preferred darker manes over lighter manes; males preferred lighter manes. **B.** Females showed a slight preference for longer manes; males significantly preferred shorter manes. **C.** *Left:* "Excess" receptive females preferred to mate with darker-maned males. *Right:* Darker-maned males more often led the response to playbacks of female roars. In **A.** and **B.**, n = number of separate tests; in **C.**, n = number of occasions/playbacks. All data from West and Packer (2002). *Photos: Top:* The arrival of the four dummies from Europe (from left to right: short dark, long blond, long black, short blond). *Middle:* Test female approaches black-maned dummy. *Bottom:* Consorting male is solicited for mating by an "excess" female.

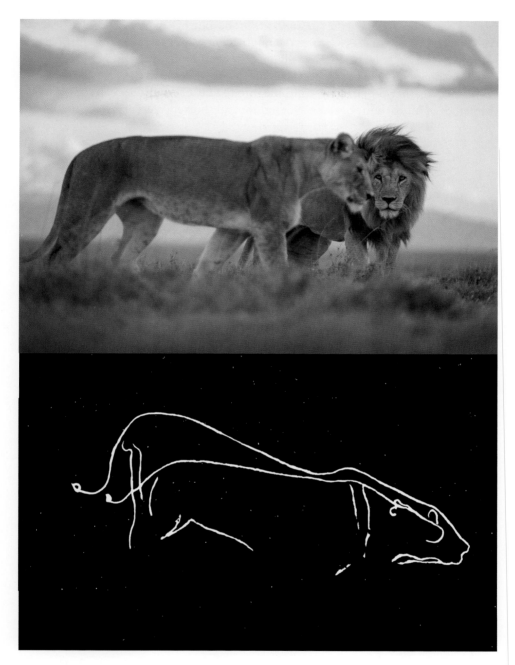

FIGURE 6.7. Consort pairs through the ages. *Top:* Consort pair in the Serengeti. *Bottom:* Consorting cave lions from the Grotte Chauvet; note the male's scrotum and lack of mane.

parts of their bodies during the summer, so the addition of a mane would have been *too* costly (Nagel et al. 2003). Perhaps as the tundra continues to thaw in this era of global warming, Russian paleontologists will unearth a mummified coalition of adult male cave lions that will help to resolve this issue. Meanwhile, with the current trajectory of warmer weather projected to continue into the next century, we can expect to see the manes of future lions become ever lighter and shorter (West and Packer 2002).

KEY POINTS

1. The onset of mane growth coincides with rising testosterone at adolescence. Manes reach full length by about four years of age and darken until about six years of age. Injury can cause the mane to fall out, and manes can also become lighter after males reach maturity.
2. Lions do not target the mane area during one-on-one fights or gang attacks, nor are wounds to the mane area more likely to cause mortality, even in age-sex classes without manes. Thus, the mane does not obviously function as a shield against wounding.
3. Black mane hair is thicker than blond hair, and black-maned males show signs of greater heat stress, including higher surface temperatures, reduced feeding in hot weather, and higher levels of sperm abnormality.
4. Mane length and coloration are both sensitive to local climate, being shorter and lighter in warmer areas or years. Darker manes are also associated with better nutrition and higher testosterone levels.
5. Dark-maned males are more likely to survive wounding, they maintain pride residence for longer, and their offspring show higher survival in most circumstances. Females prefer dark-maned males over light-maned males, whereas males avoid dark-maned and long-maned males.
6. The lion's mane appears to provide an honest indicator of a male's overall "quality," whereby females gain greater reproductive success from preferring darker-maned males, while males minimize the risks of injury by preferring lighter/shorter-maned opponents.

CHAPTER 7

FORAGING BEHAVIOR

Even strong young lions sometimes go hungry.
—PSALM 34, NEW LIVING TRANSLATION

Perhaps nothing has pervaded the popular perception of a lion pride as much as the notion of a well-oiled machine, the females working effortlessly in the common pursuit of a lone prey animal: the paragon of cooperation, the model of efficiency, the inevitable demise of the unwary prey. Lions certainly hunt together, and their hunts often do show signs of cooperation, but are they together in order to hunt? Or do they hunt together because they are already in a group? And how cooperative are they, really? At one extreme, lions in Namibia's Etosha National Park display a remarkable division of labor when hunting springbok: females employ a pincers strategy, with one or two females flanking around to the left, others flanking to the right, and the remainder moving directly forward—and the same females follow the same individual roles hunt after hunt (Stander 1992). But at the other extreme is the sight of a lone Serengeti female rising up from a sleeping group of lions and outmaneuvering her unwitting prey while her pridemates merely watch from the sidelines—and then run up to join her at the feast.

Group hunting has been the most difficult aspect of lion behavior to study: we can mimic territorial invasions using playback experiments; we can even use life-sized toys to study the males' manes—but there is no meaningful way to set up a standardized *hunt*: we can't present a standard-issue wildebeest to a particular group of lions at a specific point on the map and let them chase after it. Whereas the Etosha lions follow stereotypical hunting tactics while repeatedly hunting the same prey in a homogeneous habitat, the Serengeti lions seldom catch the same species twice in a row, and their hunting behavior is far more varied. But most problematic has been the fact that lions mostly hunt at night (figure 1.16b), especially moonless nights (figure 1.17), and they generally hunt in areas with sufficient cover to remain hidden from view (figure 9.1)—even from researchers with night-vision devices. Recent advances in satellite telemetry have made it possible to retrace the precise movements of each member of a hunting group (see Hubel et al.'s 2016 study of wild dogs in Botswana), but even here, it hasn't been possible to record the behavior of individual prey animals prior to their capture—and we focused on this topic well before satellite collars became available.

In this chapter, I present the results from our varied studies on lion feeding and hunting behavior, first, as background, second, to directly test whether lions gain sufficient benefits from group foraging to explain the origins of pride living (as traditionally assumed, e.g., Kleiman and Eisenberg 1973), third, to highlight the fact that the lions' fission-fusion social system allows them to forage solitarily when the need arises, fourth, to show the circumstances when the Serengeti lions hunted cooperatively, and, finally, to present a theoretical framework that successfully predicts the circumstances when group-living hunters *should* be most cooperative.

BACKGROUND

Lions show pronounced sex differences in hunting behavior: the smaller, faster females are adept at capturing wildebeest, zebra, gazelle, and warthog, all of which can be caught by lone females (figure 7.1). However, when pursuing the much larger and more dangerous Cape buffalo, females always hunt in groups (figure 7.2). Male lions are slower and more conspicuous than females but capturing slow-moving prey like buffalo requires *strength* rather than stealth (figure 7.3). Indeed, the lone female in figure 7.3 had spotted a lame bull buffalo, traveled half a kilometer to rouse her four resident males, then led them back to the prey and watched as they eventually pulled it down:

> **10-Jul-1984 09:00.** I find SHI, an adult female of the SH pride, as she is walking up to the four males of the BC coalition. She vigorously head-rubs several of the males then walks determinedly northwards, occasionally looking back at the males as if to make sure they follow. She moves to within 20 m of a lame bull buffalo, stops, and looks back once more at the trailing male lions, two of which catch up and continue right past her. One male leaps on the buffalo's back; the second bites hold of its hind legs. But then, a healthy bull buffalo rushes up and bashes the leg-clasping lion with his horns. The two male lions retreat, chased by the healthy bull. The healthy buffalo returns to the now-free lame buffalo and stands guard for a few minutes, snorting at the lions, before moving off. The lame buffalo is unable to follow, and two male lions jump on his back. The other two males soon pile on, and the buffalo crumples to the ground under the sheer weight of the four males. The female lion never participated, only joining at the kill after the males have suffocated their prey and opened its abdominal cavity.

Of course, females pull their weight, too, provided there are enough of them, and this observation even implies a division of labor between the sexes:

> **26-Jul-1979 09:50.** Anne Pusey and I see six females of the Munge Pride sitting within 25 m of three bull buffalo lying together in the grass. The middle buffalo sees the females, and all three get up and start to run off. Two females catch up with the trailing buffalo, as the other four females follow close behind. One female hops on the back of the buffalo but soon falls off. The buffalo runs another 100 m when all six females catch up, jump on its back, and it falls on its side. The females remain near its hindquarters, avoiding its head and legs, as the two pride males arrive. One male bites its throat; the other bites its face. The buffalo rolls on its back kicking, several females crawl onto its stomach while the remaining females keep

FIGURE 7.1. Lone females capturing wildebeest, zebra, gazelle, and warthog.

FIGURE 7.2. Group hunting of Cape buffalo.

FIGURE 7.3. Female lion watches as four males pull down a Cape buffalo. The female had led the males to the prey, which was hobbled by a bad leg.

hold of the back end, and the males remain clamped to the front end. The buffalo stops kicking, and three of the females leave to fetch the cubs as the rest of the pride starts to feed.

Figure 7.4a summarizes the group composition at 463 observed kills of the five most common prey species. Whereas female-only groups were the norm at wildebeest, zebra, warthog, and gazelle kills, nearly half of buffalo kills were made by mixed-sex groups, with all-male groups nearly as common as all-female groups. Thus, while male lions have a reputation for contributing little to their families' food supply, they sometimes provide substantial windfalls.[1]

Lions appeared to scavenge each of their major prey species at different rates (figure 7.4b), but these differences likely reflect observational bias. Although most of the scavenged carcasses were Thomson's gazelle (figure 7.4b), the vast majority of our long-term data were collected during the daytime, when cheetahs and wild dogs are active, and these two carnivores primarily capture gazelle. From our nighttime observations, we know that lions commonly scavenge wildebeest and zebra from spotted hyenas during the hours of darkness (figure 1.16c), but it was usually impossible to determine the origins of any carcass that the lions had obtained before we arrived in the morning, so lions likely scavenge these two species more often than indicated in figure 7.4b. Lions sometimes scavenge buffalo from neighboring prides, some of which may also have occurred prior to our arrival, but even if we have underestimated the proportion of buffalo carcasses scavenged by those specific individuals, the vast majority of buffalo in the Serengeti and Ngorongoro were killed by lions—and if the females caught the buffalo by themselves, they did so as groups.

HUNTING IN GROUPS—OR NOT . . .

What are the advantages of hunting in groups? Does each lion eat more meat each day if she hunts with her pridemates? Or does group foraging provide benefits from a steadier food supply that reduces her risks of starvation? Lions only feed about once every 2.33 days,[2] so a run of bad luck could potentially have serious consequences. Schaller (1972) had reported higher hunting success by female pairs and trios than by singletons, and several authors subsequently analyzed his data to predict the foraging group sizes that would maximize individual food intake rates while minimizing the variance (Caraco and Wolf 1975; Clark 1987; Giraldeau and Gillis 1988). However, Schaller's data were opportunistic—he had noted the success of any hunt he happened to observe; he hadn't set out to measure food-intake rates through time—and he also lacked the means to follow the lions at night—so his observations were necessarily limited in scope and unsuitable for estimating the precise payoffs from group foraging.

---

1 Our coverage of male hunting behavior was relatively sparse, as we mostly radio-collared females. Thus, figure 7.4a underestimates the overall proportion of all-male kills compared to all-female kills.

2 Jeannette Hanby and David Bygott conducted a series of "four day follows" of the lions in Ngorongoro Crater and Serengeti plains in the 1970s (Hanby et al. 1995). Dave Scheel and I later employed the same methodology to assess group-size specific foraging success. Combining the two datasets, the lions fed 83 times in 194 days of observation.

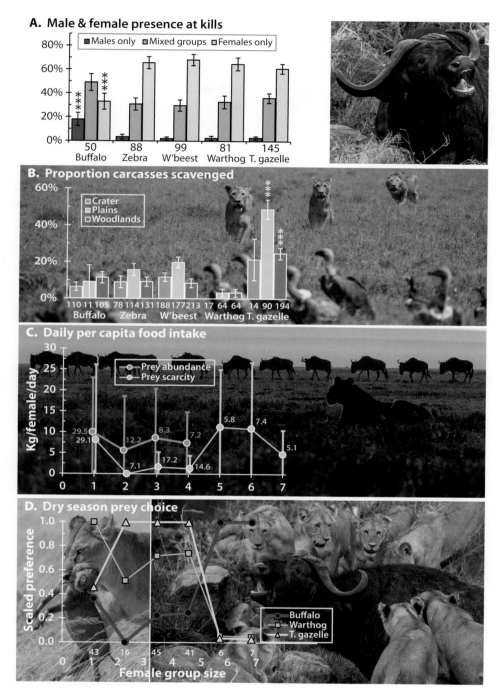

FIGURE 7.4. A. Group composition at observed kills. At buffalo kills, male-only groups (adult + subadult males) were most common ($p < 0.001$); female-only groups were least common ($p < 0.002$); and mixed-sex groups were more common than at wildebeest or zebra ($p < 0.05$). Sample size is number of kills. **B.** Proportion of carcasses of each species obtained by scavenging. Serengeti lions most often scavenged Thomson's gazelle ($p < 0.001$). Sample size is number of scavenged carcasses. **C.** Female food intake rates in the Serengeti. Intakes only varied with group size when migratory prey were scarce: females fed at higher rates when alone and in groups of five to six ($p < 0.02$), and these group sizes also showed higher variance in food intake ($p < 0.01$). Sample size is number of full days of observation. Redrawn from Packer et al. (1990). **D.** Dry season prey choice in the Serengeti. Scaled preferences calculated separately for each foraging group size: lone females mostly hunted warthog; groups of five to six females primarily hunted buffalo. Sample size is total number of hunts. Data from Scheel (1993).

For example, lone females may not be as successful as pairs during the day, but this contrast might disappear under the cover of darkness.

Thus, Dave Scheel and I conducted 143.5 full days and nights of observation in 1984–1987 to collect systematic data on per capita food intake rates in the Serengeti lions (Packer et al. 1990). Our observations consisted of a series of four-day "follows" of groups of one to seven females, starting either four days before or one day after the full moon, and using night-vision devices to observe the lions between sunset and sunrise. Large carnivores feed first on the prey's offal (liver, lungs, heart, etc.) then move on to the muscle tissue, and lions not only feed on prey that they capture themselves but also scavenge the remains of carcasses from other predators. We estimated the edible biomass of each carcass according to species and age-sex class (appendix 1) and used a scale of 1 (intact), 2 (some offal remaining), 3 (offal completely consumed), 4 (less than three-fourths of the muscle tissue remaining), 5 (less than half) and 6 (less than a quarter) to assess the amount of food obtained from each scavenged carcass. We then subtracted the amount of meat remaining when the lions lost or abandoned the remnants and apportioned the consumed biomass equally between females, as there is no dominance hierarchy among female pridemates (chapter 2).

Because of the fission-fusion nature of lion prides, females are seldom all together in the same place at the same time (chapter 1), thus a proportion of our observations on singleton females, pairs, trios, and so forth, were actually members of larger prides. In this study, we were solely interested in the effects of hunting alone or in pairs, and so on rather than from living in prides of each size. Females forage alone about a quarter of the time, regardless of pride size (see figure 7.6d), so we ended up with twice as much data on lone females as on any other group size, which was fortuitous, as we initially assumed that singletons would suffer the lowest and/or most variable feeding success. As it turned out, however, per capita feeding rates did not vary meaningfully with group size when the migratory wildebeest and zebra were abundant in their home ranges, and, when prey was scarce, solitary females fed as well as females in groups of five to six and significantly better than females in groups of two to four (figure 7.4c). Although solitaries experienced significantly higher variance in food intake than females in groups of two to four during the months of prey scarcity, their day-to-day variability in feeding success was the same as for females in groups of five to six, so feeding alone was not exceptionally risky even in the harshest time of year.

How did lone females manage to forage as successfully as groups of five to six when midsized groups performed so poorly? First consider figure 7.4d, which presents a simplified version of Dave Scheel's (1993) findings on the three most important dry season species: buffalo, warthog, and Thomson's gazelle. Dave had estimated the lions' prey preference during our 96-hour follows by dividing the total number of hunts on each prey species by the respective number of herds within a kilometer of the lions during 3,500 hourly counts. I have rescaled his findings so that the lions' "preference" for the most common prey at each group size is set at 1.0. Lone females mostly hunted warthog (which were relatively easy for singletons to capture), while groups of two to four females more often chose gazelle, and groups of five to six females primarily hunted buffalo. Buffalo carcasses can feed even the largest prides: females, cubs, subadults, males—everyone—so their per capita intakes matched that of singletons.

Dave also conducted monthly prey surveys along five 11 to 21 km transects across the long-term lion study area in the Serengeti and compared these data with his prey counts from the 96-hour-follows to estimate the lions' encounter rates with each species. Combining these figures with handling time and hunting success, he could estimate the profitability of each hunt (in terms of meat intake per hour) and thus plot the "risk-sensitive payoffs" associated with each prey species (figure 7.5). If a foraging female has the option to choose between two food items that exceed her current nutritional requirements (R), she should prefer the prey that maximizes her mean payoff while minimizing the variance (e.g., be risk averse), but if she is starving and must choose between two items that, on average, fail to meet her needs, she should choose the more variable prey on the chance that she might get lucky and obtain an unusually large payoff (e.g., be risk prone) (Stephens and Charnov 1982).

Of the five most common prey species, buffalo hunts provide the highest average payoff but the most variable outcome; Thomson's gazelle are the least profitable with the lowest variance. The four different panels in figure 7.5 contrast the payoffs associated with each prey species when migratory wildebeest and zebra are both present or absent in the study area and whether or not a pride is large enough to capture a buffalo. Large groups are predicted to prefer buffalo throughout the year (figures 7.5a, 7.5b), but whereas groups of five to six clearly preferred buffalo in the dry season (figure 7.4d), they mostly relied on wildebeest and zebra in the wet season. Figure 7.5 only considers energetic requirements, and buffalo are far more dangerous prey than either of the two migratory species, suggesting that lion foraging decisions also incorporate the potential costs of injury when safer prey are abundant.[3] In prides that are too small to catch buffalo, risk-averse females should prefer wildebeest and zebra in the wet season and warthog in the dry season (figure 7.5c, 7.5d). While the two migratory species comprise almost the entire diet when abundant in the study area, only solitaries showed a preference for warthog in the dry season (figure 7.4d); the lack of focus on warthog by females in groups of two to four may help to explain the low intake rates shown in figure 7.4c.

While the four-day follows provided our most detailed information on group-size specific food-intake rates, these data were limited to only about seven full days apiece for nearly half of the eleven group sizes plotted in figure 7.4c, so the results may largely reflect the inevitable noise from small sample sizes. However, food intake rates could also be estimated from the thousands of measurements of belly size that we collected from prides of different sizes.[4] As seen in earlier chapters, belly sizes vary consistently with habitat and season (figure 1.13a), and pride size also differed between time periods on the Serengeti plains (figure 2.8). After controlling for these ecological factors, females in intermediate-sized prides showed significantly larger residual belly sizes than in either very small or very large prides (figure 7.6a); however, as shown in figure 7.6c, entire prides only rarely forage all together, so pride size does not equal group size. However, the results clearly indicate a feeding advantage from living in

---

3 Buffalo are formidable prey. Several Crater lions died after being horned by buffalo, and a recent survey of lion skulls from Zambia found significantly higher rates of "blunt trauma" in a population that frequently fed on Cape buffalo compared to a population where lions rarely caught buffalo (Van Valkenburgh and White 2021).

4 Analyzing the data by subgroup size would have been meaningless, as pridemates could have recently reunited or separated after feeding.

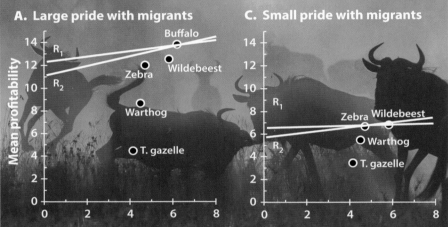
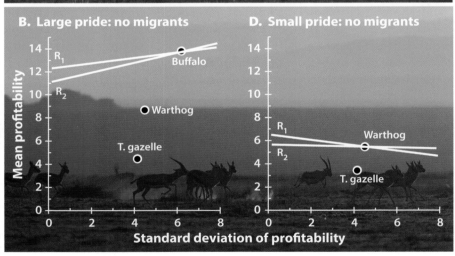

FIGURE 7.5. Risk-sensitive foraging in the Serengeti lions. Plots predict the prey species that maximize average food intake while minimizing the chances of failure (as estimated by the standard deviations). R_1 = upper limit of estimated hourly energy requirement of hunting groups; R_2 = lower limit. Data are separated by season (wildebeest and zebra are only abundant in the wet season) and pride size (small prides are unable to capture Cape buffalo). Redrawn from Scheel (1993).

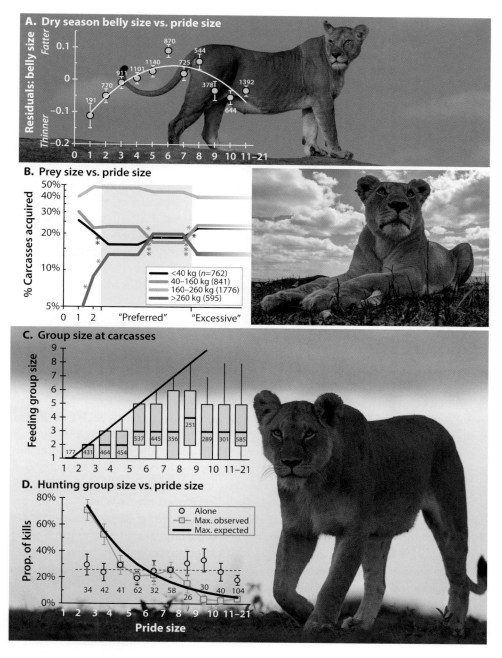

FIGURE 7.6. Food intake, prey size and grouping patterns across pride sizes. **A.** Residuals of female belly size versus pride size in the dry season. Sample sizes are the number of female measurements; residuals control for habitat, population density, periods of rapid population growth, and female age. There was no effect of pride size in the wet season, but females in intermediate pride sizes enjoyed the highest food intake during the dry season ($p < 0.0001$). **B.** Proportion of prey of each weight category versus pride size. Across all pride sizes, females obtained a relatively constant proportion of prey weighing 160 to 260 kg, but the proportion of small and large prey varied significantly with pride size. For example, prides of two females obtained a higher proportion of large prey than did solitaries, prides of three obtained a yet higher proportion, as did prides at the upper end of the preferred pride size in each habitat (see figure 5.6), but "excessively" large prides acquired a lower proportion of large prey. **C.** Feeding group sizes versus pride size. Subgroup size at carcasses reaches an asymptote at three females across all pride sizes. Data include all carcasses; diagonal line indicates maximum possible female group size for each pride size. **D.** Proportion of kills made by single females (circles) or the entire pride (squares) across pride sizes. The black line indicates the expected proportion of kills by the entire pride. Note that females hunt alone at a constant rate across all pride sizes.

intermediate-sized prides, but, as discussed in chapters 8 and 9, the lower rate of food intake in solitary females likely results from being forced to occupy marginal habitats as a consequence of competition with larger prides.

But why should food intake rates decline in the largest pride sizes? Looking across all the long-term data, prey size increases significantly with pride size but then declines again in "excessively large" prides (figure 7.6b). "Excessively large" refers to prides that exceed the preferred size-range in each habitat (figure 2.8). Because sample sizes for each prey-size category were too small to consider all possible pride sizes in each habitat, I have merged the data into broad categories for prides of four or more females according to the preferred pride sizes in each habitat, thus excessively large is twelve or more females in the Serengeti woodlands and eight or more females in the Crater and the Serengeti plains prior to 1997. While these results are not directly comparable to figure 7.4d, they clearly indicate that midsized prides (largely containing five to six females) capture the highest proportion of species like buffalo and giraffe. It is unclear why females in "excessively large" prides rely on smaller prey. Perhaps each pride range only holds a single buffalo herd, which moves elsewhere after experiencing too many lion hunts,[5] or some members of the largest prides may be forced to expand into areas with fewer buffalo.

Emphasizing the fission-fusion nature of lion prides, pridemates seldom feed all together: feeding-group size reaches an asymptote at a median of three females in prides of three or more, the upper quartile asymptotes at five females in prides of five of more, and, across all pride sizes, at least a quarter of feeding groups include only a single female (figure 7.6c). Note, though, that "feeding groups" may sometimes include individuals that arrived at the carcass well after it was captured/scavenged or only include the last stragglers after the rest of a larger group has departed. Nevertheless, the importance of solo foraging is clearly emphasized by the fact that lone females accounted for about 25 percent of 459 observed kills across all pride sizes (figure 7.6d). Remarkably, this pattern closely predicts the proportion of kills made by entire prides: assuming that each female hunts on her own with probability A_1, the probability that all pride females will be together is $1 - (1 - A_1)^n$. Therefore, only pairs and trios are expected to hunt as a unit for more than 50 percent of the time, whereas larger prides should (and do) remain intact far less often. Figure 7.6d implies that lion hunting-group decisions follow a simple rule: hunt alone with a constant probability, otherwise hunt with pridemates. Group hunting may not always be advantageous, so the fission-fusion nature of lion society allows females to forage alone whenever they need to find a quick meal, whether by catching a smaller prey animal or scavenging from other species.

BUT DO THEY COOPERATE WHEN THEY *ARE* TOGETHER?

During each hunt in our four-day follows, we recorded the distance that each female moved in pursuit of the prey (Scheel and Packer 1991). In a proportion of hunts, most females moved about the same distance toward the prey, in which case they could be labeled as "conformists." But in a sizable number of hunts, only a few females moved

---

5 Buffalo herds are known to move over 30 km in the aftermath of a successful lion kill (Sinclair 1977).

toward the prey while the rest stayed back and watched. We labeled these behaviors as "pursuing" and "refraining," respectively. Depending on the distance each female traveled compared to all the other group members, each individual could retrospectively be assigned a probability of having conformed, pursued, or refrained during that particular hunt (figure 7.7a), and the ratio of these three strategies varied according to the species of prey (figure 7.7b). For the most dangerous and demanding prey, females largely followed a pursuing or conformist role with everyone actively participating in the hunt: a buffalo is difficult to catch, and every female played a part in taking it down. However, when hunting warthog, a single female has no trouble catching the prey on her own, and the remaining group members often seem content to let just one or two group members do all the work. Thus, the lions don't always cooperate when the prey is easy to catch.

Inspired by our observations during the four-day follows, I developed a series of simple game theoretical models that asked when animals should hunt cooperatively and when they might simply let someone else do all the work (Packer and Ruttan 1988). The first consideration is whether each hunter pursues its own separate prey out of a flock, herd, or school, or if the entire group focuses on the same individual. When simultaneously pursuing separate prey animals, each group member has little temptation to refrain from the hunt, as a bystander would thereby miss a feeding opportunity and lose any potential advantages from joint action. This is especially true if hunters pursue prey that are small enough to be swallowed whole: nonhunters cannot scavenge from successful companions, and, indeed, some of the most striking examples of cooperation in nature involve group hunting of fish schools by whales and dolphins (e.g., Gazda et al. 2005). By working together to corral their prey, each predator feeds successfully, and group hunting is clearly synergistic. Ironically, we would even expect to see evidence of cooperation in cases of predators pursuing a single small prey: you can't eat if you don't join the race to dinner. It might be a bad idea to habitually forage in a group if only one group member gets to feed after every successful hunt, but if you're together for some other reason, you'd starve if you didn't join the chase.

However, if the prey animal is large enough to feed several predators, latecomers could still feed from a successful companion's kill and thereby benefit from the hunter's hard work. The advantage of arriving late to a just-killed carcass is especially clear in lions, which typically clamp down on their prey's windpipe, a process that may take several minutes to complete:

10-Oct-1978

10:00 a.m. Three females, Majani, Moss, and Mavuno, are resting near a low riverbank. Majani leaves the other two females to stalk an adult reedbuck hiding in the grass beside the river; she soon runs forward, knocks it over, and bites down on its throat. Majani continues choking her prey as Moss and Mavuno run up to the rear end of the reedbuck and tear open its abdomen. Majani snarls at her pridemates and attempts to drag it away, but the other two females hold on to the carcass and keep feeding.

10:10 The reedbuck has been dead for several minutes, but Majani keeps clenching its throat and growling at Moss and Mavuno as they finish off the offal and much of the hindquarters. Majani lets go and tries to feed at the rear, but the other two

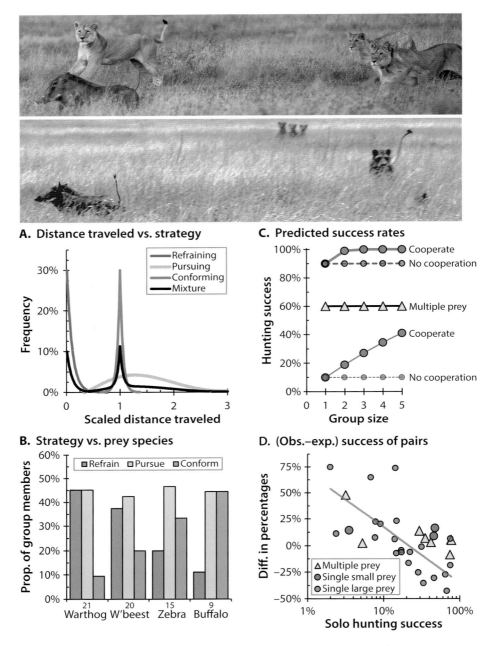

FIGURE 7.7. Group hunting. **A.** Relative distance traveled by individual females that *refrain* from active participation in the hunt (red), actively *pursue* the prey various distances (blue) or participate to a similar degree as others ("*conform*," gray). The black line represents cases where all three strategies are equally represented in the group. **B.** Proportion of lions showing each strategy while hunting each species. Sample sizes refer to the number of observed hunts. A higher proportion of group members showed restraint when hunting small prey (warthog), a greater proportion conformed during hunts of difficult prey (buffalo). Both redrawn from Scheel and Packer (1991). **C.** Expected relationship between individual hunting success and hunting-group size, if all members hunt the same prey individual simultaneously ("cooperate"), group members rely on the hunting success of a single hunter ("no cooperation"), or each group member pursues its own prey ("multiple prey"). Redrawn from Packer and Ruttan (1988). **D.** Deviation from expected hunting success of pairs plotted against solo hunting success. Positive values indicate synergy from cooperation; negative values suggest noncooperation. For single large prey, cooperation is highest where h_1 is lowest ($p < 0.001$, $n = 22$ studies). The relationship for multiple prey and single small prey is not significant. For species with $h_1 > 0.25$, the improvements in hunting success for single large prey were significantly lower than for single small prey plus multiple prey ($U = 4$, $n_1 = 8, n_2 = 7$, $p < 0.01$). Recalculated from Packer and Ruttan (1988).

females keep her off until they have eaten their fill, and Majani finally manages to feed on the remains of her kill.

Thus, group hunting in lions often involves a clear temptation to cheat.

Assuming that the initial evolutionary stages of hunting a single large prey would only involve the synchronous pursuit of the same individual but without any form of coordination or division of labor, the expected hunting success of a pair would simply be the chance that neither failed. Thus, if the success rate of a single hunter is h_1, her failure rate is $1-h_1$, the expected failure rate of a pair would be $(1-h_1)^2$, and their overall chances of catching the prey is $h_2 = 1-(1-h_1)^2$. The key aspect of this argument is that if h_1 is very low, $h_2 \equiv 2h_1$. Hence, by hunting simultaneously, two individuals could conceivably increase capture rates enough to overcome their personal costs of participating in the hunt—even in the absence of any synergistic hunting strategy. However, if h_1 is sufficiently high, a second hunter cannot augment her partner's success rate enough to cover the costs of active participation: at the limit of $h_1 = 1.0$, h_2 can only equal h_1. Further, for group hunting to be the primary driver for the evolution of sociality, the advantages of cooperation would have to be high enough to cover the costs of dividing the prey item in half—otherwise each individual should forage solitarily, and thus social species that capture single large prey would be expected to specialize on more difficult prey than do solitary species. But in our survey of 61 different combinations of predators and prey, h_1 was not significantly lower in group-living hunters than in solitary hunters (Packer and Ruttan 1988).

However, group living can evolve for many reasons besides cooperative prey capture (e.g., group territoriality, communal breeding, etc.), and group members may nevertheless hunt together because they are simultaneously confronted with the same feeding opportunity. In such cases, alternative behavioral strategies may allow a subset of individuals to exploit the hunting behavior of their cooperative companions, as seen during our observations in the four-day follows, where one female hunted on her own, while the rest of the group refrained from joining her in the pursuit (figures 7.7a and 7.7b). The identity of the "pursuing" female seemed to be conditional on who was first to spot the prey, whereas the less attentive females remained behind even after they spotted her moving forward. So, it is possible that every lion might have followed the same strategy: hunt if you are the first to spot the prey, but only join the hunt (i.e., cooperate) if the prey is sufficiently difficult to catch (and the resultant increase in hunting success exceeds the costs of participating).

Figure 7.7c shows predicted relationships between hunting success and group size for single large prey if all group members hunt simultaneously (cooperative groups). The potential improvement from cooperation is greatest when h_1 is low, whereas there is little room for improvement when h_2 is high. If the group consists entirely of individuals that follow the strategy "hunt if you are first to detect the prey, otherwise refrain from hunting," only a single individual would engage in the hunt, and group hunting success would remain unchanged with increasing group size (noncooperative groups). Hunting success for species that hunt *multiple* prey is generally reported on an individual basis (k_1), and, because all group members are expected to hunt simultaneously but independently of each other in these cases, their *individual* success rates should remain unchanged with increasing group size.

If cooperative hunting is most highly developed in situations where a single large prey is the most difficult to capture, the *observed* success of pairs should more often exceed the expectation of $h_2 = 1 - (1 - h_1)^2$ when h_1 is lowest. This prediction is broadly confirmed: *observed* values of h_2 were significantly more likely to exceed the *expected* values in twenty-two combinations of predator and prey when h_1 was smallest (figure 7.7d). In most cases, h_2 was *lower* than expected when $h_1 > 0.25$, suggesting a broad pattern of noncooperation when hunting involved easily captured prey. This pattern was not observed in species that hunt single small prey or multiple prey (and do not face a similar temptation to cheat). Considering only those species where k_1 or $h_1 > 0.25$, those taxa that capture single large prey showed a lower improvement in hunting success as pairs (h_2) than species that hunted either single or multiple small prey.

These analyses assumed that partners hunt independently of each other—for example, group-mates pursue the same prey at the same time but do not take into account each other's hunting tactics—thus there is no coordination or division of labor (see Boesch and Boesch 1989; Noë 2006). Interestingly, Stander's (1992) descriptions of the pincer movements in the Etosha lions and recent studies of chimpanzees showing a division of labor during group hunts of arboreal monkeys (Gilby et al. 2015; Samuni et al. 2018), both involve situations where h_1 is extremely low, suggesting that solving an especially difficult problem has been instrumental in the evolution of more complex hunting strategies.

To bring these findings back to the questions outlined at the beginning of this chapter, group hunting in lions is not always cooperative: some individuals may do all the work, while the remainder exploit the efforts of their companions. Collective action in lions is clearest when pursuing the most difficult prey (buffalo), but group members may parasitize a successful companion when the prey is relatively easy to catch (warthog). This pattern may also help explain why groups of two to four females fed so poorly compared to solitaries or larger groups, as they were too small to catch buffalo but perhaps too selfish to cooperate while hunting easier prey like warthog or gazelle.

But even when the lions fail to live up to their reputation as the paragon of cooperation, the fission-fusion nature of the lion's pride allows the option of foraging alone when necessary—and larger prides have the option to coalesce in order to catch buffalo. As we will see in chapter 10, buffalo are an essential dry season food item during the harshest years in the Serengeti woodlands. However, it is also clear that group hunting of buffalo cannot explain the evolution of sociality in lions: buffalo are largely absent from the diets of the Serengeti plains prides (figures 1.6 and 1.7) as well as of the smaller prides in all three study areas. Lions are social for some other reason, and their hunting behavior follows suit—they mostly hunt together because they are *already* together.

KEY POINTS

1. Female lions mostly capture prey weighing between 16 and 250 kgs, while males actively engage in prey capture of species weighing more than 400 kgs, with a sizable proportion of Cape buffalo being caught by all-male groups.

2. Per capita food intake was similar across all sizes of female hunting *groups* when wildebeest and zebra were abundant in the Serengeti study area, but, in their absence, individual payoffs in groups of two to four females were lower than for lone females or groups of five to six, as groups of two to four mostly fed on Thomson's gazelle, lone females mostly captured easy prey such as warthog, and large groups caught Cape buffalo.
3. Female hunting groups mostly take the prey species that maximizes mean food intake while minimizing the variance, although large groups prefer wildebeest and zebra rather than the more profitable Cape buffalo when the two migratory species are abundant, presumably because of lower risks of injury during prey capture.
4. Moderate-sized *prides* have the highest food intake and also take the highest proportion of large prey compared to solitary females and excessively large prides. Group size averages two to three females across all pride sizes, and females forage alone about 25 percent of the time regardless of pride size. The poor performance of solitaries likely reflects poor territory quality, but it is unclear why the largest prides take so many small prey.
5. Foraging *groups* of Serengeti females show the clearest signs of hunting cooperatively when attempting to capture large and difficult prey and the least cooperative when hunting the most easily captured prey.
6. Cooperative hunting is unlikely to explain the evolution of sociality in most species where groups typically capture only a single prey item per hunt, but if individuals live in groups for some other reason, individuals should—and do—show the clearest signs of hunting cooperatively when prey are most difficult to capture.

CHAPTER 8

INTERGROUP COMPETITION

*Who knows what true loneliness is—not the
conventional word but the naked terror?*
—JOSEPH CONRAD

Prides and coalitions engage in a constant numbers game: they play the numbers at night, when they roar, and they play the numbers when they cross paths with neighbors. Think of a sport where the odds can be altered at any point in the game. A red card in soccer means a team being down ten players to eleven for the rest of the day; a penalty in hockey puts a team down five players to six for the next few minutes. These are serious handicaps, but imagine your side being down three to five in a basketball game, or, worse, one against two in the middle of the night, and the stakes aren't just a ball thrown through a hoop, they're your life. Losing may not always mean sudden death, but it's hard to find a lion that hasn't been scarred by a member of its own kind.

Mutually assured destruction—a Cold War concept—is a good way to describe the teeth and claws of two lions poised one-on-one in a battle of wills. An awkward standoff can give an illusion of peace at a kill or a receptive female (chapters 2 and 5), but if one side can recruit a companion, it's game over; the loner has no choice but to escape. And if the pair knows their opponent from his odor, his roars, and they know that he, too, is part of a widely scattered team, they may try to use the opportunity to take him out when he's on his own. But even outnumbered, a lion still has all those claws and teeth, so it might be enough to avoid personal injury while inflicting just enough damage to let him live another day but only to count as a half or three-quarters in the fierce accounting of lion warfare. Two against one-point-five for the next six months is sweeter than any power play.

The "numbers game" dominates lion life—lions have little to fear from other species (chapter 11); their worst enemies are other lions. We've seen that the benefits from crèche formation largely derive from the mothers' greater success in protecting their cubs against infanticidal males (figure 3.7b), but infanticide is ubiquitous in leopards, tigers, and other big cats (Packer and Pusey 1984; Goodrich et al. 2008; Balme and Hunter 2013) and all of these species are militantly solitary. While group foraging confers benefits when hunting large prey, prey size largely appears to be a *permissive* factor in allowing females to be in groups rather than the driving force for the evolution of sociality (chapter 7)—a lone female can opt for smaller prey in circumstances when larger prides mostly capture Cape buffalo, and lions are hardly the perfect paragon of cooperation when hunting together.

In examining these aspects of lion life, I have tried to convey a recurrent sense that lions are social for some other reason—that they hunt together because they were already together and that mothers form crèches because they were already close to each other. So we are now ready to comprehensively explore that "some other reason"—the underlying pressure that has driven the evolution of sociality in lions: why females live in prides rather than live as solitarily as a tiger or a leopard or a jaguar, and why males defend these prides as coalitions rather than holding lone territories like most other felids. In this chapter, I will illustrate the lions' overwhelming advantages from companionship in their never-ending struggle with neighbors and strangers, whereas chapter 9 will examine how the underlying ecology of the African savannas has favored pride living in the females and how this has in turn led to coalition-formation in the males.

INTERGROUP CONFLICT

When strangers meet, the larger group of lions generally evicts the smaller, but the outcome depends on the extent of the odds. We witnessed twenty-two aggressive encounters between groups of different sizes that were otherwise perfectly matched by age and sex, and another nine cases where the sides were also equal in size. Higher odds play an increasingly decisive role until they reach three against one, when the larger group is invariably successful (figure 8.1a). But the vast majority of interpride encounters don't conveniently follow the Marquis of Queensbury rules of precisely matched weight classes: males, females, yearlings, and subadults can be present in any combination. The relatively large sample of these mixed-group encounters provides an opportunity to estimate the relative strengths of each age-sex class in a manner that most closely mimics the pattern in figure 8.1a. Figure 8.1b treats each adult male as being equal to 2.5 adult females,[1] each yearling increasing from 0 to 1.0 female equivalents between its first and second birthday, and subadult males increasing from 1.0 to 2.5 females between the ages of two to four years. But no matter how each age-sex class is weighted, the odds of winning reach near certainty at the most extreme size disparities in group size.

The photographs accompanying figure 8.1 graphically demonstrate how larger groups exploit the odds. Here, four adult females encountered a young nomadic male in the middle of their home range. Three of the females led the approach to within a few meters, then all four lay in a circle around him as he snarled and lay with his head low on the ground. The four females remained motionless for several minutes before simultaneously springing forward and attacking him. He made several desperate

---

1 As seen in many different contexts, "three" seems like a magic number for lions: "Three" is the minimum viable pride size in each habitat; pairs of nomadic males rapidly approach female trios, suggesting that they rate their individual strength as at least twice that of each female, but lone males are unwilling to approach female trios (figure 8.3c). In compiling the data in figure 8.1b, the ratio of 2.5 to 1.0 most closely mimicked the curve in figure 8.1a. Curiously, the lion ratio is similar to the well-known 3:1 rule of human combat, where three times as many attackers are presumed necessary to overpower a given number of defenders (Kress and Talmor 1999).

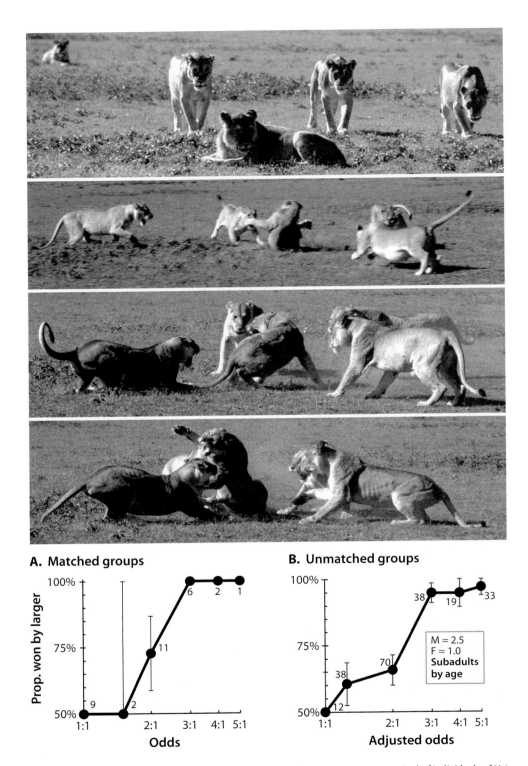

FIGURE 8.1. Outcomes of aggressive intergroup encounters where groups were comprised of individuals of (**A.**) the same age-sex class and (**B.**) different age-sex classes. In the latter, one adult male is equal to 2.5 adult females; yearlings increase from 0 to 1 female equivalent between the ages of one to two years; young males increase from 1 to 2.5 between two to four years. By convention, outcomes are set at 50 percent for encounters between equal-sized groups, as one always wins and the other loses. n = number of encounters.

contortions to try to protect himself, but they were able to bite him on the back and hind legs and left him seriously wounded.

The level of coordination typical of these gang attacks is further demonstrated by the confrontation illustrated in figure 8.2:

17-Aug-2009

08:17 Ingela Jansson finds a male named Hildur as he is chased by two males belonging to a quartet nicknamed the Killers. Hildur flees about 300 m to the north and escapes into a gully; the two males move to the east and join up with a third partner.

08:27 About a kilometer to the east, the three Killers find Hildur's companion, C-Boy, and chase him ~800 m to the northeast.

08:48 The three Killers are lying in a circle around C-Boy. C-Boy remains frozen for several minutes then tries to move, and the Killers attack. They shift around so that at least one male is always positioned to bite him on the back, thighs, or hind legs. As C-Boy twists and lunges, one male keeps a safe distance from his face and teeth, while the other males attack from behind.

08:55 C-Boy is seriously wounded. The Killers lean up against each other, rubbing their foreheads together and licking their lips. They shake out their manes and walk away.

In both sequences, the group first encircled the victim, then waited a few seconds or minutes before at least one individual lunged at the opponent's hindquarters. When the victim twisted around to defend itself, another group member lunged in from the opposite side. During attacks on dangerous prey animals like Cape buffalo, lions often just pile on—someone may try to grab it by the throat, but otherwise the goal is to knock it on its side, so its abdomen can be ripped open. But fighting another lion presents a far greater danger, the degree of cooperation is far more precise, and the consequences are far more profound than obtaining a few days' supply of meat (e.g., figures 3.4 and 3.5). Cripple one of your enemies, and the advantages can last for months; kill an enemy, and the benefits can persist for years.

On average, every lion pride encounters a neighbor or nomad about once every five days (Packer et al. 1990)—usually at night when detailed observations are virtually impossible. Yet the dynamics of intergroup interactions are highly amenable to controlled experimentation. The lion's roar is a territorial display that can be easily recorded and conveniently broadcast on a loudspeaker from a distance of 200 m of sleeping lions. The lions will invariably sit up and listen intently until the recording has ended—and a proportion of groups will then *approach* to at least the midpoint to the speaker (100 m) within the next sixty minutes.

A stranger announcing a territorial claim in the middle of a lion's home range is a direct challenge to the status quo. Males only roar after they have taken up residence in a pride (figure 8.3a). When exposed to recorded roars, nomadic males never approach the roars of resident males though they will approach roaring females (figure 8.3b)—provided they aren't outnumbered one to three (figure 8.3c). When resident males are in the middle of their territories, they always approach and almost always roar in response to the roar of a lone stranger, but residents behave like nomadic males while

FIGURE 8.2. Three members of a coalition of four males attack a member of a pair that has been caught on his own. Photos by Ingela Jansson.

outside their normal range (figure 8.3d). Males clearly recognize their companions, as they routinely join the roars of their recorded partners but never roar in unison with strangers (figure 8.3e). Whenever roaring punctuates the conclusion of an intergroup encounter, it's the winners that roar; the losers remain silent (figure 8.3f).

By broadcasting the recorded roars of intruding females, we discovered that female lions are able to calculate the odds of winning a territorial encounter solely on the basis of their opponents' vocalizations. Karen McComb (et al. 1994) played the roars of either one or three females to groups of up to eight females on days when the subjects were away from their dependent offspring. Females almost never approached the speaker when confronted with the roars of a larger group, but they almost always approached when they outnumbered the opponents by at least three to one (figure 8.4a). When the odds were one-to-one, groups of three females responded to a roaring trio the same as a lone female to a singleton: our response doesn't depend on the *number* of us; it depends on the *ratio* of us versus them.

Jon Grinnell (et al. 1995) similarly challenged resident males with recordings of one or three male intruders. Males also alter their responses according to the number of us and them, but they are far more reactive, approaching over half the time even when outnumbered one to three (figure 8.4a). A lone male approaching a male trio faces the same danger as C-Boy in his encounter with the Killers in figure 8.2. Why would any male be so reckless? When McComb (et al. 1994) tested females in the presence of their cubs, the mothers likewise always approached the speaker, regardless of the odds. Thus, mothers react to an immediate threat to their vulnerable offspring, but they are far more strategic when apart from their offspring, only directly confronting invading females when the odds are safely in their favor. Resident males behave more like mothers with cubs—so perhaps they are acting like husbands with wives and families to protect, reacting to an immediate threat with an immediate response. In fact, males are not quite as impulsive as it might seem: singleton males take an average of 45.6 ± 4.3 minutes to approach a roaring trio versus only 7.7 ± 1.0 mins to reach a roaring singleton (Grinnell et al. 1995).

Both sexes roared in the immediate aftermath of these experiments, but their roars appeared to serve different purposes. Whereas females almost always remained silent if their entire pride was together at the playback, subgroups roared half the time (figure 8.4b)—and thereby recruited distant companions in 43 percent of their post-test roaring sessions. Recall that female pridemates are almost always within earshot of roaring companions (figure 1.18b) and that subgroups frequently roared in response to distant roars during the 96-hour observations (figure 1.18c). In contrast, resident males almost always roared regardless of whether their entire coalition was present at the playback, although intact coalitions roared for slightly longer periods than did males that were apart from their companions (figure 8.4b). So, whereas females often roared to recruit distant pridemates, coalitions advertised their group size somewhat more vigorously when they were at full strength.

Although designed to test the males' and females' sense of "numeracy," Karen and Jon's studies both revealed that certain individuals consistently take the lead when approaching the speaker while others consistently lag behind. Rob Heinsohn (et al. 1995) repeatedly tested thirty-four females over a two-year period and found that four-fifths showed consistent responses to playbacks. Over half were persistent leaders,

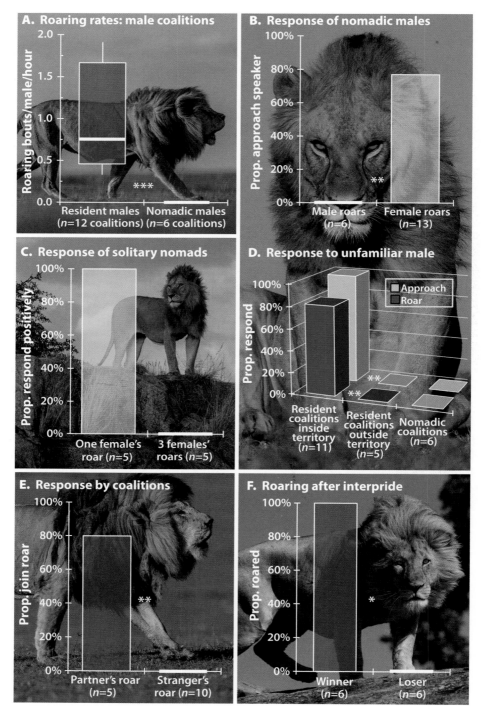

FIGURE 8.3. Roaring and territoriality. **A.** Roaring rates of resident and nomadic males. Data collected by Jon Grinnell during thirty-two nights of observation between August 1987 and May 1989. **B.** Response of nomad males to recorded roars of males and females. Redrawn from Grinnell and McComb (1996). **C.** Response of solitary nomadic males to recorded roars of one or three females. Redrawn from Grinnell and McComb (1996). **D.** Responses of resident and nomadic males to a recorded male roar; residents only responded when inside their territories. Redrawn from Grinnell and McComb (2001). **E.** Proportion of playbacks where a lone male joined the roar of the recorded male. Redrawn from Grinnell (2002). **F.** Roaring at the conclusion of same-sexed intergroup encounters. Data collected during the ninety-six-hour observations in 1984–1987.

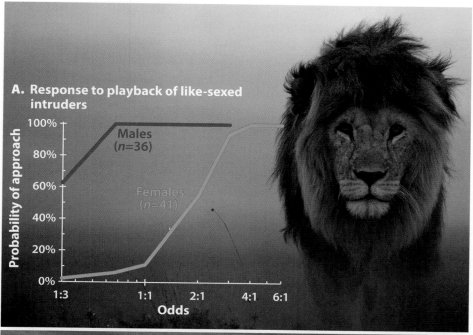

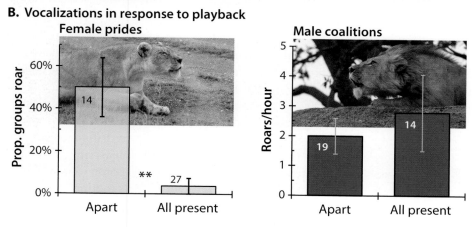

FIGURE 8.4. Group-size dependent responses to roars of like-sexed lions. **A.** Logistic curves fitted to the probability of approaching the speaker according to: (number of subjects)/(number of roaring individuals). **B.** Vocal responses to playbacks. Females: Probability of roaring when only a subset of pridemates were present (apart) or all females were present. Males: Roaring rates following playbacks when a subset of coalition partners were present (apart) or the entire coalition was present. n = number of playback experiments. All redrawn from McComb et al. (1994) and Grinnell et al. (1995).

typically approaching the speaker with head held low, but occasionally pausing and glancing backward at their companions. The remaining twelve lagged behind to varying degrees, and their behavior fell into three distinct strategies (figure 8.5a): a quarter of these females barely lagged when the odds were nearly even, but they lagged well behind the rest of their pridemates when their group safely outnumbered the roaring lions ("friends in need"). Another quarter lagged least when the odds were most favorable, while lagging longest when their pride was in the most dangerous situation ("fair-weather friends"). The remaining half lagged to a similar extent no matter the odds ("constant laggards").

When paired with another leader, an approaching leader rarely looked at her partner, but, when with a laggard, the leader often stopped and glanced backward (figure 8.5b). In a real encounter, the lead female would presumably run the greatest risk of injury, so it is surprising that the order of approach remains unchanged despite all the hesitations and backward glances—leaders lead, and laggards lag all the way to the speaker. There were no obvious physical differences between leaders and any of the broad pool of laggards—the leaders were neither older, younger, larger, or smaller than the laggards, nor did daughters show the same strategies as their mothers—and nothing obviously distinguished the fair-weather friends from the friends in need or even the constant laggards.

What could maintain this mix of strategies? Are they strategic responses to each other? Are the laggards prepared to pile on in case their leaders are attacked, the friends-in-need joining immediately when most needed, and the fair-weather friends ready to administer the coup de grâce at the tail end of a particularly chaotic melee? After all, even the slowest laggard shows up eventually: as seen in the photographs in figure 8.1, the "laggard" in the first image fully participated in the physical attack—so these strategies are not functionally equivalent to "pursue" and "refrain" as described in chapter 7, where latecomers can feast from a carcass captured by a lone hunter and contribute little or nothing to the capture of the prey. While hunting success doesn't always increase with foraging group size (figure 7.7d), the result of a territorial encounter depends entirely on overall group size (figures 8.1–8.4), so even the laggards make a meaningful contribution to the overall outcome.

This is further emphasized by the experiments that Heinsohn et al. (1996) performed on young males and females. By approximately 1.5 years of age, subadult females approach the recorded roars of adult females more often than do subadult males (figure 8.5c), and, of those that approach, young females consistently remain in the front half of the advancing group whereas young males lag behind (figure 8.5d). Although these results imply that young males are relatively indifferent to the affairs of their mothers and sisters, these boys don't just stand by during real interpride interactions: subadult males measurably contribute to their family's corporate strength (figure 8.1b).

Our roaring studies all confirm that lions can count—at least up to three—and that there are striking individual differences in the lions' responses to territorial incursions. Some lions consistently lead the charge, while others routinely bring up the rear. The leaders may be more motivated (as in the case of subadult females versus subadult males), but the leader-laggard measurements all focused on the lions' initial reactions, and their full responses during a complete gang attack could not be elicited by merely broadcasting the sounds of recorded roars. So instead of focusing solely

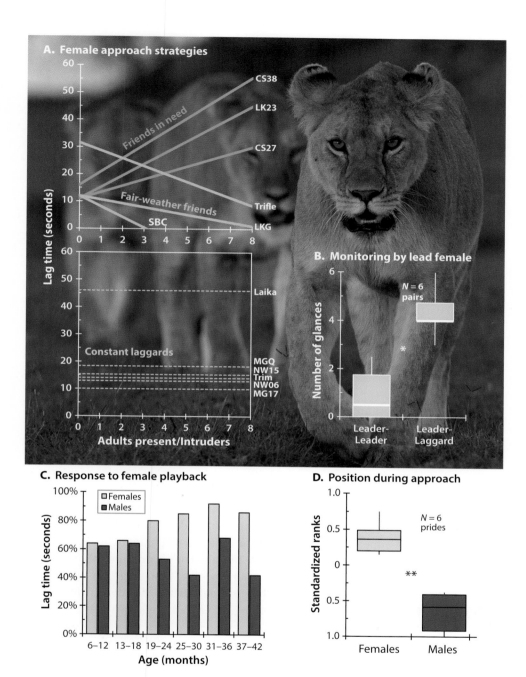

FIGURE 8.5. Leaders and laggards and the responses of subadults. **A.** Effect of the "odds" on the time that each lagging individual reached the midway point to the speaker. *Top:* Lag times of six females that lagged most when the odds were least favorable ("fair weather friends"), and those that lagged least when the odds were least favorable ("friends in need"). *Bottom:* Six females with similar lag times across all odds ("constant laggards"). **B.** Number of glances that the lead female directed at her lagging companion during a pairwise approach to the speaker, according to the companion's prior classification as a leader or a laggard. **C.** Subadult females accompanied adult females significantly more often than young males as they approached maturity ($p < 0.001$). Results based on 101 playbacks over a twenty-four-month period. **D.** Relative position of subadult females and males while approaching the speaker; females typically approached in the first half of group, males followed in the rear. n = number of prides with both males and females. **A.** and **B.** redrawn from Heinsohn and Packer (1995), **C.** and **D.** from Heinsohn et al (1996).

on the females' time to travel the first 100 m to the midpoint to the speaker, as in our original papers, it would have been more insightful if we could have measured what happens next in a real encounter.

Recall the sequence of events in figure 8.1 where the laggard joined her three companions in encircling the intruder, and all four lunged synchronously during the actual attack. Perhaps by arriving asynchronously, the females were better able to configure their final attack formation: arriving in a bunch might have led to some confusion as to where to position themselves with respect to each other. The sequence in figure 8.2 further emphasizes the importance of attacking the opponent from behind. Perhaps the final lion to arrive plays the role of executioner while the early arrivals draw the victim's attention from the opposite direction. Finally, the staggered approaches might serve to spread out the group in such a way so as to display its size during its advance toward the opponent(s). Perhaps the sight of an expanded formation of oncoming lions more frequently inspires an inferior group to turn tail and run before any physical attack takes place.

Unfortunately, testing these scenarios would have been impossible without staging actual battles between two sets of combatants or by employing state-of-the-art robotics, neither of which would have been ethical or feasible.

BLOOD ON THE TRACKS

Look closely at almost any lion and you'll find signs of past battles (see figure 6.4a). Lions are wounded at least once every eight to twenty months, depending on sex and social group size (figure 8.6a). Wounding rates in females *seem* to be relatively constant across pride sizes, and wounding rates in singleton males *appear* to be lower than in coalition males, but our measures are misleading. We did not attempt to rate the severity of each wound—just whether or not the individual had been obviously wounded. While pridemates claw at each other during squabbles at kills, and coalition partners spar over access to receptive females (see chapters 2 and 5; figure 6.3), any resultant wounds are usually superficial. But solitary females and singleton males can only be wounded by rival groups, so their wounds are more serious. Consequently, wounded solitaries of each sex suffer substantially higher mortality than do group-living lions (figures 8.6b and 8.6c), and these tabulations almost certainly underestimate their true mortality rates as an unknown proportion of wounded singletons must have died *before* their wounds could be recorded.

The most important lesson from figure 8.6 is the vulnerability of solitary females: once wounded, they are far more likely to die than are unwounded solitaries. Wounded members of a pair do not suffer a similar risk, and pair members are just as likely to survive wounding as are members of even the largest prides. Thus, as long as a wounded female has even one companion capable of foraging—or of standing guard— while she is recovering, she will often survive her convalescence whereas a wounded solitary is nearly always doomed. For resident males, mortality rates are highest for singletons, regardless of observed wounding (figure 8.6c); wounded singleton males suffer substantially higher mortality than unwounded singletons, but this difference is not statistically significant.

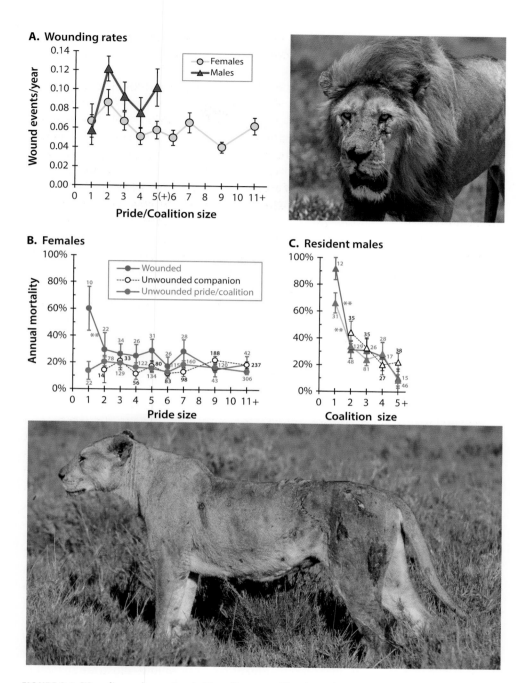

FIGURE 8.6. Wounding and mortality. **A.** Wounding rates of females and males in prides/coalitions of each size. Rates are wounds per female-year or male-year; error bars are determined by the number of years and assume one wounding event per animal per year. Data are pooled for prides of eight to ten and eleven to twenty-one females and for coalitions of five to nine males. Annual mortality rates for wounded individuals, unwounded companions and prides/coalitions where no one was wounded for (**B.**) females and (**C.**) males. Mortality rates are significantly higher for wounded solitary females than for unwounded solitaries. For males, mortality rates are higher for both wounded and unwounded singletons compared to wounded/unwounded members of pairs. n = number of individuals.

The wounding data provide compelling evidence of the hazards of a solitary life in a world of prides and coalitions. When Thomas Hobbes described the state of mankind during wartime as being nasty, brutish, and short, he could have been talking about the life of a lone lion. In fact, his full quote concludes with "worst of all, continual fear, and danger of violent death; and the life of man *solitary*, poor, nasty, brutish and short." A world of territorial prides is indeed a world at war: solitary females are vulnerable to the whims of neighboring groups, being forced to surrender any desirable piece of real estate at any time, and a solitary female is far more likely to come into contact with another pride than is any member of a larger pride (figure 8.7a). Being at the mercy of their larger neighbors, solitary females are often forced to seek refuge in suboptimal habitats, as illustrated by one of our ninety-six-hour observations:

13-Apr-1987
19:18 The solitary female, SBG, rests on a ridgetop in the middle of her wet season home range, an area usually devoid of prey. The migrating wildebeest are currently so abundant that she has scavenged four half-eaten carcasses from hyenas the past three days. Six females from the PN pride suddenly surround SBG in the darkness, she sees them in time and escapes to the east as the PN females roar from the ridgetop.
19:39 SBG has fled 1.11 km to the northeast. She roars and listens intently, but there is no response in any direction. She roars a total of 25 times over the next five hours; none of which elicit any response.

14-Apr-1987
07:15 SBG rests on a ridgetop 3.9 km north of the site of the chase.

By roaring and listening so frequently, SBG not only staked claim to another area that was temporarily rich with prey, but she also confirmed the absence of any other lions in the vicinity.

Intergroup encounters are more common at kills, and larger carcasses attract more frequent interpride interactions (figure 8.7b). However, the ownership rule described in figure 2.5 not only applies to pridemates but even to members of rival prides. If lions kill a large prey near their territorial boundary, the neighboring pride will sometimes join them at the feast—and though the first pride may snarl more vigorously than usual, each feeding lion will defend its site at the carcass rather than abandon the kill to chase away a trespasser. However, if the carcass is nearly finished, hungry latecomers may supplant the now-sated hunters, and, after the carcass has been completely consumed, the larger pride will chase away any lingering intruders.

The following observations from the ninety-six-hour follows illustrate several of these points. These again involve the solitary female, SBG.

18-Sep-1986 21:58. SBG catches an adult zebra in a clearing 20 km north of her usual dry season range and outside the northern boundary of the long-term study area. After she finishes throttling her prey, she opens up its abdomen and feeds on its entrails, but frequently lifts up her head to look around and listen. After about 15 mins, several hyenas appear, and SBG is increasingly alert, feeding more quickly

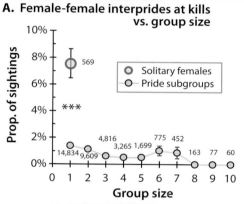
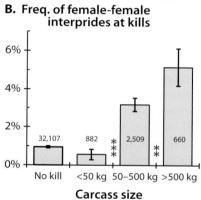

FIGURE 8.7. Frequency of female-female interpride interactions. **A.** Solitary females contact neighboring prides more often than pride females in any group size. *Top:* a playful interaction between two solitary females before returning to their respective home ranges. **B.** Interpride encounters are more common at larger carcasses, with the largest carcasses being the most attractive to neighboring females. *n* = number of sightings.

and peering into the distance. The hyenas start to vocalize, and SBG immediately runs to a nearby kopje as an adult male lion chases the hyenas from the zebra carcass. He is soon joined by a second male. They both start feeding, and neither makes any attempt to locate SBG. SBG's belly size was 3.5 before capturing the zebra, and 2.5 when she fled.

19-Sep-1986 21:00. SBG reaches an area with abundant wildebeest where she stops and listens to distant roars. A few minutes later, she hears a disturbance about half a kilometer to the northwest and walks determinedly to a group of three unknown females that are feeding on a wildebeest. The pride females snarl but continue feeding as SBG joins them on the carcass. She feeds for about 20 minutes then leaves the three females at their kill and walks further west.

20-Sep-1986 04:00. Relocating SBG an hour after she had crossed a ravine, Dave Scheel finds her as she leaves another half-eaten carcass being consumed by another unknown pride. Her belly size is now a 2.0.

SBG readily abandoned her zebra after she had extracted a decent meal, and she didn't hesitate to join two sets of pride females where she ate enough to maintain her belly size over the next few days. The essential point here is that the lion's *territorial* behavior is primarily focused on the land itself—not on individual food items. If strangers catch dinner in your dining room, your first priority is to eat while there's still food on the table—hyenas or yet more lions may steal your carcass while you're chasing away the trespassers. But, if you find a stranger snooping around your pantry, chase her away immediately. As the African folktale puts it, the roaring lion is telling the world, "This is my land. *My land. My land.* Mine, mine, mine, mine."

A solitary female literally has no land she can truly call her own. She can occupy an area if no one else wants it, but she is forever bouncing around the nooks and crannies between larger prides, living like a ghost in a world of danger. Forced to live in suboptimal habitat and with no help to rear her cubs, a solitary has little chance of recruiting an eventual pridemate. Thus, the solitary state is essentially a terminal condition: solitary females suffer an annual "extinction" rate of approximately 25 percent (figure 8.8a). Although pairs are generally more successful than solitaries, they risk shrinking to a single individual by the following year, and trios risk shrinking to a pair (figure 8.8a). Thus, over the entire study period, 80 percent of prides that were either founded by a solitary or ever shrank to a lone female eventually went extinct, and about a quarter of the prides that shrank to two to three females also went extinct (figure 8.8c). By way of comparison, the fate of lone males is even more stark: in the next year, a singleton has a 70 percent chance of vanishing and pairs and trios both face a 40 percent annual chance of shrinking to the next-smaller size (figure 8.8b).

WAR GAMES OF THE SEXES

High population densities inevitably increase the risks of direct confrontation between neighbors, and, even if you live in a large pride, more neighbors can mean more problems. Anna Mosser measured how pride females fared when surrounded by differing

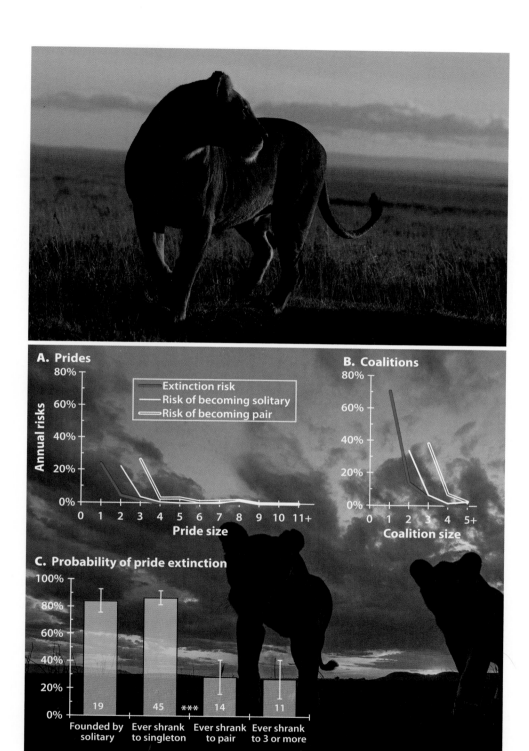

FIGURE 8.8. Extinction risks of prides and coalitions. Annual probability of extinction and/or reduction to one to two individuals for (**A.**) prides and (**B.**) coalitions. **C.** Proportion of prides that ever went extinct if they shrank to one to three females. Prides almost always go extinct if they are reduced to a single female; n = number of prides.

numbers of neighbors (Mosser and Packer 2009). Females with the most adult neighbors suffered significantly lower reproductive rates than females in relatively empty areas (figure 8.9a), and, as would be expected from the need to protect themselves from the greater risks of intergroup interactions, females in crowded areas are less likely to spend time alone (figure 8.9b). Anna's analysis included a major surprise: although female wounding and mortality rates also increased in crowded areas, these patterns largely resulted from the greater numbers of neighboring resident *males* (figure 8.9c). So, we must add another hazard to a female's life, not just the us-and-them of female-female competition, but the battle between us-and-them-and-*their husbands*.

We had long ago gained hints that males could influence the outcome of female-female competition, as in the following observation:

26-Jul-1980 08:55. Three females and two resident males from the SE pride are walking near two females from the SH pride. Two females and one male from the SE pride start chasing the SH females. The male passes the SE females and catches up to the SH females where he seems to realize that he is also resident in the SH pride. The SH females stop, look at the male, then start advancing toward the SE females. As soon as the male starts to follow, the three lions chase the SE females back to the starting point.

In this case, the rather slow-witted co-resident male was recruited by both sides, and his alternating allegiance determined the outcome of each chase. But now consider this observation from thirty-three years later where the male coalition was only resident in one of the two prides:

4-Apr-2013 07:00. Daniel Rosengren finds two females from the BF pride, BF18 and BF45, and BF45's two 6-week-old cubs resting 50 m from five females and two resident males from the KB pride. The males, S#94 and S#99, belong to the quartet called the Killers. The five KB females start chasing the two BF females, and the two BF cubs dive beneath a bush. The Killers follow behind the KB females for the first few hundred meters, but the females stop and watch as the two males pass them by and continue to pursue the BF females for another kilometer. BF18 and BF45 split up, and the males focus on BF18, chasing her another few hundred meters. Exhausted, BF18 slows down. The males catch up and bite her flanks and spine (see figure 8.9d). She collapses, paralyzed from the waist down. The Killers lie down nearby. Unable to eat, BF18 dies on the 12th of April; BF45's two cubs are never seen again.

In the following months, the KB females expanded their range eastward into areas that were once the BF pride's exclusive territory. Here, the males again followed the cues from the females in their pride, but they were not co-resident with the BF females. The BF females were competitors of the Killers' wives; by killing BF18, the Killers enhanced their reproductive success by improving conditions for the mothers of their own cubs.

As Machiavellian and idiosyncratic as this observation may seem, Schaller (1972) also saw an adult male stalk and kill a female from a neighboring pride, and the Killers

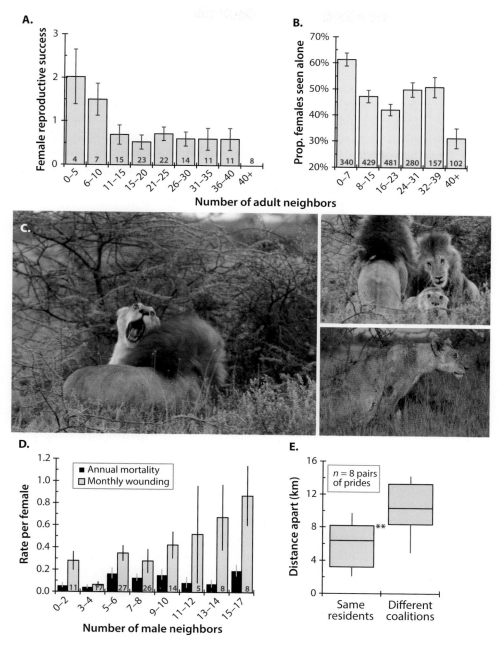

FIGURE 8.9. Impacts of adjacent prides on females. **A.** Per capita female reproductive success declines significantly with increasing numbers of adult neighbors; *n* = number of prides. **B.** Females spend significantly less time alone with increasing numbers of neighbors; *n* = number of females. **C.** Crippling attack on the female BF18 by two resident males from the KB pride. **D.** Female wounding and mortality rates increase significantly with the number of neighboring adult males; *n* = number of prides. **E.** Adjacent prides remain significantly closer together when they share the same male coalition; data are restricted to co-residencies of at least one year preceded/followed by a total of three years where neither pride contained a co-resident coalition, and each pride was located at least twenty times when coalitions were shared and twenty times when not. A.–C. redrawn from Mosser and Packer (2009).

were called "the killers" because of two prior cases where they were found next to the freshly killed carcasses of females from neighboring prides. This basic conflict also helps explain why SBG was so quick to abandon her zebra carcass to the two males (above)—yet readily joined two groups of females at their kills—and why adjacent prides remain significantly closer together when they share a co-resident male coalition compared to when they have different resident males (figure 8.9e). Long-term studies of chimpanzees have found that when males kill females from neighboring communities their wives subsequently enjoy higher reproductive success (Williams et al. 2004) and male leopards are known to kill young females that might compete with the adult females within their territory (Balme and Hunter 2013). Anna's statistical analysis suggests this behavior is *routine* in lions (figure 8.9c), and, indeed, both the lower reproductive rates and greater reluctance to forage alone in crowded areas (figures 8.9a and 8.9b) are measured by the combined number of neighboring females plus males—males aren't just mating partners, they are also their wives' best allies. Though it has been convenient to think of females as the sole competitors of other females, females can also rely on their husbands to do at least some of their dirty work for them.

INTERGENERATIONAL-INTERGROUP COMPETITION

Competition isn't just between neighbors, disputing territorial boundaries in the here and now. Successful prides produce cohorts of dispersing daughters that carry the battle into the next generation. But the ranging patterns of dispersing subadults reveal an underlying tension between mothers and maturing offspring—a tension that is seen in female leopards as well as lions. Leopards show the typical pattern of a solitary species: the mother cedes part of her original range to her maturing daughter, and the two subsequently defend adjacent territories (figure 8.10). Although female lions also cede part of their territory, the changes involve whole groups: a maternal *pride* surrenders a portion of its range to a maturing *cohort* of daughters (figure 8.11). This process may be repeated with each successive set of daughters until the immediate area becomes so packed that it's difficult for a dispersing cohort to integrate into the territorial network. For example, the TR pride gave rise to a number of daughter prides during the 1990s and 2000s, including the YT pride, which settled along the eastern boundary of their mothers' range in 1999. In July 2002, a cohort of five three-year-old females left the TR pride to form the TT pride. A few months later, we attached a GPS collar to one of the TT females, TR64. The TT females briefly moved well to the east of their natal range, threading their way through the territories of several prides, but they soon returned home, and then made at least six separate forays beyond the northern limits of the study area before eventually managing to carve out a range between their mother's pride and the YT pride (figure 8.12).[2]

But while most descendant prides stay close to home, others venture forth to settle new lands, spreading the family beyond the limits of their mothers' territories. The nested nature of river confluences in the Serengeti and the geography of the Crater

---

2 Similarly, although female leopards often settle adjacent to their mothers (as in figure 8.10), a proportion of daughters in crowded populations are forced to disperse farther distances from their mothers (Naude et al. 2020).

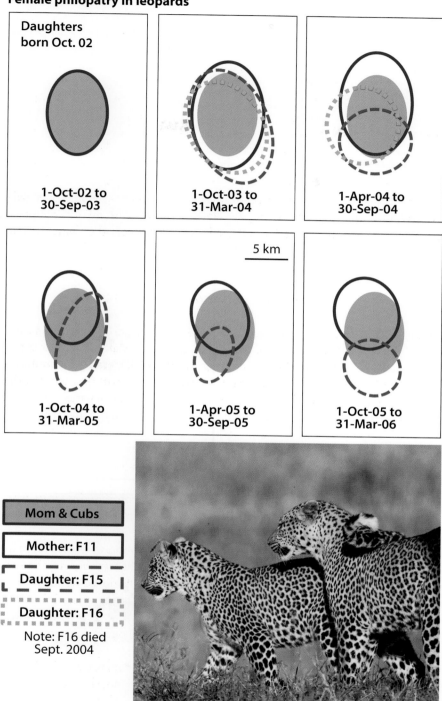

FIGURE 8.10. Ranging patterns of a female leopard and her two daughters in South Africa's Phinda Private Reserve. Panels show the approximate home range of each female over successive six-month intervals, retaining the mother's original range in orange. Redrawn (and simplified) from Fattebert et al. 2015. Photo from the Serengeti.

185

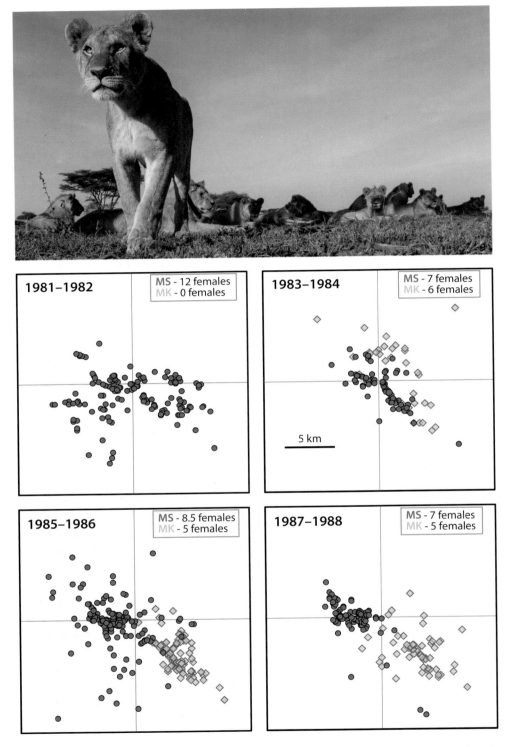

FIGURE 8.11. Ranging patterns of adult females from the MS Pride (solid circles) and their maturing daughters in the MK pride (diamonds). Data are plotted over two-year intervals.

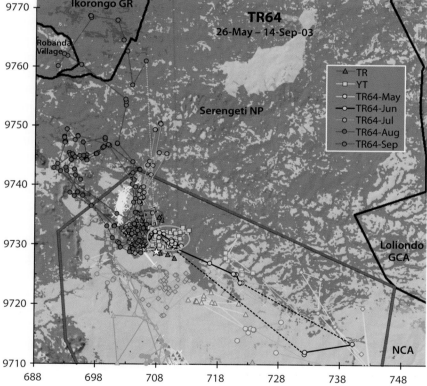

FIGURE 8.12. Movements of a dispersing female cohort. Colored lines indicate movements of a GPS-collared female, TR64; solid lines connect locations measured at approximately six-hour intervals; dotted lines link locations more than twenty-four hours apart (owing to occasional transmission failure). The star indicates TR64's location on 24-Nov-2003, the first resighting after her collar failed on 14-Sep-2003. Dark and light blue markers are contemporaneous locations of VHF-collared females from the TR and YT prides, respectively, and ovals indicate their dry season core areas. Gray markers similarly mark sightings of VHF-collared females in the remaining prides in the northern half of the study area.

floor provide a natural pattern of subdivision across the entire landscape. Thus, lion territories are typically clustered in a series of *neighborhoods*, each of which is occupied by one to five prides that mostly descend from the same matriline. As shown below, most daughter prides remain immediately adjacent to their mothers' territory and only shift away from that immediate location under certain circumstances (see figures 8.12 and 8.15). The largest pride in each neighborhood centers its territory around the most valuable landscape features (river confluences in the Serengeti woodlands, kopjes and marshes on the Serengeti plains, and freshwater springs in the Crater) and generates cohorts of dispersing daughters. Descendant prides remain close to their mothers for the first five years after dispersal (figure 8.13a), especially if they are surrounded by larger numbers of unrelated neighbors (figure 8.13b) or if their neighborhood consists of higher-quality habitat (figure 8.13c), where "quality" is measured by the per capita reproductive rate of each 1 km^2 cell in the long-term Serengeti study area (figure 9.2d).

Figure 8.14 summarizes these patterns in the Ngorongoro Crater. Hydrology divides the Crater floor into four distinct quadrats, each containing a river or spring, and the total number of females in each quadrat has remained relatively stable for decades (figure 8.14a). The richest quadrat of the Crater (in the northeast) has given rise to fourteen daughter prides, of which ten remained in their natal neighborhood, while the remaining four moved elsewhere (figure 8.14b). In contrast, the least productive quadrat (the southeast) has given rise to only four daughter groups, half of which moved to a different neighborhood. At any one time, each quadrat has contained a single "dominant" pride of about five females that generated the smaller prides in each neighborhood (figure 8.14c). Looking across the entire Serengeti and Ngorongoro study area, dominant prides generate more daughter prides in neighborhoods with the greatest mismatch between the typical size of the largest pride and the total number of females that can be supported in those areas (figure 8.15a). Across all neighborhoods, the average size of the dominant pride ranged from 2.0 to 11.12 females (median 5.9), whereas the average number of females occupying each neighborhood totaled 4.7 to 16.0 (median = 8.8). Thus, dominant prides give rise to more remaining daughter prides in neighborhoods with more available "space" for additional females.

Of greatest interest here are the eighteen of eighty-eight (20%) of daughter prides that move to new neighborhoods. These "emigrants" are significantly larger than the cohorts that remain in their natal neighborhoods, and emigrants move to neighborhoods where the largest pride is significantly smaller than their mothers' pride (figure 8.15b). Adding together every female in the neighborhood (the dominant pride plus all the "satellite" prides), emigrants move to neighborhoods with significantly fewer females than in their natal neighborhoods and with fewer females than in the other adjacent neighborhoods (figure 8.15c). Finally, larger cohorts are not only more likely to emigrate to new neighborhoods, but they are also more likely than smaller cohorts to themselves spawn yet more prides in their first ten years of existence, especially if their largest neighbor contains only three females or less (figure 8.15d). Thus, descendant prides leave their natal neighborhoods when they can become dominant prides in new areas, spreading their mothers' lineage across the broader landscape.

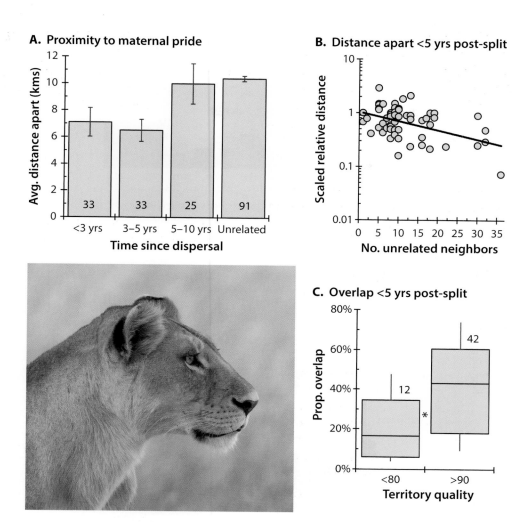

FIGURE 8.13. Spacing between related prides. **A.** Descendant prides remain significantly closer to their mothers for the first five years post-dispersal. **B.** Recently dispersed prides stay significantly closer to their mothers in areas with more unrelated neighbors; "scaled relative distance" = distance to mothers)/(average distance to other prides. **C.** Territories of recently dispersed prides overlap more with their mothers' in high-quality habitats. **D.** Three females confront a young female (second from right) from a descendant pride that had dispersed twenty-five months earlier; the young female was unharmed. **A.–C.** redrawn from VanderWaal et al. (2009).

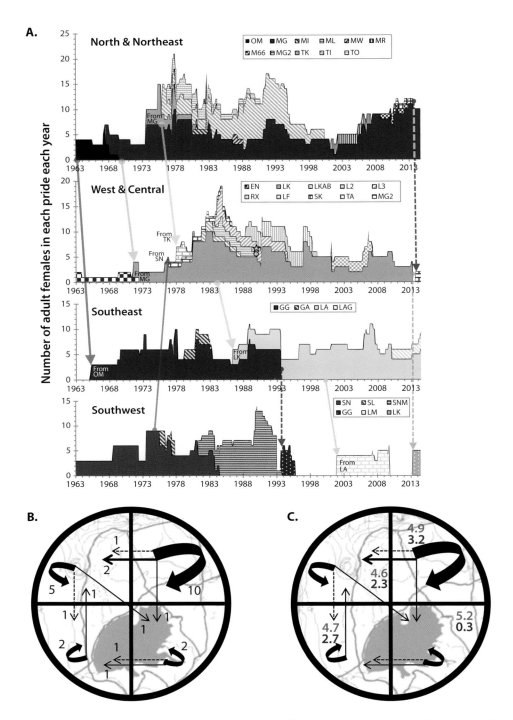

FIGURE 8.14. Lion neighborhoods in Ngorongoro Crater. **A.** Pride composition, formation, and dispersal in each neighborhood. Stacked height of filled area indicates number of adult females in each additional pride; dotted arrows indicate preexisting prides that moved to new neighborhoods; solid arrows are prides that shifted at natal dispersal. Yellow star and circle indicate a pride merger between two solitaries (in west central in 1990). **B.** Summary of prides generated in each neighborhood/quadrat. Dotted arrows: preexisting prides that eventually shifted neighborhoods. Solid arrows: prides that shifted at natal dispersal. Curved arrows: prides that remained in their natal neighborhoods. **C.** Average number of females in each neighborhood. Bold red numbers: average size of the "dominant pride" in each neighborhood (e.g., the pride that produced descendant prides). Brown numbers: average total females in all the remaining prides in each neighborhood.

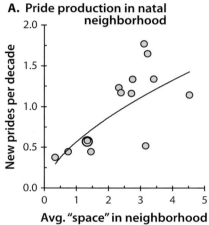
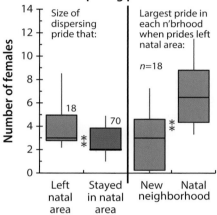
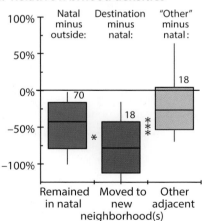
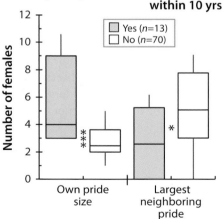

FIGURE 8.15. Neighborhoods, emigration, and breeding success of new prides. **A.** Dominant prides produce significantly more descendant prides in neighborhoods with more available "space." "Space" is the average number of females in a neighborhood minus the average size of the dominant pride. **B.** Pride size and natal emigration. *Left:* Emigrant prides are larger than prides that remain in their natal neighborhoods. *Right:* Emigrant prides move to neighborhoods where the dominant pride is smaller than their mother's pride. **C.** Emigration depends on relative population density. Emigrating prides move to neighborhoods with lower population densities compared to their natal neighborhoods (*middle*) than could have been obtained by nonemigrating prides (*left*); destination neighborhoods also offer a greater reduction in population density than adjacent neighborhoods (*right*). **D.** Pride size and pride production. *Left:* Successful prides are larger than prides that fail to produce descendant prides. *Right:* Successful prides live in neighborhoods with smaller rivals than do unsuccessful prides. n = prides.

GROUPS BEGETTING GROUPS—OF BOTH SEXES

Daughter prides originate from maternal prides, but that's only part of the story. Daughter prides have dads, too, and coalitions have to have come from somewhere. How does the formation of new a social group relate to the size of its mothers' pride and paternal coalition? Here, I only consider the origins of groups with a realistic chance of begetting yet more groups in the next generation. For a new pride to be successful, the minimum size is three or more daughters, as solitaries and pairs are seldom able to thrive. Similarly, "viable" coalitions of sons must contain at least two males, as solitaries have little chance of successful reproduction.

As would be expected, larger prides more frequently give rise to viable daughter prides than do small prides (figure 8.16a), but the production of viable descendant coalitions, on the other hand, reaches a peak in midsized prides (figure 8.16c), perhaps because the largest prides are vulnerable to more frequent male takeovers (figure 3.9a), and sons are forced to disperse before they are old enough to survive (figure 3.2b). Thus, the relative production of prides and coalitions shows a U-shaped relationship with maternal pride size: solitary females never produce viable groups of either sex, whereas midsized prides produce more groups of sons than of daughters, and the largest prides produce more daughter groups than sons (figure 8.16e). The pattern for resident males is quite different, as singleton males produce relatively more descendant prides than coalitions, while the larger coalitions show the opposite (figure 8.16b, 8.16d, 8.16f). Singleton males are almost always ousted before their offspring have reached maturity (figure 2.12a), and dispersing subadult females have higher survival than subadult males (figure 3.2b), leaving more descendant prides than descendant coalitions. In contrast, large coalitions remain resident long enough to allow groups of sons to remain together until they can take over their first new pride (figure 2.12a).

We have now seen where new groups come from and why. As prides grow in size, they produce more and more descendant groups, first with a bias toward sons, but eventually begetting more and more daughter groups as their pride reaches an upper limit in size. For males, solitaries tend to give rise to groups of daughters, whereas the largest coalitions are best able to spawn intact coalitions of sons. So, to the largest the spoils, mothers and daughters, fathers and sons. And how big is your pride? Well, it's almost entirely an accident of birth. If you were born in the northeast quadrat of the Crater or the right part of the Serengeti woodlands, you were born to rule. From the land, all wealth stems.

KEY POINTS

1. Lions of both sexes advertise their status as territory holders by roaring, generally allowing neighbors to remain well apart from each other. But when close contact does occur, territorial encounters sometimes involve gang attacks whereby a single-sex group encircles a like-sexed intruder, biting it from behind and inflicting serious wounds.
2. In aggressive encounters between like-sexed groups, larger groups have an advantage against smaller groups and always win when outnumbering opponents

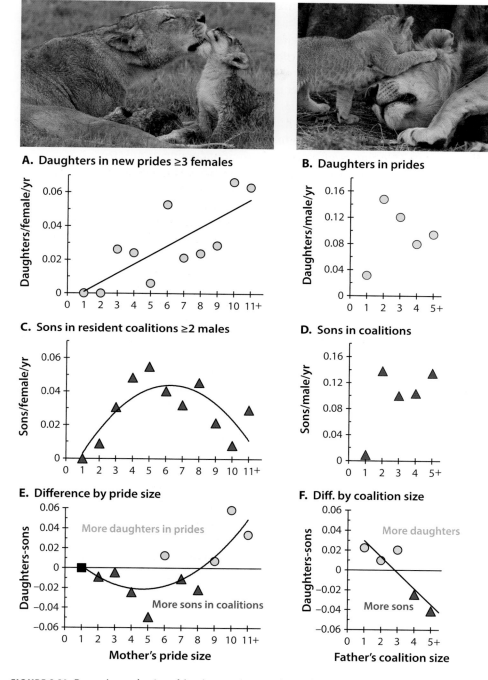

FIGURE 8.16. Per capita production of daughters and sons in descendant prides and coalitions large enough to reproduce successfully. **A.** The largest prides give rise to the most females in viable descendant prides. **B.** Coalitions of two to nine males generate more daughters in viable prides than do singleton males. **C.** Females in midsize prides produce significantly more sons in viable descendant coalitions. **D.** Coalitions of two to nine males produce more sons in viable coalitions than do singleton males. **E.** Midsize prides produce more viable sons than daughters whereas the largest prides produce more viable daughters. **F.** Smaller coalitions produce relatively more viable daughters while larger coalitions produce more viable sons.

by a ratio of at least 3 to 1. In mixed-sex groups each adult male adds the equivalent of about 2.5 females to the group's corporate strength. The willingness of all-female groups to approach recorded female roars closely mirrors the odds, whereas males will approach recorded males even when outnumbered, though such approaches are far more cautious.
3. Although every group member participates in the final stages of a gang attack, individuals differ in their initial responses to a simulated territorial incursion, with some regularly leading while the remainder lag behind. These alternatives may either allow groupmates to sequentially encircle their foes or enable the opponents to more accurately assess their oncoming group size.
4. Compared to pride-living females, solitary females are more likely to come into close contact with neighboring prides, and solitary females are also more likely to die shortly after being wounded. Small prides risk shrinking to even smaller sizes and are likely to go extinct if they ever shrink down to a single female. Solitary females and singleton resident males both suffer higher annual mortality than their group-living counterparts.
5. Female fitness is adversely affected by greater numbers of neighboring adults, particularly neighboring resident males. Females with more neighbors spend less time by themselves, and prides that share a co-resident coalition remain shorter distances apart than do prides with different resident coalitions, presumably because of the lower risks of attack on themselves and their cubs, as males will attack and kill females from adjacent prides.
6. Maternal prides cede portions of their territories to dispersing daughters, who remain nearby in "neighborhoods" that are bounded by geographical features that limit access to water or shelter. The maternal pride dominates its neighborhood as long as it remains larger than any of its daughters' prides. Large dispersing cohorts sometimes take over adjacent neighborhoods when they outnumber the largest pride in the new area.
7. Although larger prides and coalitions both generate more descendant offspring *groups*, the ratio of daughter prides to descendant coalitions varies with the size of the parental group. Pride-size specific recruitment results in a higher ratio of daughter cohorts dispersing from large maternal prides, but only large coalitions and midsized prides typically experience male tenure lengths that are long enough to assure the survival of sons rather than just daughters, thus resulting in relatively more descendent coalitions.

CHAPTER 9

THE EVOLUTION OF LION SOCIALITY

> A tribe including many members who, from possessing in a high degree the spirit of patriotism, fidelity, obedience, courage, and sympathy, were always ready to aid one another, and to sacrifice themselves for the common good, would be victorious over most other tribes; and this would be natural selection.
> **—CHARLES DARWIN**

Group living has arisen repeatedly in birds and mammals, yet among the felids there is only one truly social species: the lion. We have seen how pervasively lions work together in raising their cubs, hunting their prey, and defending their territory. But while crèche formation protects against infanticidal males, other cat species are just as infanticidal as lions. Though lions do cooperate to catch certain prey like Cape buffalo, singleton hunters gain similar food intake rates as cooperative groups, and other felids can also capture relatively large prey, for example, 20 kg lynx catching 100 kg reindeer, despite remaining militantly solitary. The advantages from group territoriality, on the other hand, are compelling: in the face of competition with a neighboring group, a lone female has little chance of controlling an exclusive area where she can successfully raise her offspring, and a lone male has little chance to gain or maintain pride residency long enough to ensure his cubs' survival.

But if group territoriality offers the most plausible explanation for lion sociality, we must now address the obvious question: Why lions? Other felids are territorial, and they jealously guard their turf as lone individuals. What's so special about being a lion? For the answer we must focus on the savanna landscapes where lions mostly likely evolved.

THE THREE RULES OF LION REAL ESTATE

As the apex predator of the African savanna, lions can live wherever they choose. Except for the hazards of being trampled underfoot by a herd of angry elephants or thumped by a herd of Cape buffalo, lions pretty much get to do whatever they want wherever they want to do it. So, given all that inviting landscape—open grasslands, shady woodlands, flowing springs, the privacy of thick brush, and the commanding view from a granite dome—why do they divide their time between certain locations rather than others? At the scale of a pride "neighborhood," lions clearly follow the pattern of source-sink dynamics as originally envisioned by Pulliam (1988): the northeast quadrat of the Crater produced an excess of prides that dispersed to the other three quadrats, while the southwest quadrat was repeatedly recolonized by prides that originated from other parts of the Crater (figure 8.14b). But what makes some areas more reproductively valuable than others? Can we relate these differences to specific landscape features?

There really are three rules of lion real estate. We've already seen the first rule: a female needs a good spot to hide her newborn cubs (figure 4.1). The second rule is a steady source of water. There are only a limited number of freshwater springs, and, after the end of the rainy season, surface water persists in only a few marshes and river confluences, thus it is easy to locate the major water points (see figure 9.2a). The third rule, a reliable place to catch dinner, is trickier. Mapping the food supply across the landscape requires careful consideration, as the food doesn't want to be eaten, but the lions obviously succeed often enough to be able to survive and sometimes even to thrive.

Grant Hopcraft (et al. 2005) characterized the landscape features surrounding hundreds of lion kills in the Serengeti. The results varied somewhat by habitat and season, but, in general, the radio-collared females caught more prey than expected in the immediate vicinity of river courses (figure 9.1a) and in areas with abundant woody vegetation (figure 9.1b): lions frequently ambush thirsty prey, and they need to conceal themselves to be successful. Kills were also common within the "view-shed" of a kopje or other highpoint (figure 9.1c). View-shed refers to the area visible from a particular spot, so when lions are being kopje cats, they are often surveilling the evening's hunting prospects. Finally, kills were common within 100 m of an erosion embankment that was tall enough to provide cover in at least one direction; these include "erosion terraces" on the open plains (figure 9.1d) and banks of bare earth near river courses. The landscape also predicts where lions scavenge more carcasses, particularly within view-sheds (that lion looking up at the sky isn't contemplating the weather, she's scanning for vultures) and in open areas where cheetahs and hyenas can be seen catching their prey.

If you're going to do battle over the exclusive rights to a particular piece of property, make sure it's worth the fight. If you're a lion, you should work hardest to guard the best denning sites (Rule #1), the most persistent waterholes (Rule #2), and the spots where you are most successful in obtaining your prey (Rule #3). To understand how these translate into a true measure of real estate value, Anna Mosser (et al. 2009) merged a comprehensive set of habitat maps with forty years of long-term data on the individual reproductive performance of every female in the study. Anna mapped out rainfall patterns, the course of each river, and the location of all the dry season kills (figure 9.2a); she then added the distribution of woody vegetation, kopjes, and wet season kills (figure 9.2b). After subdividing each pride's history into two-year chunks (the time it takes to rear a set of cubs), she found the locations where females persistently lived at the highest densities (figure 9.2c), consistently gave birth to the most cubs (figure 9.2d) and the most surviving offspring (figure 2e).[1] By converting these totals

---

[1] Reproductive rates in figures 9.2d and 9.2e are "relative" to the most successful pride each two-year time-step so as to control for year-to-year variations in cub production/survival and long-term changes in population density

FIGURE 9.1. Location of kill sites. Lions capture more prey than expected near (**A.**) rivers and (**B.**) areas with extensive woody vegetation. Blue circles/triangles indicate the ratio of observed divided by expected. Redrawn from Hopcraft et al. (2012). **C.** Lions resting atop a kopje with an extensive viewshed. **D.** Large cub sitting beside an erosion terrace on the Serengeti plains.

into the number of descendants per female per year for every square kilometer in the study area (figure 9.2f), she derived a fine-grained map of lion property values over the full forty-year time span.

A single landscape feature stood out: the junction of two tributaries. Besides the dense vegetation required to make a good den and the extra turbulence necessary to gouge out a persistent waterhole, a typical confluence also serves as a funnel, hemming in prey that are reluctant to cross either drainage line (figure 9.3). So, if your territory contains enough confluences, you effectively live in a "source" area for lions, and your pride will be likely to spawn daughter prides and coalitions of sons that disperse elsewhere (figure 9.2f). But if you live too far from the nearest confluence, your home is essentially a "sink" where your pride, and any pride that replaces you, will be doomed to reproductive futility (figure 9.2f).

Now look at the range map for the twenty-eight study prides in the Serengeti in 2007–2008 (figure 9.4a). Keep in mind that these are year-round ranging patterns, and that most prides shift southward during the rainy season and northward during the dry season (figure 1.9). Ranges are generally contiguous in the high real estate–value region at the northwest corner of the study area, while pride ranges are often spaced apart from each other in the poor-quality southeast. Looking separately at each Serengeti habitat, larger prides in the woodlands have significantly larger ranges than smaller woodlands prides (figure 9.4b), but ranges on the plains are similar across all pride sizes (figure 9.4c). This contrast likely derives from the geography of the two regions: woodlands prides can annex incrementally more acreage of high-quality habitat as they grow larger, whereas plains prides rely on drainage systems that are spaced too widely to be spanned by a single pride, no matter how large. The greater flexibility of pride ranges in the woodlands is illustrated by figure 1.18b: neighboring woodlands prides move closer together during the dry season, whereas plains prides remain similar distances apart throughout the year.

After assessing the "quality" of each 1 km^2 grid cell, Anna found that the *absolute* number of adult females and the *relative* number of resident males significantly influence the quality of each pride's overall territory in each two-year time-step. However, whereas an *increase* in pride size results in a clear improvement in territory quality ($p < 0.05$), a *drop* in pride size does not significantly reduce quality, indicating that female recruitment is particularly important to adding high-quality habitat that they will retain for a period of time even after shrinking in size at some later date. These effects translate to the equivalent of a growing pride moving approximately 115 m closer to a confluence with each additional female, but a shrinking pride only shifting about 58 m farther away. On the other hand, by having a resident coalition that is larger than their neighbors' coalitions, females shift the equivalent of approximately 160 m closer to a river confluence for each extra male, while prides with a smaller resident coalition move about 275 m farther away.

The case study in figure 9.5a illustrates these patterns. The GU pride was formed by two dispersing females that initially remained close to their mothers' pride range, but, as they set out on their own, they occupied marginal habitat until they recruited

---

(Mosser et al. 2009). However, absolute reproductive rates were used in figure 9.2f to determine whether each 1 km^2 area was a "source" or "sink."

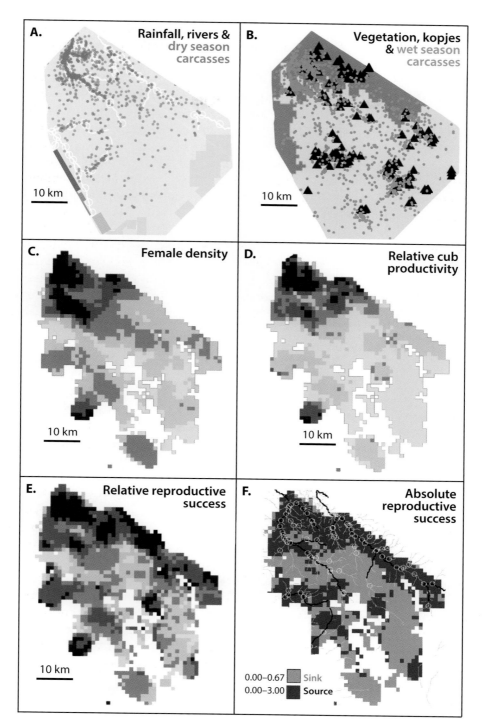

FIGURE 9.2. Lion real estate in the Serengeti study area. **A.** Rainfall, major rivers and location of 1,278 dry season kills (orange dots). Dry season rainfall ranges 94 to 232 mm (light to darker shading); white circles indicate river confluences. **B.** Vegetation, kopjes (black triangles), and locations of 1,278 wet season kills (green dots). Darker coloration indicates woodlands, lighter areas open grasslands. **C.** Female density. Density ranges from less than 0.03 to greater than 0.17 (light to darker shading; empty areas are in white). **D.** Relative cub productivity per female per two-year timestep; values range from less than 0.1 to greater than 0.5. **E.** Relative reproductive success per female per two-year period; values range from less than 0.1 to greater than 0.5. **F.** Areas where absolute reproductive rates exceed (dark gray) or fall below (light gray) replacement rates. All redrawn from Mosser et al. (2009).

199

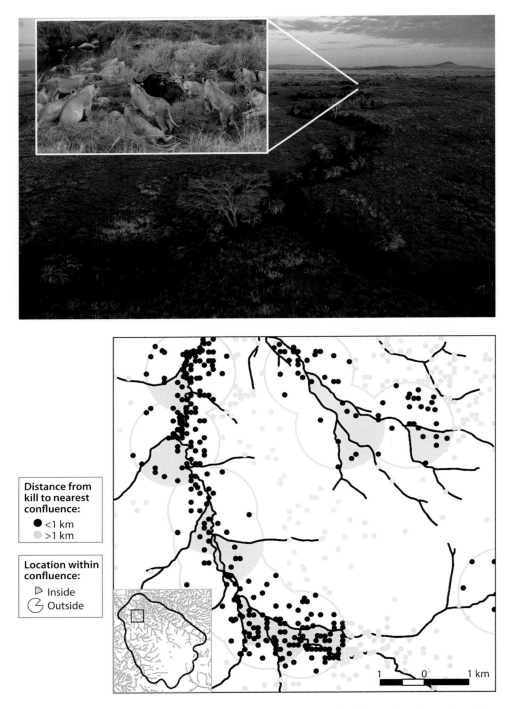

FIGURE 9.3. River confluences and hunting success in the Serengeti. *Map:* Black dots indicate kills made within 1 km of a confluence; shaded areas are between converging tributaries. Significantly more kills were made between converging tributaries than expected by chance. The map only illustrates 100 km² of the overall study area.

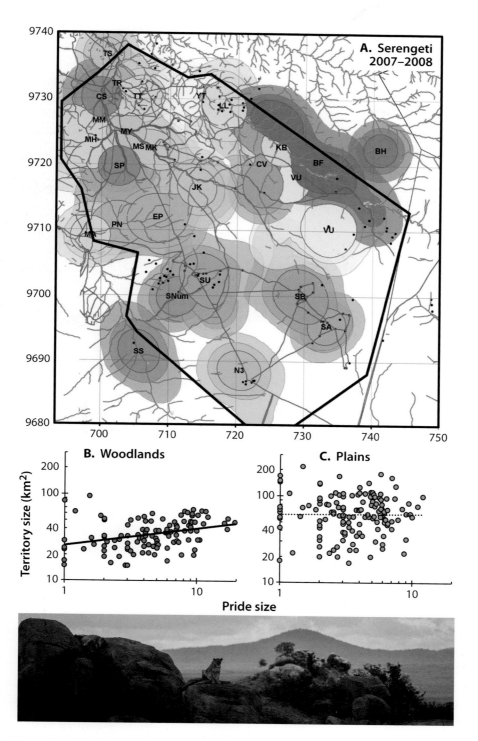

FIGURE 9.4. Pride territories in the Serengeti. **A.** Kernel density estimates of lion ranging patterns over two years of observation (2007–2008). Outer circles enclose 95 percent–use areas, middle circles are 75 percent–use areas, and inner circles are 50 percent areas. The 75 percent kernels provide the best estimates of territorial boundaries, and these increase with pride size in (**B.**) the woodlands but not in (**C.**) the plains. Graphs redrawn from Mosser and Packer (2009).

daughters of their own. As their pride grew, they acquired higher-quality areas, and they also attracted a relatively large male coalition, further increasing the quality of their territory. However, when they were taken over by a relatively small coalition in 2001, they lost access to the best parts of their territory. Territorial competition is further illustrated by figures 9.5b and 9.5c: three adjacent prides shared a 30 km^2 overlap area in 1991–1992.[2] The smallest pride went extinct by 1993, and the larger surviving pride claimed 21 km^2 of the overlap area compared to only 9 km^2 claimed by the smaller surviving pride.

In summary, some areas are far more valuable than others. Richer areas are conducive to more successful reproduction, and larger prides not only gain more land in competition with their neighbors, but they also control the best real estate. Thus, the rich get richer in manifold ways.

THE LANDSCAPE OF FEMALE SOCIALITY

Lions consistently cooperate when confronted by trespassers; they track the odds of winning intergroup encounters and recruit distant pridemates when outnumbered. Larger prides successfully defend larger territories in the woodlands and control higher-quality territories in both the woodlands and plains. Daughters are routinely recruited into their mothers' prides and actively engage in territorial defense from an early age. Group territorial species in insects (Adams 1990), birds (Brown 1970), fish (Clifton 1990), and mammals (Waser 1981; Bowles 2009) show fundamental similarities to lions: larger groups win intergroup contests (e.g., Carlson 1986; Cheney 1987; Adams 2001; Wilson et al. 2014) and possess higher-quality territories (Woolfenden and Fitzpatrick 1984), and their offspring benefit from territorial inheritance (Lindstrom 1986), particularly in heterogeneous landscapes (Stacey and Ligon 1987). Grouping has long been considered beneficial in species that rely on heterogeneous resources (e.g., Crook 1964; Wilson 1975; Bradbury and Vehrencamp 1976), and David Macdonald (1983) developed the Resource Dispersion Hypothesis (RDH) to explain how a patchy landscape could support additional individuals at virtually no cost to a territory owner or breeding pair in circumstances when each individual relies on access to a succession of hotspots through time, and each patch contains enough food to meet the requirements of multiple individuals (also see Kruuk and Macdonald 1985).

The RDH was subsequently modified to include the advantages of joint defense of high-value patches (Johnson and Macdonald 2003) and thus links to a second hypothesis that follows from classic foraging theory: at high population densities, intruder pressures become so high that a lone individual can benefit by recruiting a partner to share the costs of territorial defense (Davies and Houston 1981) and a saturated habitat promotes recruitment of maturing offspring into the natal group (Gaston 1978; Koenig et al. 1992; Emlen 1994). Unlike other *Panthera* species, lions live almost exclusively in savanna habitat, a landscape characterized not only by pronounced habitat heterogeneity (Pickett et al. 2003; Boulain et al. 2007; Levick and Rogers 2008) but also by

---

2 The discontinuity in the KB pride territory in figure 9.5b was due to the separation of their wet and dry season ranges in 1991–1992; Serengeti prides typically shift their ranges to the south and east each wet season (figure 1.9).

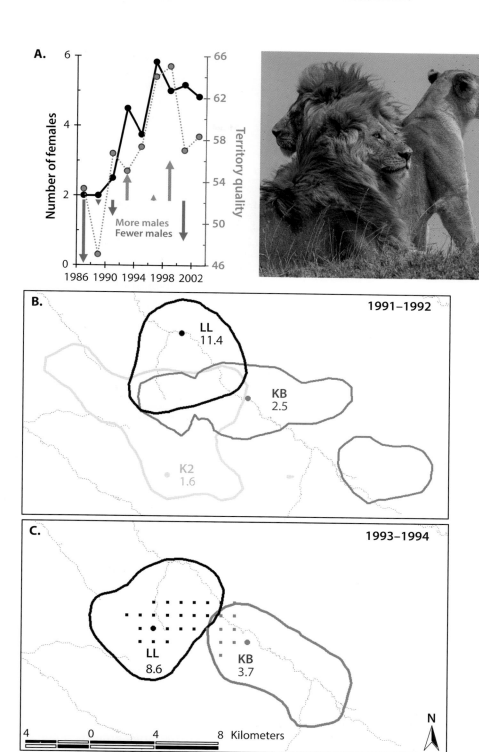

FIGURE 9.5. Territory dynamics. **A.** Changes through time in territory quality (green) of the GU pride with concurrent changes in pride size (black) and resident coalition (arrows). Arrow lengths indicate the extent to which the resident coalition was larger or smaller than the neighboring coalitions. **B.** Territorial boundaries of three adjacent prides in 1991–1992; numbers are pride size, squares indicate 1 km² overlap areas, circles designate the center of each territory. The territory "center" is the mean north-south and east-west coordinate, weighted by the number of individuals in each sighting. **C.** Boundaries of the two remaining pride ranges in 1993–1994. Darker squares were annexed by the LL pride, lighter squares by the KB pride. All redrawn from Mosser and Packer (2009).

203

TABLE 9.1. *Typical population densities (individuals/100 km^2) of the five Panthera species*

Species	1st quartile	Median	3rd quartile	n
Lion	6.36	9.93	14.16	34
Leopard	2.68	4.50	7.92	48
Tiger	2.00	3.74	5.09	18
Jaguar	0.75	1.39	2.38	40
Snow leopard	0.74	0.87	1.07	5

Sources/Notes: Values for lions, tigers and jaguars are based on expected densities in the absence of anthropogenic effects (Jędrzejewski et al. 2018; Harihar et al. 2018; Packer et al. 2013); values for leopards are from areas with minimal human impacts (Balme et al. 2010, 2019; Chapman and Balme 2010; Grey et al. 2013; Havmøller et al. 2019; Strampelli et al, 2019; G. Balme, unpublished); values for snow leopards are from large-scale surveys (McCarthy et al. 2010; Chetri et al. 2019). n = number of study areas.

high levels of primary productivity (Sinclair 1979a). Lions typically live at more than 2.5 times higher densities than tigers and jaguars and over twice as high as leopards (table 9.1), so population density is also likely to be an important additional factor for explaining the evolution of group territoriality in lions (Packer 1986).

To formally test whether landscape heterogeneity and/or population density could have favored group living, Anna Mosser and Margaret Kosmala developed a spatially explicit agent-based computer simulation (Mosser et al. 2015). Their model allows female "agents" to move across a landscape of 60 × 40 1-km^2 cells and to produce daughters according to the amount of resources they are able to accumulate in competition with other agents. The simulation landscape contains a total of seventy hotspots surrounded by low-quality habitat. In a highly peaked landscape, the hotspots are far more valuable than adjacent areas; in a low-peaked landscape, the hotspots are only slightly more valuable (figure 9.6a). Hotspots are either distributed in the same way as the river confluences in the Serengeti study area or are scattered randomly across the landscape (figure 9.6b). To avoid starvation, agents must obtain a minimum amount of resources at each time-step, but they can only consume a finite amount of food per day. Thus, as territories expand, benefits increase asymptotically, whereas the costs of territorial defense grow ever larger, and only a limited range of territory sizes provides sufficient benefits to exceed the costs (figure 9.6c). *Group* territoriality requires an area rich enough to meet the needs of multiple group members, and, if the territory provides enough resources, the group is able to grow further by recruiting adult daughters (figure 9.6c).

Neighbors only contest individual grid cells that are close enough to the midpoint of each pride's territory to provide a net benefit despite the costs of added defense (see Crofoot and Gilby 2012). When two groups fight, each pride experiences a 5 percent chance that one group member is killed. For unequal group sizes, the probability of the larger group winning is equal to $N_L/(N_L + N_S)$ where N_L is the size of the larger group and N_S is the size of the smaller group. But if the odds of the larger group winning are greater than 55 percent, the smaller group withdraws without a fight. Thus, larger groups expand the size and/or quality of their territories at the expense of smaller neighbors. Individuals mature and die according to the typical age-specific mortality for females in the Serengeti and Ngorongoro Crater.

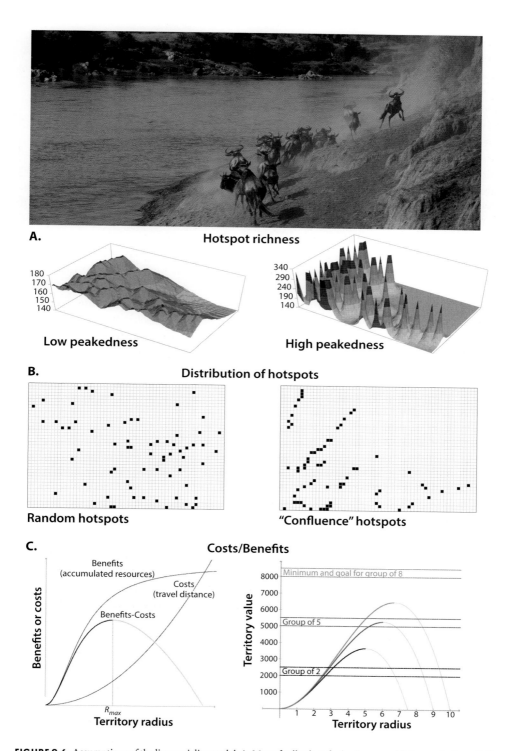

FIGURE 9.6. Assumptions of the lion sociality model. **A.** Map of cell values for landscapes with low peakedness (*left*) or high peakedness (*right*). **B.** Hotspots can either be distributed randomly across the landscape (*left*) or to mimic the spacing of confluences in the Serengeti (*right*). **C.** Changes in the costs and benefits of territoriality with increasing territory size (*left*). Requirements for groups to benefit from joint territorial defense (*right*). Daughters can only be recruited if the territory provides sufficient resources.

At the start of the simulation, all mothers are solitary and force their daughters to disperse at maturity regardless of the resources available in their territories. A mutant strategy is introduced into the population that allows daughters to remain with the mother, provided that the feeding requirements of the resultant pride would not exceed the available resources in that territory. If multiple daughters are forced to disperse, the cohort remains together as a new pride and first tries to establish a territory in the grid cells immediately adjacent to their mothers.

In testing the model, we varied two key variables. To assess the importance of the RDH, we altered the "peaked-ness" of the landscape by either concentrating the hotspots' values within smaller areas or spreading them out over broader areas. To assess the importance of population density, we increased the average value of each grid cell, with higher values allowing for greater numbers of agents within the landscape and, thus, more frequent encounters between neighbors. We ran each simulation for five hundred yearly time-steps with twenty replicates for all possible combinations of peaked-ness and landscape value.

Figure 9.7 shows a simplified version of the model containing two hotspots of low peakedness. The initial population is entirely solitary with the exception of a single social "mutant." The social female recruits her first daughter (figure 9.7a), then the mother and daughter move closer to the center of the hotspot and give birth to two cubs (figure 9.7b). The pride grows larger, expanding its territory and monopolizing most of the local hotspot (figure 9.7c). To keep from growing too large, the pride forces a lone daughter to disperse (figure 9.7d). The pride subsequently produces four sets of dispersing daughters, two of which establish territories next to their mothers, while the other two cohorts shift to the second hotspot (figure 9.7e). Although the social strategy eventually spreads across the landscape (figure 9.7f), the trait does not result in every pride being actively social: eight of twenty-two prides comprise a single female with the genetic potential to be social, but they each occupy a territory with insufficient resources to support two or more adults.

Figures 9.7g and 9.7h summarize the conditions favoring the evolution of lion sociality. If hotspots are *randomly* distributed, greater landscape heterogeneity (hotspot peakedness) and higher population density (average cell value) both favor group living, as sociality never evolves in homogeneous landscapes or at low densities (figure 9.7g). In contrast, the *clustered*-hotspot pattern generated by river confluences promotes sociality in two divergent circumstances (figure 9.7h): first, at low population densities in highly heterogeneous landscapes, lions can only survive at the hotspot clusters; the rest of the grid is essentially uninhabitable, so competition is intense and highly localized. Second, at most levels of heterogeneity in the richest habitats, lions can survive anywhere on the grid, but intruder pressure is so high that groups are better able to defend their territories than can solitaries. Thus, habitat heterogeneity mostly serves to increase population densities at the hotspots, so the two factors often promote sociality in tandem.

Our landscape model is obviously an oversimplification, not least because we only considered female-female interactions, whereas *resident males* affect cub survival (chapters 3 and 4) and recruitment of daughters (figure 5.12), and *neighboring males* affect female mortality and ranging patterns (figure 8.9). But in the broadest sense, the model makes intuitive sense in that tigers, leopards, and jaguars mostly live at

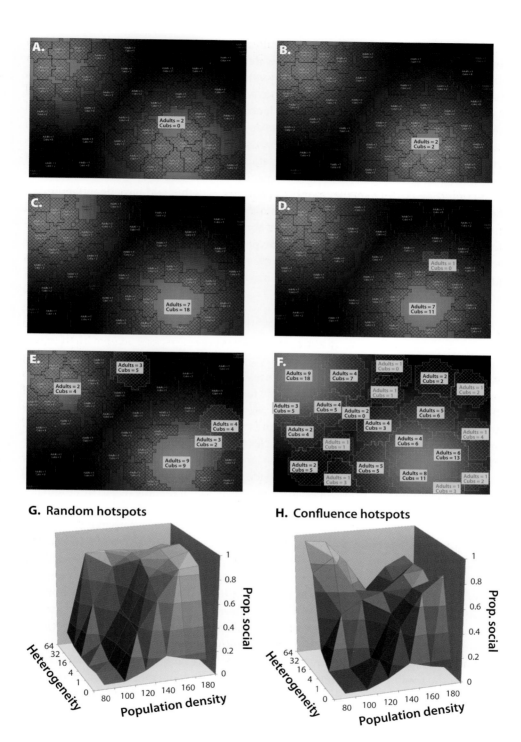

FIGURE 9.7. **A.–F.** Representative time-steps of the sociality model on a landscape with two hotspots (highlighted in bright green). Black cross-hatching shows the territories of solitary agents. Yellow cross-hatching shows the social agents; yellow boxes report the number of mothers and cubs in each pride; green lettering highlights "social" agents that are currently solitary. **G.–H.** Proportion of agents in the final population that carry the social gene, where less than 50 percent indicates an advantage to the solitary strategy and more than 50 percent indicates an advantage to the social strategy, for (**G.**) random and (**H.**) clustered hotspot distributions. Redrawn from Mosser et al. (2015).

lower densities in habitats that are generally more homogeneous (forests rather than savannas). Where lions do live at very low densities (e.g., deserts such as the Namib and the Kalahari), hotspots are extremely valuable as most of the landscape is uninhabitable. However, leopards also live in savannas, yet savanna-dwelling leopards are as solitary as their forest cousins. Where leopards and lions co-occur, leopards live at somewhat lower population densities than do lions, but more important is the intense aggression that lions direct toward leopards (see chapter 11), which would make it virtually impossible for leopards to form stable groups. Interestingly, saber-toothed cats like *Smilodon* were the largest felid species of their era, and fossil evidence suggests that they sometimes survived broken bones, which could have been facilitated by injured animals being able to feed from their groupmates' kills (Heald 1989)[3] and the relative abundance of *Smilodon* bones at the La Brea tar seeps suggests that they foraged in groups (Carbone et al. 2009). So perhaps we should add yet another factor to the evolution of group territoriality: body size. Only the largest felines can develop group territoriality, as they prevent smaller sympatric species from enjoying the same advantages of group living.

PRIDE SIZE REVISITED

What causes pride size to differ from one area to the next? One possible explanation is prey size, as a large carcass allows multiple females to feed from the same kill (Packer 1986). In all three of the long-term study areas, females typically capture prey large enough to feed multiple individuals (figures 1.6 and 1.7), and larger prides capture a greater proportion of the largest prey species (figure 7.6b). In contrast, lions in the Gir Forest of India typically capture prey that provide less than 50 kg of edible biomass compared to the approximately 100 kg provided by an adult wildebeest (see figure 9.12a), perhaps explaining why prides of Asian lions typically contain only two to three females (Jhala et al. 2019). Depending on availability, lions throughout Africa prefer similar-sized prey as in the Serengeti and Ngorongoro, ranging from wildebeest up to Cape buffalo (Hayward and Kerley 2005). But pride sizes range from two to three females in both the Kunene region of Namibia (Stander et al. 2018) and Samburu-Isiolo region of northern Kenya (Bhalla 2017), 5 females in Hwange (Mbizah et al. 2019), and six to eight females in both Okavango (Kotze et al. 2018) and Kruger (Ferreira and Funston 2010). Further, carcass size hardly varies across our three long-term study areas during the wet season, whereas Crater lions typically capture the largest prey during the dry season yet form the smallest prides in any of the three study areas (figure 9.8).

What about population density? Across Africa, lions vary widely in population density with the small prides of desert lions in Namibia and northern Kenya living at by far the lowest densities and the large prides of Kruger/Okavango lions living at densities similar to the Serengeti woodlands and Ngorongoro Crater. Consistent with this overall pattern, prides frequently undergo fragmentation in small, fenced reserves in South Africa (see chapter 12) where managers limit their lion populations by removing

---

3 We observed several cases where lions survived for over a year after losing the use of a back leg, and one male even survived several months after losing the use of a foreleg.

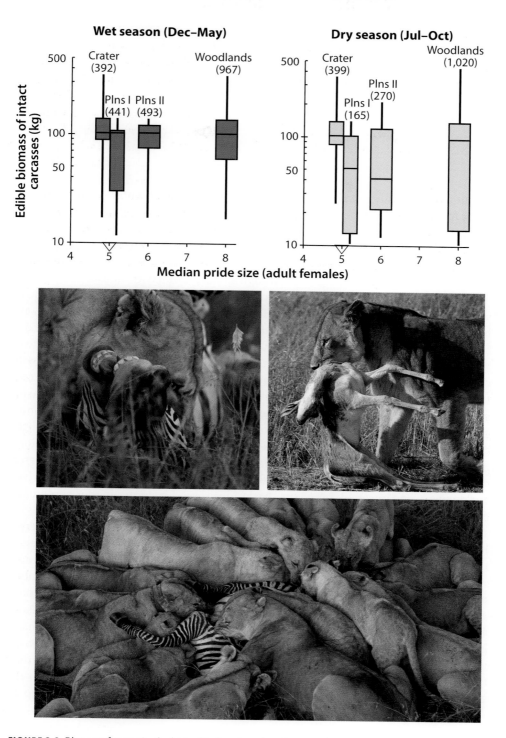

FIGURE 9.8. Biomass of prey animals obtained by lions in each study area each season. "Edible biomass" is estimated from the edible fraction of an intact carcass (see appendix); no attempt was made to correct for meat missing from scavenged carcasses. Median pride size was five females both in the plains prides in 1966–1996 (Plains I) and in the Crater, but pride size on the plains averaged six females in 1997–2015 (Plains II).

TABLE 9.2. *Population density, landscape heterogeneity, and pride size in the long-term study areas*

Habitat	Density: Females/100 km²	Landscape heterogeneity	Median pride size	Expected pride size
Ngorongoro Crater	11.2	Low	5	Small
Serengeti Plains 66–98	3.3	Low	5	Small
Serengeti Plains 00–15	5.7	Medium	6	Medium
Serengeti Woodlands	10.4	High	8	Large

all but a single pride (McEvoy 2019). Thus, instead of the one pride containing five to six females as would be typical for the habitat, the pride splits into multiple small prides of one to three females. On the other hand, lion density was far lower in the Serengeti plains in 1966–1998 than most years in the Ngorongoro Crater, yet pride size was virtually identical between the two habitats (figure 2.8). Nevertheless, population density may still influence pride size through an interaction with landscape heterogeneity (figure 9.9a). Table 9.2 summarizes the observed relationship between these three variables in the long-term study areas. The Crater floor and the Serengeti plains are both characterized by relatively few landscape features (figure 9.9b and 9.9c), and our sociality model predicts that population density will have little effect on pride size in relatively homogeneous landscapes (figure 9.9a). Female density on the plains nearly doubled after the expansion of taller grasses into shortgrass habitat (chapter 10), and the increased vegetative cover expanded the number of sites suitable for successful prey capture, possibly increasing the heterogeneity of the plains habitat. Thus, the combined increase in density and heterogeneity on the plains may explain the subsequent increase in pride size. Woodlands prides have a similar diet as the Crater lions (figures 1.6 and 1.7) and a similar population density, but the woodlands study area contains a greater diversity of vegetative cover (figure 9.9d) as well as numerous river confluences, and our model predicts that prides should be larger in more heterogeneous habitats, particularly at high population densities (figure 9.9a).

With three variables and only four data points, the relationship in table 9.2 doesn't hold much statistical power, but the landscape variables certainly make more sense than prey size in explaining why the largest prides were found in the woodlands and why the Crater prides were so small. A rigorous test would require a far larger number of study populations and a single biologically relevant measure of landscape heterogeneity that could be applied across all of Africa. Previous attempts to use the size and distribution of prey herds as measures of resource heterogeneity at Hwange have had only limited success, other than to suggest that, first, herd size at waterholes may help explain maximum foraging group size (Valeix et al. 2012) and, second, distance between waterholes better explains lion home range size than does pride size (Mbizah, et al. 2019).[4] But whereas we could readily use our real estate map of

---

[4] We also found no correlation between pride size and territory size on the Serengeti plains (figure 9.4c), and it is noteworthy that of the three components of group territoriality that formed the basis of our landscape model (the numerical advantages of larger group size in winning interpride encounters, the ability of larger prides to control larger territory sizes, and the recruitment of daughters into the maternal pride), the evolution of sociality was *least* sensitive to gaining larger territory size (Mosser et al. 2015). Habitat *quality* mattered more than total area.

FIGURE 9.9. Landscapes and pride size. **A.** Predicted effects of population density and landscape heterogeneity on pride size in a confluence-clustered landscape. Model outputs were rescaled to match the observed range of pride sizes across the long-term study areas. Representative prides and habitat in the (**B.**) Ngorongoro Crater; (**C.**) Serengeti plains; and (**D.**) Serengeti woodlands.

lion fitness (figure 9.2) as a template for measuring landscape heterogeneity in the Serengeti, comparable data are not yet available from any other lion population—and some sites would be virtually impossible to characterize: Botswana's Okavango Delta, for example, undergoes annual variations in flooding that alter the landscape more rapidly than the two-plus years required by mothers to recruit daughters into their prides. Thus, the Delta prides are perpetually chasing a moving combination of landscape heterogeneity and population density.

THE ORIGINS OF MALE SOCIALITY

When a female leopard or tiger successfully rears a litter of two sons, the dispersing brothers go their separate ways. But when a female lion raises a pair of sons, the brothers disperse together as a coalition—and even if a female only manages to rear a single son, chances are that he'll seek out a coalition partner once he has left home. Male lions are at least as gregarious as females—providing as striking a contrast with tigers, leopards, jaguars as do female lions. How did *male* sociality arise in lions and why? The easy answer is that not only do lions have significantly larger litters than any of the solitary big cat species (figure 4.14e), thus increasing the chances that maturing sons will have ready-made partners, but lion prides also produce cohorts of similarly aged sons, and larger cohorts produce larger male coalitions (figure 5.5c). In fact, *solitary* female lions only ever produced three sons that eventually gained residence in our long-term study areas: all three started out as singletons (figure 9.10a), and they all teamed up with unrelated partners to gain residence as members of a pair. Prides containing only two females also typically produced singleton males, and only prides of three or more females routinely produced coalitions of two or more sons. The pressure to form coalitions is so great that over three-fourths of resident males that had dispersed from their natal prides as singletons soon teamed up with other singletons or pairs, and a quarter of dispersing pairs formed trios with an unrelated third (figure 9.10b).

Despite the pressure to find enough partners to gain (figure 2.11) and maintain (figure 2.12) residency, it is somewhat surprising that coalition partners are only rarely found all together in the same place at the same time. Like females, coalition partners show fission-fusion group patterns, and, whereas pairs are typically found together, members of trios and larger are typically found apart from each other (figure 9.10c). By spreading themselves out, the resident coalitions are presumably better able to detect trespassing males, at which point they can recruit their partners to expel the invader(s) (figures 8.3d, 8.3e, and 8.4b). A resident male from a pair or a very large coalition is slightly, but significantly, less likely to be found apart from his companions than were males from trios and quartets (figure 9.10d), possibly because pairs face the greatest risks of meeting larger coalitions, whereas quintets and larger don't need to spread themselves so thinly across their territories. However, we never attached more than one radio collar on any male coalition, so we lack data on how individual coalition partners distribute themselves around their joint territory.

Presumably, territorial males defend territories in order to prevent rival males from mating with the females within their domains. When a male jaguar, leopard, or tiger

FIGURE 9.10. Origins, additions, and associations of coalition partners. **A.** Effect of maternal pride size on the number of related males in each resident coalition. Solitary females only produced single surviving sons, female pairs produced equal numbers of singletons and male pairs, and larger prides routinely produced male pairs, trios, or larger. n = number of cohorts that became resident. **B.** Proportion of cohorts that teamed up with outside males before becoming resident. **C.** Male group size does not vary significantly with coalition size, and, except for pairs, partners are seldom all found together. **D.** Individual members of trios and quartets were significantly more likely to be seen apart from the remainder of their coalition than were members of pairs or coalitions of five to nine males. Data in **C.** and **D.** include observations when the male was seen alone as well as when he was the only coalition member present in a group with females and/or cubs.

establishes a territory, his range overlaps the ranges of several females so that he can likely father most or all of their offspring. When male lions gain residence, the coalition's range overlaps the pride range, and the residents do indeed father all of the females' cubs—at least in the Serengeti (chapter 2). However, paternity tests in Namibia's Etosha National Park revealed that 41 percent of thirty-four cubs were fathered by males from adjacent prides (Lyke et al. 2013), and exclusive access of a pride by only one male coalition is the exception rather than the rule in the Gir Forest of western India (Chakrabarti and Jhala 2019). Why does breeding exclusivity of the resident coalition vary from one lion population to another?

Figure 9.11 presents a simplified schematic representation of male and female ranging patterns in solitary felids and in lions of the Serengeti, Etosha, and the Gir Forest. Each idealized male in a typical solitary species maintains exclusive access to his females, and male coalitions in the Serengeti are known to be successful in fathering all the cubs in their prides. However, female lions in Etosha range over such wide areas (500–1,000 km^2) that two *singleton* males fathered only three of twelve cubs, and a *co-resident* trio fathered six of eleven cubs in two adjacent prides, although the three pairs that resided in single prides did manage to sire all eleven offspring with their respective females (Lyke et al. 2013). The smaller coalitions in the Gir Forest (four singletons, five pairs, and one trio) all shared residence with one or more other coalition, whereas the sole quartet in the Gir Forest managed to monopolize their prides. Thus, I have presented the mating system in Etosha as being characterized by female home ranges that are often too large for resident coalitions to monopolize and illustrated the Gir prides as often having multiple resident coalitions.

The situation in Etosha seems intuitive: females cover too large of an area for singleton males to be able to monopolize successfully or for a trio to be safely co-resident in adjacent prides, but, even here, the resident pairs fathered all the cubs in their respective prides. Pride ranges in the Gir Forest are about the same size as in the Serengeti woodlands, yet resident pairs and trios in the Gir are much less successful in monopolizing their prides. What can explain this pattern? Resident males must divide their time between patrolling for strangers, monitoring their females, protecting their cubs, and finding food. Though males may be as competent as—or even better than—females in catching prey as large as a Cape buffalo (figure 7.3), they primarily rely on the females to capture zebra, wildebeest, or any of the smaller prey species, and scavenging from the females will be most profitable when the prey is large enough to feed a mixed-sex group. As described above, typical prey size is far smaller in Gir than in the Serengeti or Ngorongoro (figure 9.12a), and the extent to which males associate with females depends on the presence and size of captured prey (Chakrabarti et al. 2021): males in all three areas are least likely to associate with females in the absence of a carcass, and the degree of association at a carcass depends on the size of the prey (figure 9.12b). When Gir lions feed on carcasses as large as in the Serengeti and Ngorongoro, male-female associations are almost identical in all three areas, but Gir males associate far less with females at small carcasses. The pattern in Gir appears to extend to situations where kills are absent, resulting in far lower levels of overall male-female association than in the Serengeti or Ngorongoro Crater. Indeed, the higher level of male-association in the Crater males even in the absence of a kill (also see

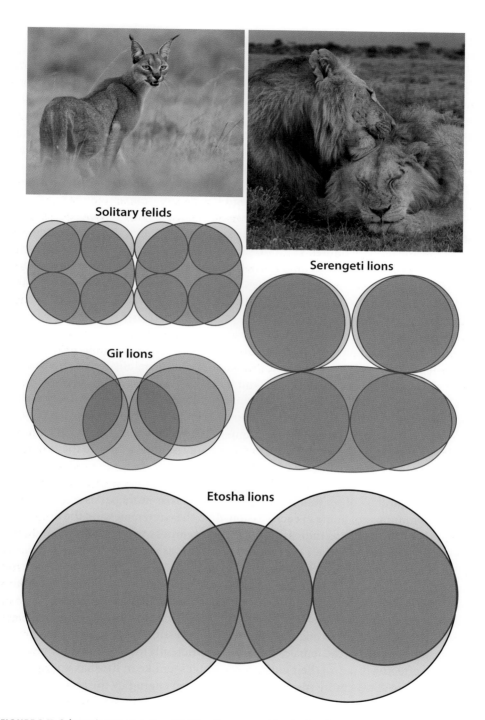

FIGURE 9.11. Schematic representation of felid mating systems. Females in solitary species occupy nonoverlapping home ranges; the home range of each male overlaps the ranges of multiple females. In the Serengeti, each lion pride has a separate home range; the ranges of resident coalitions overlap that of their pride(s); larger coalitions maintain residence in multiple prides. In Etosha, female home ranges are too large for a single coalition to cover the entire area, preventing the males from fathering all the offspring. In Gir, male and female home ranges overlap, but males spend so little time with "their" females that a second coalition may hold residence in the same pride at the same time.

figure 3.7a) may result from the males anticipating that the next carcass will be large enough to scavenge a decent meal from the females.

The degree to which resident males associate with females likely results from three factors: first, the males' efforts to join the females at a particular location and to remain with them until the females move off. Males generally join female groups more often than vice versa, whereas females are more likely to depart from the males (figure 9.12c). This pattern is most striking at small and midsized prey (which are primarily captured by females [figure 7.4a] and scavenged by males [figure 2.5a]), although not at the largest carcasses, where females have the opportunity to feed from kills made by the males (figure 7.4a). Second, the fission-fusion nature of lion prides allows females to distribute themselves across their pride ranges, and female subgroups in the Serengeti are spaced more closely together during the dry season (when females rely more on ambush hunting along watercourses) than during the wet season (when they more often feed in open areas: compare figures 9.2a and 9.2b) and are also closer together than female subgroups at Gir (where artificial waterholes are abundant and prey availability does not change seasonally) (figure 9.13a). When not directly associating with their females, resident males in Gir are significantly farther away from their females than are Serengeti males at any time of year (figure 9.13a), presumably because of a greater need for the Gir males to forage on their own rather than scavenge from females.

Third, distant pridemates communicate by roaring, thus providing opportunities for resident males to locate their pride females. Males generally roar more than females, although this varies by season and location (figure 9.13b). In the Serengeti, females roar at similar rates as the males during the wet season (when wildebeest and zebra are abundant), but females reduce their roaring in the dry season (when prey are small and scarce), whereas the males continue to roar at the same rate. In Gir, prey size remains constant throughout the year, and their prey is more similar to the size of dry season prey in Serengeti, during which time the females' roaring rates are virtually identical in the two habitats. In contrast, the males at Gir roar at similar rates as in the Serengeti, regardless of season. However, Gir males roar more persistently (roaring for a greater number of bouts in each session) than the Gir females and the dry season Serengeti males—perhaps indicating a greater eagerness to find mates that do not necessarily want to share their consistently small carcasses.

Taken together these findings suggest that small prey size and greater dispersion among pridemates prevents the Gir males from spending as much time with females as do the males in Serengeti and Ngorongoro, and, consequently, renders the Gir males less able to prevent their females from interacting with rival coalitions. Thus, the ability of Serengeti/Ngorongoro coalitions to successfully monopolize their prides likely results from, first, the females remaining sufficiently localized that the resident males can effectively patrol their entire pride ranges (unlike in Etosha), and, second, larger prey size allows more frequent intersexual scavenging by the resident males, thereby reducing the females' risk of exposure to outside coalitions. However, the male quartet at Gir nevertheless managed to monopolize its prides, and, even in Etosha, male pairs fathered every cub in their prides, whereas singleton males only fathered 25 percent of cubs. Thus it is notable that although sharing a pride with a second male at Etosha would have resulted in a 50 percent reduction in individual reproductive success, the payoff of pair formation still exceeded the 25 percent paternity rate for

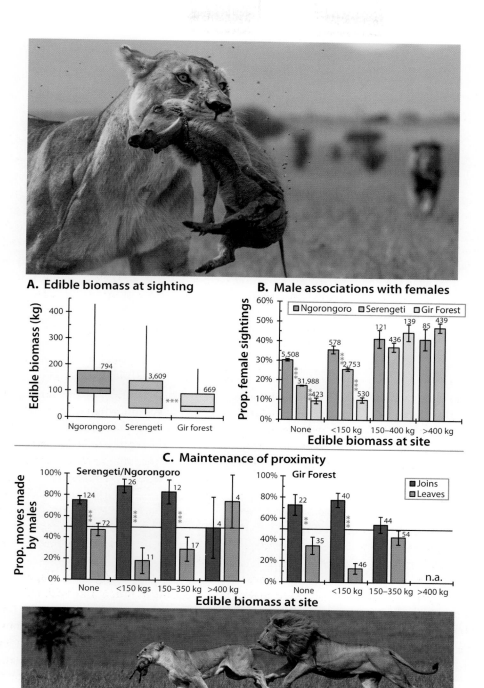

FIGURE 9.12. Comparisons of prey size, male-female associations, and roaring between the Serengeti/Ngorongoro Crater and the Gir Forest, India. **A.** Females in the Crater and the Serengeti typically obtain over twice as much edible biomass per feeding event as in Gir. Biomass for captured prey is calculated as in the appendix, while correcting for meat remaining from scavenged carcasses and summing the total biomass at multiple kills; n = number of female sightings. **B.** Females associate with males at significantly higher rates in the two Tanzanian populations except when feeding at larger carcasses. Data exclude sightings of consort pairs. n = number of female sightings. **C.** Males primarily maintained proximity with females in most circumstances, as they joined females more often than vice versa and were less likely to leave the females, although this contrast disappeared at the largest carcasses. Data from the Serengeti and Ngorongoro were merged as sample sizes from the Crater were small, and trends were identical. n = number of sightings where single-sexed subgroups joined or left opposite-sexed subgroups.

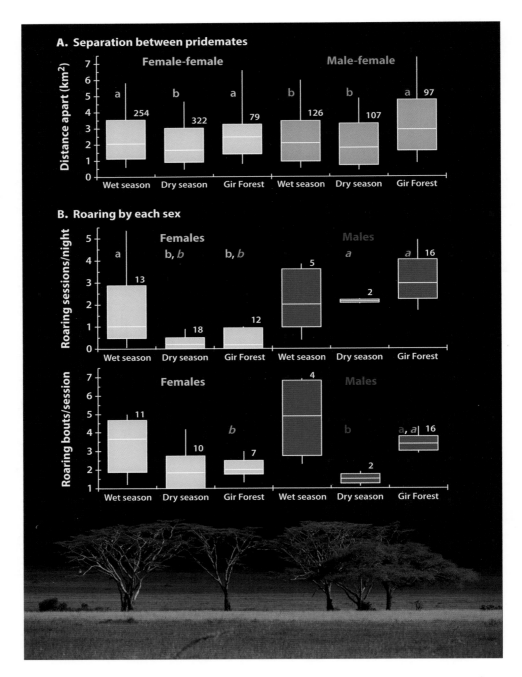

FIGURE 9.13. Separation of subgroups and roaring in the Serengeti and Gir Forest. Letters designate significant differences in female-female (tan), male-female (orange) and male-male/female (brown) comparisons, respectively. **A.** Female subgroups were closest together during the dry season in the Serengeti, whereas resident males were farthest from their females in Gir. Data from the Serengeti are restricted to distances between radio-collared pridemates that were located on the same day; data from Gir are observations where multiple subgroups were located on the same day. n = number of paired sightings. **B.** Roaring in single-sexed groups in the Serengeti and Gir. *Top:* Serengeti females roar more in the wet season than in the dry season and Gir. Males roar at similar rates in each situation but more than females in the dry season in Serengeti and in Gir (as indicated by italicized letters). *Bottom:* Females roar for a similar number of bouts, regardless of situation. Gir males roar longer than both Gir females and Serengeti males in the dry season. n = number of observation periods of at least one full night on females and a half-night on males. All redrawn from Chakrabarti et al. (2021).

singletons—even without any of the other advantages of coalition formation outlined in chapter 2.[5]

At some level, the advantages of male sociality derive from the clustered nature of lion females: by holding a territory large enough to overlap an entire pride, a cooperating pair would fare better than either would alone—a mutant ancestral male lion that formed a coalition with a second male would easily invade a population of ancestral male proto-lions that followed the tiger-leopard strategy of nonoverlapping individual territories.

THE LION: THE PARAGON OF GROUP LIVING

In *The Descent of Man* (1871), Darwin discussed altruistic and self-sacrificial behavior in humans, "He who was ready to sacrifice his life . . . rather than betray his comrades, would often leave no offspring to inherit his noble nature." Darwin then argued that such behavior might be beneficial to warring groups: "A tribe including many members who . . . were always ready to give aid to each other and sacrifice themselves for the common good, would be victorious over most other tribes; and this would be natural selection." We could easily substitute "pride" or "coalition" for "tribe" and frame an equally compelling description of lion behavior.

Females band together to defend valuable pieces of real estate; they coalesce to hunt a buffalo and split up when they need to hunt alone. Mothers form tight-knit nursery groups to protect their cubs from a variety of dangers, with the option to forage alone while a babysitter remains on duty. Maternal prides populate their neighborhoods with descendant prides—and daughters sometimes expand their mothers' empire to adjacent neighborhoods. By giving rise to coalitions of sons, prides can extend their genetic legacy even wider. But prides and coalitions can be too large—super prides are prone to fragmentation, resort to taking smaller prey, and suffer from more frequent male takeovers; super coalitions are less well coordinated and prone to fragmentation.

The pride is egalitarian rather than despotic, thus the advantages from group living are so profoundly mutualistic that females treat their pridemates as necessary partners rather than rivals—they respond to the roars of short-handed pridemates when outnumbered in territorial encounters, they gently rebuff the suckling attempts of pridemates' cubs while accepting them as future partners for their own offspring, and they encircle trespassers before administering a potential coup de grâce. Yet they don't always work together when hunting, they can be selfish when hogging a carcass, cubs frequently steal their aunties' milk, and consorting males jealously guard receptive females.

One of the longest-running debates in evolutionary biology is the relative importance of individual selection versus group selection. Darwin's fundamental principles of natural selection were based on the differential survival and reproduction of individuals, but as his writings on tribal warfare made clear—and has been more formally explored by Traulsen and Nowak (2006), Bowles (2009) and Bowles and Gintis

---

5 We did not have any singleton males when we performed our paternity testing, but I wouldn't have been surprised if they had not successfully fathered all the cubs in their prides—especially on the open plains, where female home ranges are much larger than in the woodlands.

(2011)—the unit of selection can be the social group, especially where groups have the power to annihilate each other. In these cases, group benefits are so strongly mutualistic that there is little conflict between individual selection and group selection. The selfish individual that undermines its group will be killed along with everyone else. However, this doesn't mean that everything should always be sweetness and light: individual selection will still favor a certain degree of selfishness, provided these petty disputes remain petty, and the group's ability to do battle against its enemies is not compromised. And, indeed, one of the great fascinations of watching lions for so many years has been the way that their behavior varies so markedly from one context to the next: lions can be utterly selfish at the dinner table or in the bedroom, but when the time comes to pull down the buffalo or to defend hearth and home, their self-interests align toward getting enough food during a particularly harsh dry season or repelling neighboring groups from a valuable stretch of real estate.

The very strong advantages of cooperation likely explain why females accept their pridemates as true partners rather than risk injury from competing with each other for higher reproductive status—egalitarianism resulting from a perpetual truce between individuals that might otherwise cripple a necessary companion. The competitive spirit is still there—all those ear notches come from a determination to maintain access to a particularly desirable spot at a carcass—but the intensity of aggression is held in check. Males get more agitated when competing against their coalition partners for access to a receptive female, but once the consortship is clearly established, they respect each other's ownership. Males are certainly more hierarchical than the females, but pairs gain strong mutualistic advantages from group membership, and the larger coalitions may require kinship either to accept a strong degree of reproductive skew—being compensated via inclusive fitness effects from the success of a close relative—or to wait in a reproductive queue until their higher-ranking companion burns out and they can take their turn as the most successful breeder in the group.

Brian Bertram (1976) originally attributed several of the more interesting aspects of lion sociality to kin selection—including communal nursing and reduced competition between coalition partners. But what is the role of kinship in these situations? Mothers are more likely to nurse the cubs of their first-order kin, but the underlying pressure to form a crèche derives from mutualistic advantages of collective cub defense. As for male coalitions, unrelated partners are behaviorally indistinguishable from related partners—except that related partners show greater skew in mating activity and reproductive success than do unrelated partners. However, this is all driven by the fact that daughters are recruited into their mothers' prides, and cohorts of daughters mostly consist of sisters and cousins. Natal philopatry inevitably leads to kin groups, and even though we can identify mutualistic advantages from group formation, these will still confer inclusive fitness effects, and there must be considerable proximal benefits from interacting with the same individuals since birth—and from living in the same familiar area where they grew up—and our simulation model highlighted that sociality was far more likely to evolve when daughters joined their mothers' prides. Thus, kinship is integral to lion sociality, even if we think of lions as living in a mutualistic society that necessarily involves an enlarged family of like-sexed siblings and cousins that may squabble at the dinner table but truly respect each other and rise to each other's aid when the time comes to do something grand.

I cited Darwin's famous quotation at the beginning of this section because I have long found the parallels between lions and humans—and even chimpanzees—to be so compelling. Warfare appears to be a human universal (Bowles 2009), and boundary disputes and gang attacks between neighboring communities are ubiquitous in male chimpanzees (Wilson et al. 2014).[6] Warfare plays a central role in Bowles and Gintis's (2011) models of human evolution, based on the high levels of mortality suffered by the losers—but their work emphasizes the importance of within-group punishment to provide a favorable environment for higher-level cooperation, primarily through various types of "social control." Boehm (2011) summarized extensive evidence from hunter-gatherer societies that noncooperative individuals were subjected to group-level shunning, shaming, ridicule, expulsion, and even assassination. However, lions consistently dominate smaller groups in interpride encounters and work together to surround intruders and protect their cubs from potentially infanticidal males despite showing no signs of intragroup punishment. When two females hold back during a hunt then keep the successful hunter from feeding (as Moss and Mavuno did with Majani, see chapter 7), all is forgotten (or even forgiven?) by the next time the three companions do anything together. Lions aren't vindictive, thus they are able to maintain a warlike society without any obvious form of policing or social control. The threats from neighboring prides and nomadic males arise almost daily, so perhaps there's no need to remind pridemates to keep in line.

The contrast between the lions' intragroup squabbling and intergroup warfare fits within a framework of classifying aggression into two broad categories: reactive versus proactive. A lion is *reacting* in anger or frustration when a female lashes out at the approach of a pridemate to her spot at a kill or when a male snarls at a coalition partner that has come too close to his consort. But lions are *being proactive* when they stalk their prey with intent to kill, as is a male when he is killing the cubs of his predecessors or when a group of lions approaches the recorded roars of a simulated intruder or encircles a trespasser. These two responses are triggered by different neurological pathways and reflect contrasting selective pressures (reviewed by Wrangham 2018).

Although direct neurological assessments have not been conducted in lions, Jessica Burkhart has tested the responses of captive lions to elevated levels of oxytocin, a peptide hormone that plays an important role in prosocial and affiliative behavior. Working at a wildlife sanctuary in Dinokeng, South Africa, Jessica used an atomizer to quickly administer oxytocin to every member of a captive group then presented them with experimental challenges (Burkhart et al. 2022). In one set of experiments, a 15 × 40 cm frozen blood "popsicle" was placed in the group's half-hectare enclosure. As with wild lions, the first animal to reach the food item behaved possessively by threatening any companion that came too close, but oxytocin had virtually no effect on how near the other lions came to the "owner" and their rates of aggression were little changed. In contrast, when Jessica played recorded roars of an unfamiliar individual just outside their enclosure, the oxytocin-treated lions moved closer together and seemed less concerned by the outside threat, as they remained silent,

---

6 Whereas chimps, *Pan troglodytes*, are group territorial, their closest relatives, *Pan paniscus*, are much less so—and, interestingly, their contrasting habitat preferences mirror those of lions versus tigers: chimps more often live in savanna habitats, whereas bonobos more generally live in rain forests (Hohmann et al. 2010).

whereas the "control lions" almost always roared in response (figure 9.14) (Burkhart et al. 2022). These contrasting results suggested that reactive aggression (during feeding competition) versus proactive aggression (in territorial defense) may involve similarly divergent neurological processes as have been described in other mammalian species (Tulogdi et al. 2010, 2015).

Regardless of the precise physiological mechanism, lions are far more tolerant of conspecifics than are their solitary cousins. Whereas all the big cat species are territorial, only the lion allows members of its own kind to remain within its territory, and even if lions express anger and frustration with each other when feeding together at a kill, they come together day after day and feed together again and again. They presumably elicit a burst of oxytocin every time they rub their heads together, every time the reunite after a few days apart—they must in some deep-seated way be programmed to live together, work together, and solve together the challenges posed by their own kind.

KEY POINTS

1. Female reproductive rates are highest in areas closest to river confluences as these provide more continuous access to water, safer den sites for hiding cubs, and better opportunities for capturing prey. Larger prides control larger and higher-quality areas.
2. Agent-based simulation models suggest that group territoriality can drive the evolution of sociality under particular combinations of population density and habitat heterogeneity. Lions are likely to be the only social cat species because savanna habitats more closely meet these conditions than do the relatively homogeneous forested or otherwise low-density habitats preferred by the other species in the genus *Panthera*.
3. Average pride size is not strongly influenced by typical prey size in the long-term study areas but instead appears to be influenced by contrasting levels of landscape heterogeneity and lion population density, with the largest prides being found in the highly heterogeneous, high-density Serengeti woodlands.
4. Male sociality ultimately arises from female sociality: large prides generate cohorts of male relatives, thus forcing singleton males to find unrelated companions. In the Tanzanian study areas, resident coalitions typically maintain exclusive access to their pride females. But extrapride males often sire offspring in populations where resident males spend considerable time apart from the pride females.
5. Lions show many hallmarks of group selection, as social groups compete against each other for valuable real estate, and reproductive success depends so strongly on group size. These factors may have contributed to the low intensity of within-pride competition for food and mates compared to the severity of interpride conflicts.

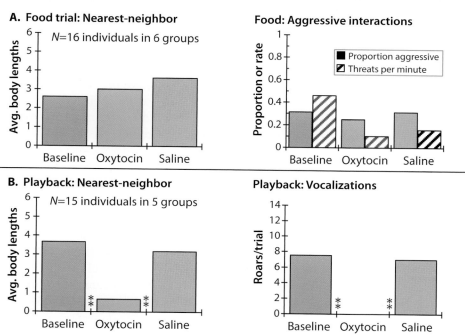

FIGURE 9.14. Effects of oxytocin on lion behavior. **A.** Food-object trials. There was no significant difference across treatments in either the distance between neighbors (*left*) or aggressiveness in the presence of the food object (*right*). **B.** Roar-playback trials. Lions remained in significantly closer proximity to their nearest neighbor post-oxytocin (*left*) and showed a significant drop in territorial vocalizations (*right*). Vertical bars are 95 percent confidence intervals. Data from Burkhart et al. (2022). *Photograph:* A field assistant uses a piece of meat to draw a captive lion close to the fence so that Jessica Burkhart can administer a measured dose of oxytocin with an atomizer.

CHAPTER 10

POPULATION REGULATION

Population, when unchecked, increases in a geometrical ratio.
—THOMAS MALTHUS

Ecologists seek to understand the forces that shape the abundance and distribution of organisms. Why are there so many lions in one area and so few in another? Why does the population grow or shrink from one year to the next? If the size of a population is limited by its food supply, it is said to be subject to "bottom-up" regulation. If a population is held below its carrying capacity by predation or disease, it is sensitive to "top-down" processes. Over the past half century, the Serengeti study area has held as few as 100 lions and as many as 350 individuals; the Crater population has varied from a low of 10 to a peak of 124. Both populations responded to changes in food availability, and both have suffered from lethal outbreaks of infectious disease. But the two populations experience such striking contrasts in food supply (the Great Migration in the Serengeti versus the Great Meat Locker of the Crater) and in their susceptibility to infectious disease (the Crater lions lead the Serengeti five die-offs to one) that we need to treat their respective histories like two different case studies.

SERENGETI: TRACKING LIONS IN A CHANGING ENVIRONMENT

When I first arrived in the Serengeti, the ecosystem had only just recovered from the Great Rinderpest Epizootic that had swept across Africa in the 1890s, killing 80 to 90 percent of all ruminants (Sinclair 1979b). The virus was finally eliminated from the ecosystem in the early 1960s, and the wildebeest population grew from a few hundred thousand to 1.4 million by 1977, while the Cape buffalo doubled in numbers and Thomson's gazelle increased by about 20 percent, meanwhile zebra (which are unaffected by rinderpest) remained relatively constant (figure 10.1). The massive increase in at least three of their five major prey species[1] allowed the lions to double their numbers between the mid-1960s and mid-1970s (figure 10.2). But whereas prey populations showed no further growth after the 1970s, the woodlands

---

1 Warthog are also susceptible to rinderpest, and their numbers must likewise have increased from the early 1960s, but no long-term data are available.

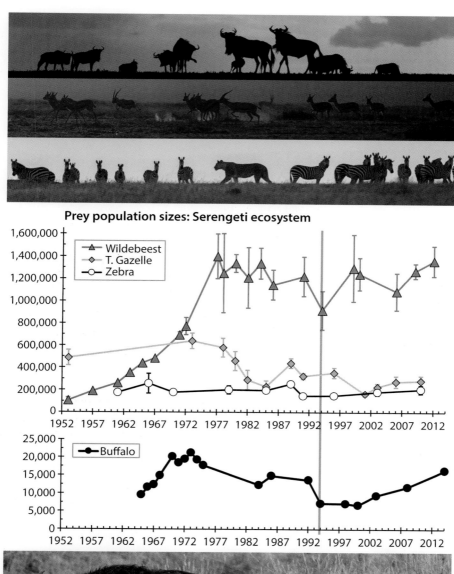

FIGURE 10.1. Serengeti herbivore population sizes. *Top:* Estimated total populations of migratory wildebeest, Thomson's gazelle, and zebra. *Bottom:* Total count of Cape buffalo in the central Serengeti. Red line highlights the drought in 1993. Redrawn and updated from Packer, Hilborn, et al. (2005).

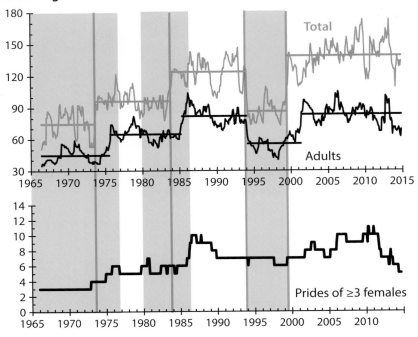
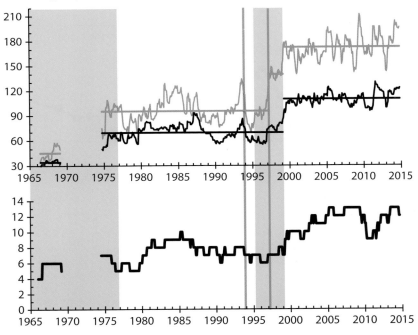

FIGURE 10.2. Lion population sizes each month in the Serengeti in the (**A.**) woodlands and (**B.**) plains. *Top:* Blue lines include all individuals; black lines indicate lions over two years of age. Horizontal lines indicate time periods when population sizes were statistically indistinguishable but different from adjacent periods. *Bottom:* Number of prides containing at least three females each month. Pale green blocks highlight time spans when the populations were presumed to be below the local carrying capacity; dark green lines demarcate the first year within each time span with favorable dry season rainfall. Red line shows the CDV die-off in 1994. Redrawn and updated from Packer, Hilborn, et al. (2005).

lion subpopulation suddenly grew a second time in the 1980s and the plains lions soared in the 1990s (Packer et al. 2005).

I had no idea what had triggered these later changes, and I was puzzled why each subpopulation had remained so stable before it suddenly jumped to a new level and remained so constant for the following decade, so I contacted Tony Sinclair, who had initiated, maintained, and coordinated the ecological monitoring of the Serengeti herbivores, grasslands, and woody vegetation since the mid-1960s (Sinclair and Norton-Griffiths 1979). Tony was quick to point out how certain changes in the local landscape might have improved the lions' *accessibility* to their prey. The tenfold increase in wildebeest had physically altered the park. Enormous numbers of wildebeest have the same effect as a vast lawn mower that consumes most of the fuel for a potential wildfire the next dry season (figure 10.3a). Pre-recovery, there was plenty of uneaten grass, and fires regularly burned most of the park, but the recovering wildebeest created enough firebreaks by the early 1980s to drastically reduce the extent of grassfire across the Serengeti (figure 10.3b). Seedlings and young trees were less often incinerated by the annual conflagrations (figure 10.3c), and more trees meant more cover for lions (photo inset to figure 10.3), improving their hunting success and thereby allowing the woodlands lion population to grow again. On the other hand, shallow soils and low rainfall inhibit tree growth on the Serengeti plains, so the plains portion of the study area remained largely unchanged until a severe drought in 1993–1994 killed large numbers of wildebeest and buffalo (figure 10.1) and caused the wildebeest to "skip" a substantial proportion of the tall and intermediate-grass plains. The altered grazing pattern allowed the grasses to grow tall enough to conceal hunting lions in a larger portion of the plains study area (figure 10.3d).

Thus, the elimination of rinderpest had three significant consequences. First, the massive increase in ruminant prey directly supported a near doubling of the overall lion population. Second, the lawn-mowing service of the fully recovered wildebeest population reduced the extent of wildfire enough to increase prey accessibility for the lions living in the regenerating woodlands. Third, the first severe post-recovery drought shifted the wildebeest far enough away to leave the taller-grass plains unmown and thereby increase the hunting success of unseen lions. But while these changes can account for the direction and approximate timing of the subpopulation shifts, we still need to explain the stop-start pattern of decade-long stability punctuated by sudden change. Why didn't the woodlands subpopulation grow in lockstep with the expanding prey populations during the 1960s and 1970s or as gradually as the trees in the 1980s?

The answer lies in the vulnerability of the Serengeti lions to the year-to-year whims of nature: their most important prey species come and go, and no two years are exactly the same. In a typical year, the wildebeest enter the long-term study area shortly after the onset of the rains in November and leave to the south and east of the study area by mid-December (figure 1.4). As the rains taper off in May, the migration returns to the center of the park, moving north and west, well beyond the territories of our study animals. As we've seen, the dry season (July–October) is the critical period for food limitation (Fig 1.13a) and cub mortality (figures 1.14b and 4.16b), but some years are harsher than others. Weather in East Africa is strongly affected by the "El Niño–Southern Oscillation," and the Southern Oscillation Index (SOI) is measured by the difference in barometric pressure between Tahiti and Darwin, Australia. Dry season rainfall in the Serengeti correlates

FIGURE 10.3. Wildebeest, fire, and the regeneration of woody vegetation in the Serengeti woodlands. **A.** The extent of wildfire is inversely related to the size of the wildebeest population. **B.** Wildfire reached a low point in the late 1970s and early 1980s. **C.** Population growth rates of acacias in the Serengeti woodlands peaked in the early 1980s. Green band indicates time period when the woodlands lions experienced the greatest increase in prey accessibility. Redrawn from Packer, Hilborn, et al. (2005). **D.** Long-term changes in grass height along three transects on the Serengeti plains. The extent of tall grass increased after 1994 in two of the three transects. Redrawn from Packer, Hilborn, et al. (2005).

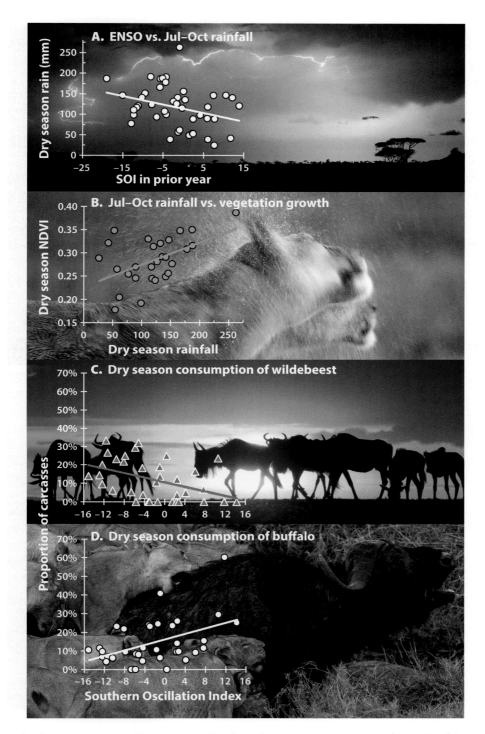

FIGURE 10.4. Dry season rainfall, vegetation, and lion diet in the Serengeti. **A.** Dry season rainfall is highest following years with the lowest Southern Oscillation Index. **B.** The Normalized Difference Vegetative Index is highest during dry seasons with higher levels of rainfall; redrawn from Sinclair et al. (2013). Proportion of (**C.**) wildebeest and (**D.**) buffalo carcasses in the lions' dry season diet varies significantly with SOI in the previous year. Data only include dry seasons (July–October) with at least ten observed carcasses.

with the average SOI from February to June during the prior year (figure 10.4a). With sufficient rain, grass grows in the ordinarily parched months of July–October (figure 10.4b), attracting the wildebeest to our part of the ecosystem (figure 10.4c)—whereas the woodlands prides rely most on Cape buffalo during the driest years (figure 10.4d).

Figure 10.5a summarizes the effects of dry season rainfall on the diet of the study lions: in years with wetter dry seasons, the lions feed more on wildebeest and zebra and less on Thomson's gazelle. In the driest dry seasons, the lions feed more on buffalo, and, when fewer zebra are available, lions consume more warthog and buffalo. During the recovery of the wildebeest population and the changes in vegetation (represented by the green rectangles in figure 10.2), the lions were much thinner than usual (figure 10.5b), despite their subpopulations falling below the rising carrying capacity. Although the ongoing ecological changes may have gradually improved the *potential* food supply/accessibility in each study area, the *actual* food was out of reach to the north or west. Despite a continuous improvement over a broad scale, our study lions couldn't benefit directly until the first good year of dry season rainfall: a wetter dry season means more wildebeest and zebra in the local lions' diet (figure 10.5a) and directly translates into higher cub survival (figure 10.5c). The first good rainfall years after each major ecological change (1973, 1984, and 1999 in the woodlands; 1997 on the plains) provided the necessary conditions for virtually every pride to produce surviving cubs and to ultimately generate large cohorts of dispersing daughters (bottom of figure 10.2). Thus, most years were so challenging that prides could only maintain their numbers, but the occasional good year enabled most females to breed successfully, and a new crop of maturing daughter prides subsequently worked their way into the territorial framework. These were the times when groups produced new groups—inevitably creating a coarse-grained pattern of population change. If the lions were solitary, we would have seen a more tapered response, rising like the slopes of a sand dune, but instead the pattern of growth was as jagged as a pile of stones.

Finally, although we had a reasonable understanding of why and when each Serengeti subpopulation grew, we had no way of predicting precisely how many lions each area *should* have held—there were simply too many unmeasurables. Each Serengeti subpopulation was clearly sensitive to the local food supply, but the lions survive each dry season on a scattering of a half-dozen resident prey species and varying numbers of migrants that pass through for varying periods of time—and if things get too harsh, they may even abandon their territories for a few days or weeks in search of the migration (as described for SBG in chapter 8). None of these could have been measured with any precision, so we had no way to estimate local carrying capacities from one year to the next.

TOP-DOWN POPULATION REGULATION: PREDATION, INFANTICIDE, AND DISEASE

Predation and Infanticide

Lion cubs are vulnerable to leopards, spotted hyenas, cheetahs, and neighboring females, but predation on cubs appears to be far too rare to limit the population as a whole (chapter 4). Infanticide by incoming males, however, has profound pride-level

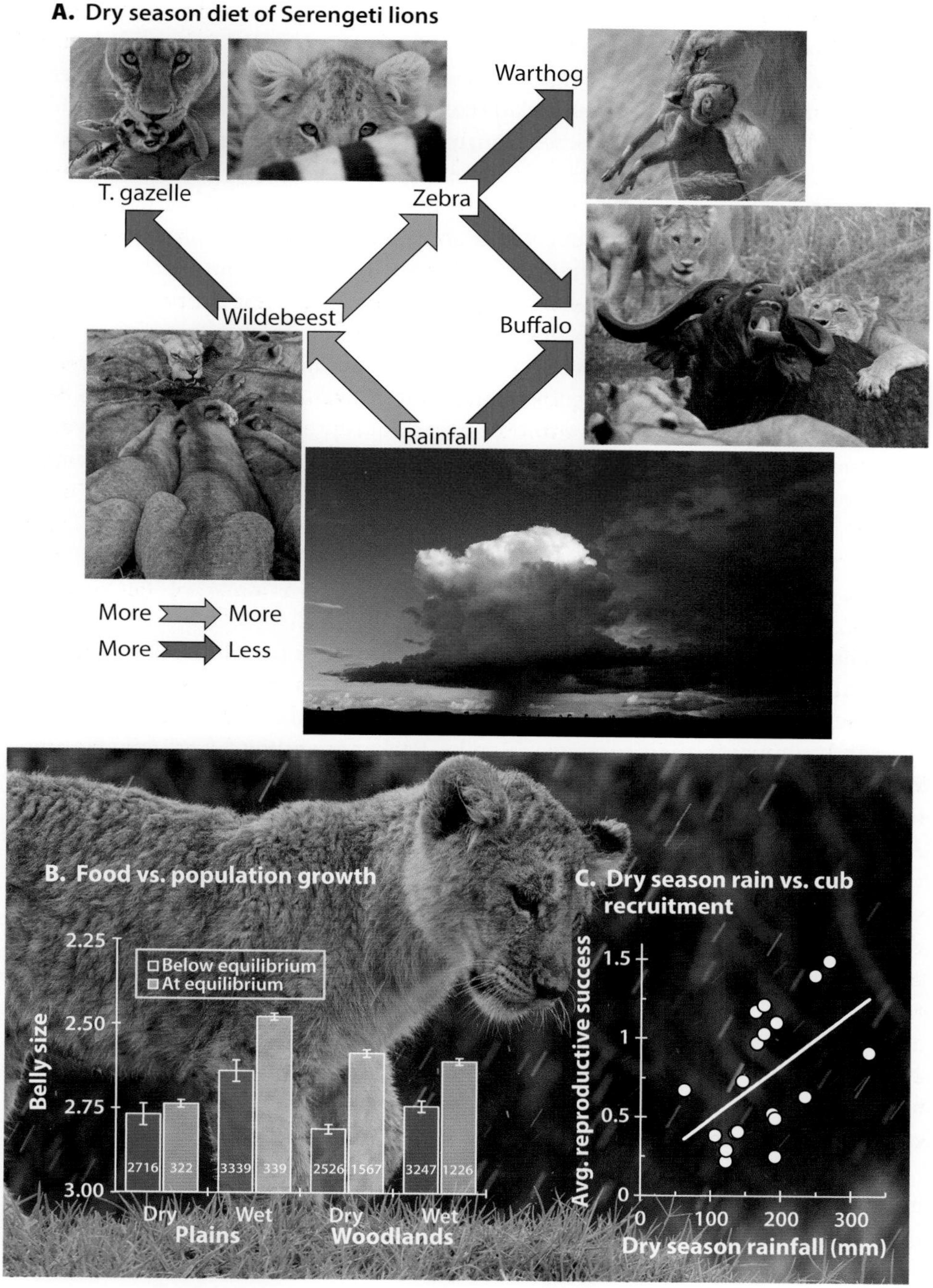

FIGURE 10.5. **A.** Linkages between dry season rainfall and diet. Solid arrows indicate statistically significant relationships. **B.** Serengeti lions were significantly thinner during most periods when their local subpopulations had not yet increased in response to vegetation changes. **C.** Cub recruitment rates of Serengeti females increase with dry season rainfall; redrawn from Mosser (2008).

effects, and synchronous takeovers can have larger-scale impacts. Varying numbers of nomadic males pass through the Serengeti study area each wet season while heading toward the migratory herds (see figure 1.10), particularly in years with above-average rainfall (Borrego et al. 2018). These include cohorts of young animals and once-resident males that have recently been evicted from other parts of the Serengeti. Between 1974 and 1989, the number of nomadic males entering the study area increased to a similar extent as the growing number of study prides but subsequently declined (figure 10.6a). The initial growth presumably reflected an overall increase in lion numbers throughout the rest of the woodlands (figure 10.2a), while the subsequent decline likely resulted from the increasing levels of human-lion conflict around the periphery of the park (as discussed in chapter 12). In years with greater numbers of entering nomads, a larger proportion of Serengeti study prides experience male takeovers (figure 10.6b), and resident cubs and yearlings suffer significantly lower survival (figure 10.6c). Thus, the study population is sensitive to the successful reproduction of prides living in other parts of the ecosystem—and may even benefit from the anthropogenic removal of nomadic males (chapter 12).

Disease

A bright red stripe in figure 10.2 marks the severe impacts of an outbreak of infectious disease that killed nearly 40 percent of the Serengeti lions and left several of the survivors with permanent neurological damage, such as facial twitches (photos in figure 10.7) and tremors in their forepaws (Roelke-Parker et al. 1996). Tanzania National Park's chief veterinarian, Melody Roelke, collected samples from nearly a hundred lions and sent blood and tissue samples to the US and Europe for laboratory analysis. A few weeks later, an American pathologist, Linda Munson, detected cellular sites of viral multiplication ("inclusion bodies") in the epididymis of a young male that had suffered convulsive seizures just before death. The inclusion bodies were diagnostic clinical signs for canine distemper virus (CDV), a highly infectious morbillivirus that is closely related to rinderpest and measles, and the virologist Max Appel's antibody tests confirmed that nearly every lion had recently been infected with CDV. Appel also tested all the lion samples that we had collected for our genetic surveys in the 1980s and early 1990s, and, by measuring the proportion of infected animals of each age each year, we were able to construct "age-prevalence curves" that revealed the precise timing of the CDV outbreak.

The die-off began when the virus entered the Serengeti study area in Dec. 1993 and persisted until September 1994 (figure 10.7a). Samples from other parts of the ecosystem later confirmed that the virus crossed the Kenyan border to the Maasai Mara Reserve in August 1994 and to the Maswa Game Reserve by November 1994 (Roelke-Parker et al. 1996). Appel's serological survey revealed that the Serengeti study population had also been exposed to CDV in 1981 as well as in a possible outbreak in 1977 (figure 10.7a), and the lions had additionally experienced periodic outbreaks of calicivirus, parvovirus, and coronavirus (Packer et al. 1999). Survivors enjoy lifelong immunity to CDV and long-term immunity to the other three viruses, and the extent of each epizootic closely adhered to the predictions of "herd immunity": the higher

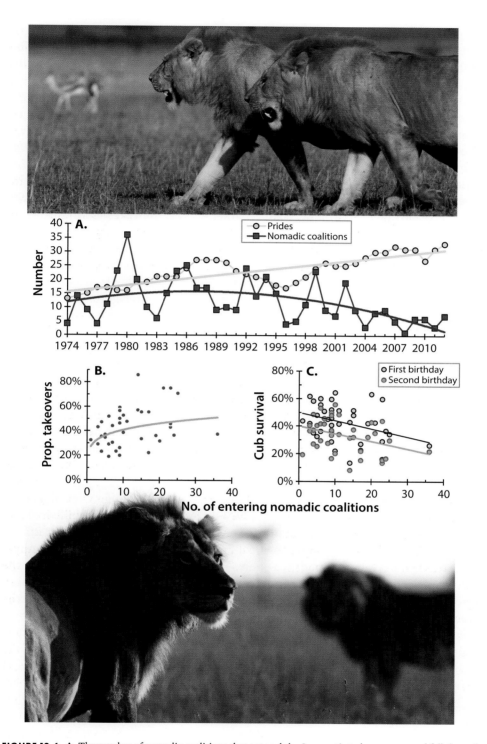

FIGURE 10.6. A. The number of nomadic coalitions that entered the Serengeti study area rose and fell through time while the number of resident prides increased. **B.** The proportion of study prides experiencing male takeovers was highest in years with the most nomadic coalitions. **C.** The proportion of cubs surviving to their first and second birthdays declined with increasing nomadic coalitions. All redrawn from Borrego et al. (2018).

the proportion of unexposed hosts at the start of the outbreak, the more extensive the spread of the virus through the lion population (figure 10.7b). Few lions born before 1981 were still alive by 1994, thus the vast majority of the population was susceptible to CDV in 1994, and over 95 percent became infected.

Besides its unprecedented breadth, the 1994 CDV outbreak was also unique in severity: no other virus was associated with a measurable decline in survival in the Serengeti (figure 10.7c). Indeed, the discovery of the earlier CDV outbreaks came as a complete surprise. Although one or two animals developed facial twitches in 1981, there was no detectable increase in mortality in either 1977 or 1981 (figure 10.7c).[2] When we first reported on the 1994 CDV outbreak, we attributed the contrasting severity in the successive outbreaks to a possible strain difference: perhaps the 1994 strain was particularly virulent (Roelke-Parker et al. 1996). This hypothesis proved difficult to evaluate as CDV infections typically run their course in a week or so, and, while Melody had been on hand to extract blood samples from animals that were obviously sick in 1993–1994, our earlier sampling had only begun in 1984, and, even in later years, we felt no urgency to sample large numbers of animals that appeared to be healthy. A second lethal CDV outbreak struck the lions in Ngorongoro Crater in 2001 (Munson et al. 2008), but we were only able to sample a handful of survivors after they had all cleared the virus. Thus, even though the study populations experienced at least seven separate CDV outbreaks from the 1970s onward, we only managed to collect viral sequences from the 1993–1994 outbreak in the Serengeti.

Working around this shortcoming, Weckworth, Davis, Roelke-Parker, et al. (2020) analyzed complete genomes of various different CDV strains from North America, Asia, and Africa that were known to produce differing severity of clinical distemper (e.g., symptoms of respiratory, enteric, and neurological disease) and found several mutations in the 1993–1994 Serengeti strain that increased viral replication, immune suppression, and onward transmission. More transmissible strains of SARS-CoV-2 (the virus causing COVID-19) are also known to inflict higher mortality (Davies, Jarvis, et al. 2021), so it seems reasonable to suppose that the Serengeti lions had been exposed to a particularly "hot" strain in 1993–1994. In addition, Nikolin et al. (2017) found evidence of two contrasting strains of CDV in the Serengeti: one that mostly circulated in canids (e.g., domestic dogs, bat-eared foxes) while the second transmitted more readily between noncanids (e.g., lions, spotted hyenas)—thus raising the possibility that the "silent" outbreaks in the lions could have been caused by canid strains, whereas the lethal die-offs might have been caused by noncanid strains. However, Weckworth, Davis, Dubovi, et al. (2020) found that the canid and noncanid strains from 1993–1994 differed by only thirteen mutations and had therefore likely diverged

---

2 While there were no obvious signs of serious illness or elevation in mortality during the outbreaks in the Serengeti in 1976, 1981, 1998, or 2007, Brandell et al. (2020) found similar changes in ranging behavior in these silent outbreaks as in the 1993–1994 outbreak, with adjacent prides showing comparable reductions in territory size and territorial overlap during the outbreak years compared to other years. The most likely cause of the lion prides remaining farther apart may have been a general lethargy and, presumably, an unwillingness to confront neighboring prides in any sort of territorial dispute. So, although CDV may not have caused visible signs of "distemper" in these years, the virus may still have influenced the lions' health.

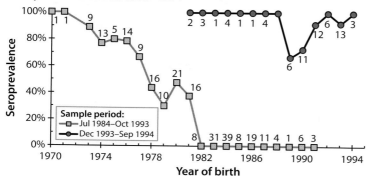
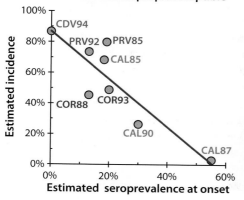

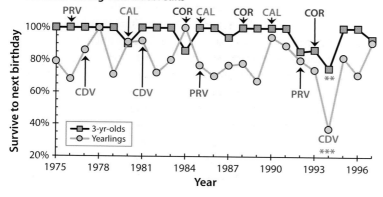

FIGURE 10.7. Viruses of the Serengeti. **A.** Only sampled animals born between 1970 and 1981 were seropositive for CDV between July 1984 and October 1993, whereas lions of all ages were seropositive between December 1993 and September 1994. **B.** The estimated proportion of newly infected individuals was lower when a higher proportion of the population had previously been exposed to the same virus. **C.** Survival of yearlings and three-year-olds was significantly lower in 1994 than in any other year, whereas there was no similar effect for calicivirus (CAL), coronavirus (COR), parvovirus (PRV), or either of the earlier CDV outbreaks. All redrawn from Packer, et al. (1999).

approximately eleven months prior to the onset of the outbreak,[3] thus these were not two circulating strains of independent origins but a descendant set of viral lineages that derived from a single spillover from domestic dogs (see figure 12.6). Given the evolutionary convergence in geographically isolated strains of CDV that are all known to infect lions (Weckworth, Davis, Roelke-Parker et al. 2020), why wouldn't a noncanid strain have arisen in each new outbreak in domestic dogs?

If strain differences do not provide the answer, what else might explain the severity of the lethal outbreaks? Guiserix et al. (2008) suggested that CDV might have been endemic in the Serengeti prior to 1981 (rather than briefly sweeping through the population in 1977 and again in 1981) and then disappeared completely for a dozen years before returning in 1993–1994. In this scenario, only young cubs would have been vulnerable to infection during the endemic years prior to 1982, and hence clinical signs would have largely gone unnoticed as all the adults would have been survivors of childhood infection. If this were true, we should have seen an increase in cub survival after the virus somehow disappeared by 1982 (as cubs would no longer be exposed to CDV each year), but there was no detectable difference in cub mortality between 1975–1981 and 1982–1992 (figure 10.7b). Further, with more contemporaneous sampling from the 1980s onward, the pattern of CDV prevalence has clearly remained episodic, with only a few outbreak years punctuating longer disease-free periods (figures 10.9 and 12.2), so there is no evidence that CDV has ever remained endemic in the lion population for any significant length of time.

Could differing disease outcomes across successive outbreaks instead result from variations in the lions' environment? Animals rarely die from CDV alone; the virus is so immunosuppressive that a co-infection with a second pathogen can be devastating. But it wasn't until the 2001 die-off in the Crater that Linda Munson identified the fatal co-factor for both epizootics. Serological tests confirmed that even the youngest Crater victims had been exposed to CDV in 2001, so the virus had struck within a few weeks of our blood and tissue sampling. The mortality rate was again close to 40 percent, and the histologic lesions and depleted lymphocytes again provided clear evidence of CDV-related pathogenesis (Munson et al. 2008)—but while collecting samples in the field, we had seen various overt physical symptoms that initially led us to think that the Crater lions were suffering from something else entirely.

In 1962, the Crater lion population had famously dropped from an estimated seventy-five lions to less than a dozen (Fosbrooke 1963). Henry Fosbrooke described a "plague" of biting *Stomoxys* flies that erupted after months of heavy rain ended the drought of 1960–1961. Covered with thousands of Stomoxys, some of the lions started climbing trees on the Crater walls to escape the onslaught. The lions became lethargic and developed a series of circular skin lesions that Fosbrooke thought were caused by the flies, and he also thought that the flies literally bled the lions to death. The Crater lions were again covered by Stomoxys flies in 2001 and again developed large festering skin lesions. Richard Kock led our investigation on the Crater floor, collecting dozens of Stomoxys flies, taking biopsies from the skin lesions, drawing blood and lymphocytes, and performing necropsies.

---

3 By comparison, SARS-CoV-2, like CDV, is a single-stranded RNA virus and rapidly accumulates mutations that accelerate its spread (Maher et al. 2022).

It turned out that the lions were anemic, but this wasn't caused by a loss of blood to the Stomoxys flies but instead by intravascular hemolysis caused by tick-borne parasites. The lions were so lethargic that they expended little effort to shoo away the biting flies—so the wall-to-wall coverage with biting flies was a *sign* of illness rather than the cause. Their skin lesions showed no trace of vector-borne infectious agents; the open sores instead resulted from the lions obsessively scratching themselves in easily reached spots on their flanks and sides—which could have simply reflected a dermatological reaction to tick-borne disease.

In the end, it was the combination of drought and flood in the Crater in 1961–1962 that provided the most plausible explanation for the unusual severity of the CDV outbreaks in 1993–1994 and 2000–2001. A drought in 1993 had weakened the Serengeti buffalo (as measured by the consistency of their bone marrow) to such an extent that the lions were able to capture an unprecedented number of buffalo, and a drought in 2000 had a similar effect in the Crater (figure 10.8). Starvation is also strongly immunosuppressive, and malnourished *herbivores* typically suffer from tick infestations (e.g., Sinclair 1977) and, hence, associated outbreaks of tick-borne disease. Catching so many weakened buffalo had been irresistible to the lions, but the buffalo carcasses exposed the lions to large numbers of disease-bearing ticks. Linda Munson saw clear signs of parasitic infection in tissue samples from the Crater lions, and her protégé, Karen Terio, found that the fatal CDV outbreaks in both the Serengeti and the Crater coincided with exceptionally high levels of infection by Babesia, a tick-borne blood parasite that the lions normally carry at low levels (figures 10.9a, 10.9b). Prides that suffered severe mortality during the two die-offs typically showed higher antibody titers for CDV (figure 10.9c), but while almost all the lions were infected with CDV, the prides that were most heavily infected with Babesia showed the highest mortality (figure 10.9c). Indeed, two prides were infected with CDV without being exposed to high levels of Babesia, and neither suffered elevated mortality, while a third unharmed pride was exposed to Babesia but not CDV.

Neither CDV nor Babesia was enough by itself, but the combination of the two pathogens was lethal, and it took a drought to line them up: droughts render herbivores more susceptible to tick infections, and CDV infects a variety of carnivore species, each of which would be more likely to contact the other at one of the few remaining waterholes during drought, so it is not surprising that the two diseases might sometimes flare up together.[4]

A final virus that merits discussion is the lion version of feline immunodeficiency virus (FIV_{Ple}) that was first identified in the Serengeti by Olmstead et al. in 1992. The human equivalent is HIV, which causes AIDS, and every lion in the Serengeti and Ngorongoro is infected with FIV by the time they reach sexual maturity (Brown et al. 1994; Packer et al. 1999). We had never detected any evidence that FIV was harmful to lions until a fine-grained analysis suggested that, of the three viral clades in the Serengeti, lions carrying the B-clade suffered 28 percent mortality during the 1994 CDV outbreak compared to 52 percent that carried either the A or C clades; of the

---

[4] Henry Fosbrooke died in 1996, so I was unable to ask if any lion samples had been collected during the 1962 die-off. The East African Veterinary Services in Nairobi once housed any colonial-era samples that would have been collected in Tanganyika, but all their materials had been discarded a few years earlier.

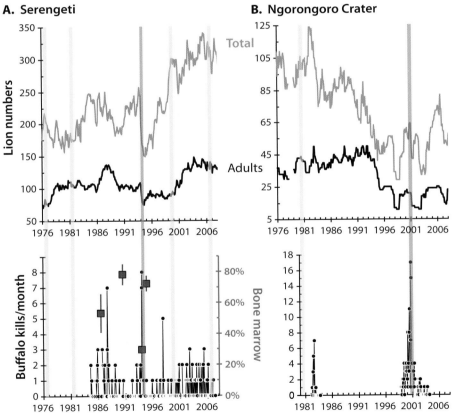

FIGURE 10.8. CDV outbreaks and buffalo consumption in (**A**) the Serengeti and (**B**) the Ngorongoro Crater. *Top:* Black lines show the number of adults more than four years of age); blue lines show total population sizes. Gray bars indicate timing of "silent" outbreaks that were only detected retrospectively by serology; red bars show fatal outbreaks. *Bottom:* Number of buffalo carcasses in the diet of each lion population each month (black dots) and bone marrow fat scores (red squares) of Serengeti buffalo carcasses. Data are restricted to years with comparable levels of search effort; staff were only stationed full-time in the Crater in 1982–1983 and 1999–2004. Redrawn from Munson et al. (2008).

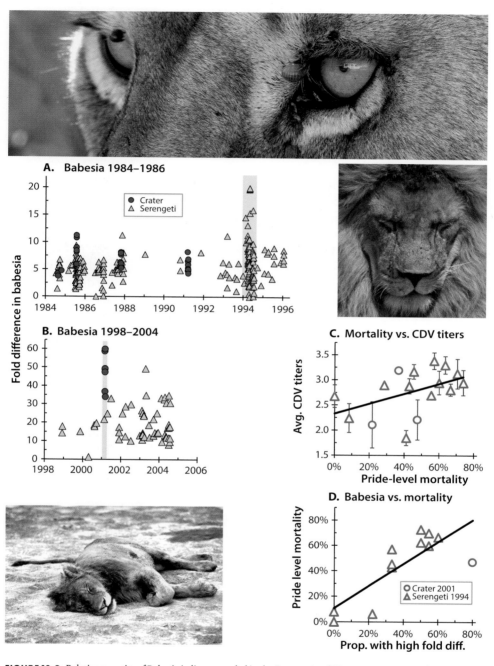

FIGURE 10.9. Relative quantity of Babesia in lions sampled in the Serengeti and Ngorongoro Crater as determined by real-time polymerase chain reaction (PCR) from (**A**) red blood cell pellets collected in 1984–1996 and (**B**) whole blood in all subsequent samples. The relative quantity of hemoparasite DNA was calculated as [average threshold PCR cycle]/[hemoglobin concentration] and expressed as the fold difference greater than the sample with the smallest quantity of DNA. Levels of Babesia infection were significantly higher in (1) the two fatal outbreaks, (2) the Crater, and (3) assays performed on whole blood ($n = 344$ samples). **C.** Relationship between pride-level mortality rates and CDV titer level during the fatal outbreaks. **D.** Pride-level effects of hemoparasitemia on mortality rates during the fatal outbreaks in the Crater and Serengeti. Prides with a greater proportion of individuals showing high hemoparasitemia suffered significantly higher mortality rates. All redrawn from Munson et al. (2008).

seventy-three lions that were known to be co-infected with CDV and Babesia, the B-clade lions suffered 38 percent mortality versus 62 percent with the A or C clades (Troyer et al. 2011). Outside of this extreme immunological challenge, we could find no evidence that the lions suffered serious harm from the virus, and there is no hint that FIV significantly shortens the lions' lifespans or limits their population size in either habitat (Packer et al. 1999).

NGORONGORO CRATER: DISEASE VERSUS DENSITY

The Crater lion population has consistently shown an exceptional vulnerability to infectious disease, with the 1962 and 2001 die-offs being the most obvious (Kissui and Packer 2004). But something else besides CDV/Babesia must also have afflicted the Crater lions, as three other die-offs with unusually high levels of adult mortality over a two-to-four-month period occurred in 1994, 1997, and 2007. We have no diagnostic data for any of these events except for the age distribution of the victims: the mortality rate for prime-aged lions was the same during each of these die-offs as seen in the Serengeti in 1994 and the Crater in 2001 (figure 10.10a). The 1994 Crater die-off may have been caused by CDV/Babesia, as the 1993 drought parched the entire region, and CDV was known to have entered the Ngorongoro Conservation Area by 1994. The 1997 Crater die-off, however, must have been caused by a different disease, as almost all the animals older than four years of age would have carried antibodies against the virus in 2001, but mortality rates in 2001 were the same for older animals as for younger individuals. We will explore why the Crater lions are so vulnerable to infection at the end of this chapter.

After the dramatic die-off in 1962, the Crater lions recovered rapidly, growing nearly ten-fold over the next dozen years (figure 10.10b). The population remained close to one hundred individuals until the early 1980s then went into decline, followed by the die-offs that repeatedly knocked the population down below thirty to fifty animals in the 1990s and 2000s. As expected, if the Crater lion population is primarily food limited: the lions were significantly thinner at higher population sizes (figure 10.11a). Excluding the outbreak years, the Crater lions showed clear signs of density dependence: the population grew fastest when small and declined when large (figure 10.11b). Note, though, that food and disease were not the only factors affecting population growth. With such a small number of prides, coincident takeovers will inevitably lower cub survival in the entire population, and, whereas the Crater population almost always grew in years without any male replacements, the population mostly declined in years when multiple prides suffered takeovers (figure 10.11c). However, takeover rates did not vary with population size, and thus the upper limit on population size is ultimately set by the local food supply.

Prey species remain abundant all year round on the Crater floor (figure 1.3), and local prey abundance is far easier to track. Wildebeest, zebra, and buffalo account for about 90 percent of the Crater lions' diet (figure 1.6), and their numbers have been counted most years since the 1960s (figure 10.12a). Over twenty thousand wildebeest occupied the Crater floor in the dry season of 1964 but subsequently declined to about five thousand by the mid-1990s (Estes et al. 2006) and then

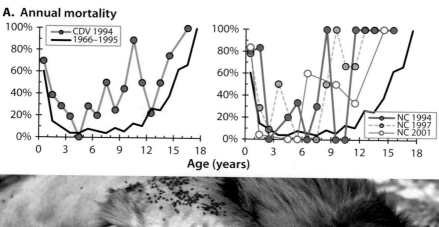

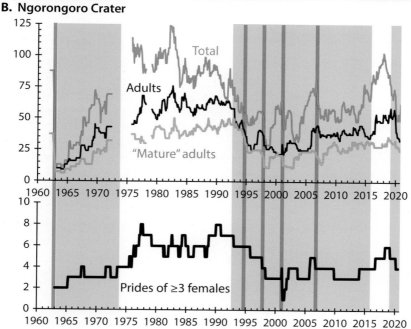

FIGURE 10.10. Disease and death in the Ngorongoro Crater. **A.** Age-specific annual mortality during known and presumed disease outbreaks. *Left:* Red line indicates mortality during the 1994 CDV outbreak in the Serengeti. *Right:* Red lines indicate mortality in the Crater during undiagnosed outbreaks in 1994 and 1997 and the 2001 CDV outbreak. Black lines in each graph show the background age-specific mortality of the Serengeti and Ngorongoro lions in 1966–1995. **B.** *Top:* Lion population size each month in the Crater. Blue lines include all individuals; black lines indicate lions more than two years of age, brown lines are adults more than four years of age. *Bottom:* Number of prides containing at least three females each month. Pale green blocks highlight time spans when the population was presumed to be below the local carrying capacity; vertical red lines indicate disease outbreaks. All redrawn from Kissui and Packer (2004).

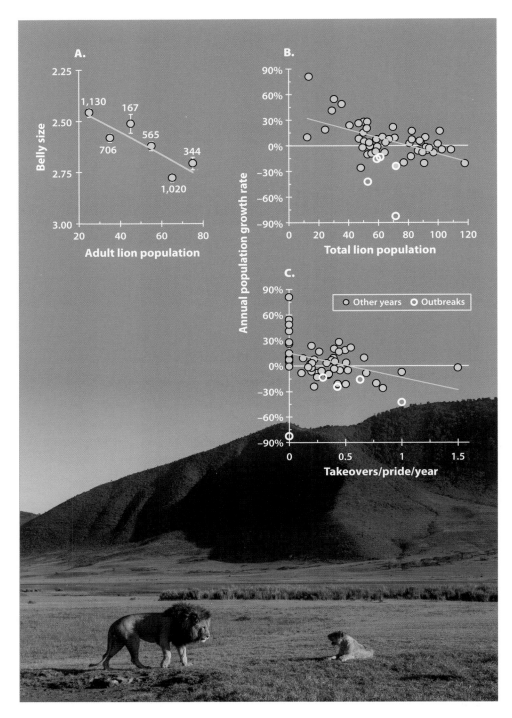

FIGURE 10.11. Density dependence in the Crater. **A.** Crater lions are significantly thinner when the adult population is larger. n = number of individual measurements. **B.** Annual rate of population change in the Crater lions is significantly lower at larger population sizes. Data from outbreak years are indicated by white circles. Updated and redrawn from Kissui and Packer (2004). **C.** Population growth is significantly lower in years followed by greater numbers of male takeovers per pride. Outbreak years are again in white.

doubled to about ten thousand animals by the 2010s. Cape buffalo only returned to the Crater floor in any numbers in 1973 (having presumably been eradicated by rinderpest at some point) and increased to about five thousand by 2011. Zebra declined from about six thousand in the 1960s to three thousand in the 1990s but subsequently recovered. All three species showed signs of density dependence, with populations growing at low densities and declining at higher densities (see figure 11.3a).

In contrast to the Serengeti, the Crater has shown no lion-friendly changes in vegetation—no increase in tree cover or grass height—so variation in overall prey *abundance* is the only factor that would be expected to affect the lions' food intake. We can simplify the food supply by considering each prey species in terms of "wildebeest equivalents" (Kissui and Packer 2004): one buffalo weighs the same as 3.33 wildebeest, and a zebra weighs as much as 1.3 wildebeest. Though less plentiful, each buffalo provides far more food than a wildebeest, so the long-term decline in wildebeest numbers was ameliorated to a considerable extent by the growth in the buffalo population, and the Crater has held between twenty thousand and thirty thousand wildebeest equivalents over most of the past fifty years, reaching a low ebb in the 1990s and recovering by the 2010s (figure 10.12b).

In terms of biomass available to each lion, conditions following the 1963 die-off were truly exceptional: the wildebeest population was at an all-time high, while the Crater lions were at their all-time low. Per capita food availability fell rapidly as the lion population recovered, reaching a minimum in the 1970s, and increasing again before reaching a plateau of about five hundred wildebeest equivalents per lion by the late 1990s (figure 10.12b). The growth rate of the Crater lion population depended strongly on the dry season food supply: the population grew fastest in years when more food was available per lion—even when excluding the exceptional year of 1964 (figure 10.12c). Prey availability does not differ by season (an average of 420 ± 58 wildebeest equivalents per lion in the dry season versus 441 ± 40 in the wet season), as the departing buffalo are replaced by incoming wildebeest each dry season and vice versa in the wet season, and there is virtually no correlation between wet and dry season totals in the same year ($r^2 = 0.041$, $n = 34$ years). Lion population growth did not vary significantly with food supply during the wet season, and lion belly size does not vary seasonally in the Crater, so year-to-year variations in food intake must somehow be more critical for cub survival during the dry season months or hunting success might be more sensitive to prey abundance each dry season, perhaps owing to a limited number of ambush sites.

Whatever the reason, the regression line in figure 10.12c crosses the x-axis at 303 wildebeest equivalents per lion: the lion population grew at higher levels of food availability and declined at lower levels. If we calculate the intercepts for the three major prey species (figure 11.3a), we get: 10,554 wildebeest, 4,051 zebra, and 2,665 buffalo. Add these together, and we get 24,685 wildebeest equivalents, suggesting that at an overall equilibrium of prey, the Crater could support 24,685/303 = 81.3 lions. Looking solely at lion numbers, the regression of population growth versus population size in figure 10.11b crosses the x-axis at 79.3 lions. Thus, it seems reasonable to suggest that, on average, the Ngorongoro Crater should contain about 80 lions.

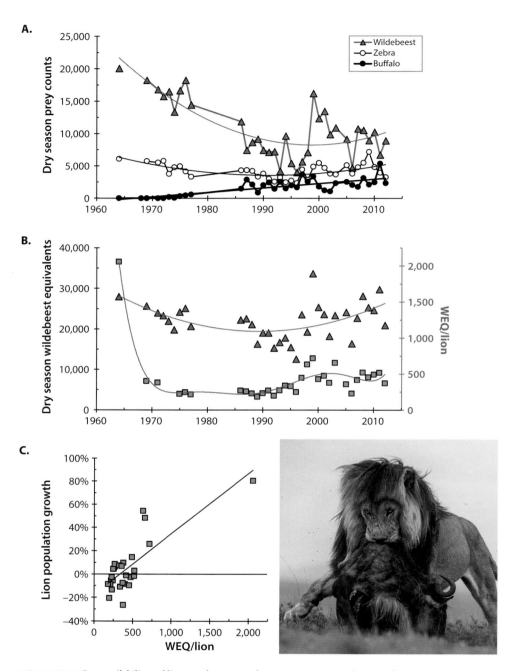

FIGURE 10.12. Prey availability and lion population growth. **A.** Dry season ground counts of the Crater lions' primary prey species (see Estes et al. 2006; Moehlman et al. 2020). **B.** The number of dry season "wildebeest equivalents" (WEQ) and wildebeest equivalents per lion through time. **C.** Lion populations grew faster in years with more wildebeest equivalents per lion during the dry season; the relationship is highly significant both with and without the outlier.

KEEPING IT IN THE FAMILY

The Ngorongoro Crater is an island of rich savanna habitat surrounded by dense forest to the east and dry grasslands to the west. The Serengeti lion population is about 60 km to the west, thus there is very little genetic exchange with outsiders and little incentive for male coalitions to leave. Over 80 percent of the resident males in the Crater were known to have been born on the 250 km^2 Crater floor, which stands in stark contrast with the 33 percent of resident males that were born within the 2,000 km^2 Serengeti study area (figure 10.13a). Not only did Crater males stay far closer to home, but a far higher proportion became resident in their natal pride (figure 10.13b), resulting in nearly ten times as many Crater cubs being fathered by "natal males" as in the Serengeti (figure 10.13c). The Crater's rich food supply allows prides to produce coalitions of four to six males that prevent most singleton males and pairs from establishing residence on the Crater floor. Outside coalitions have only successfully immigrated into the Crater on nine occasions between 1963 and 2020, and the newcomers entered when there were typically only three to six resident males on the entire Crater floor (versus six to eleven in all the other months) (figure 10.13d).

Our genetic assays from 1984–1987 showed that the Crater lions already had lower genetic diversity and higher homozygosity than their Serengeti counterparts (O'Brien et al. 1987), but we were never able to perform paternity tests with our limited number of samples. Given that the entire population descended from the handful of survivors of the 1962 die-off, the low levels of male dispersal thereafter, and the fact that two of the founding prides from 1962 eventually went extinct, leaving all the Crater prides as descendants of a single pride (see figure 2.14) and most of the resident coalitions descended from a single paternal coalition (figure 2.16), it seems reasonable to assume that the population has been subject to considerable levels of inbreeding over much of the past half century, and, indeed, Crater males had significantly higher levels of sperm abnormalities (Wildt et al. 1987) and lower levels of sperm motility (Brown et al. 1991).

If we simply dichotomize the cubs born in the Crater either as the offspring of two Crater-born parents or as "outcrossed" by virtue of having an immigrant father, we can compare various measures of cub survival. As mentioned in chapter 4, females with swollen udders and suckling stains sometimes lose their unseen cubs. Litters of two Crater parents were significantly more likely to be "lost" than were "outcrossed" litters (figure 10.13e). Considering only those cubs that survived long enough to be observed, there was no significant difference in initial litter size (figure 10.13f), but cubs of two Crater parents again had significantly higher mortality before their first birthday (figure 10.13g). The origins of the immigrant males were mostly unknown, so we were unable to estimate their extent of genetic dissimilarity to the Crater females, but these trends are consistent with the possibility that the Crater cubs suffered lower viability as a result of their genetic isolation. On the other hand, it is also possible that the immigrant males may have been more effective in protecting their offspring in some way.

Inbreeding is known to increase susceptibility to infectious disease (e.g., Acevedo-Whitehouse et al. 2003; Reid et al. 2003; Dorman et al. 2004), and resistance to bovine tuberculosis was successfully restored by translocating sixteen individuals into an inbred population of eighty lions in South Africa's Hluhluwe-iMfolozi Park (Trinkel et al. 2011) (see figure 12.7). Looking at the history of the Crater lions from this

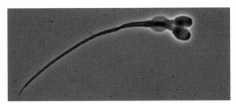
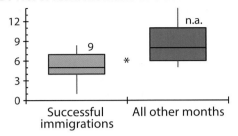
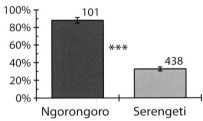
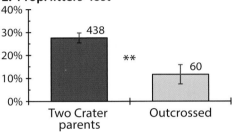
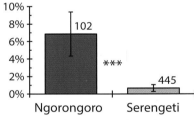
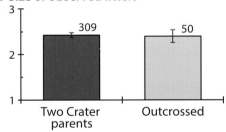
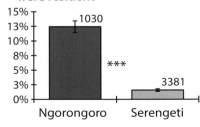
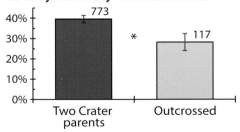

FIGURE 10.13. Male dispersal and inbreeding. **A.** Almost all resident males in the Crater were born on the Crater floor, whereas the majority of resident males in the Serengeti were born outside the study area. n = number of resident males. **B.** Crater coalitions were more likely to reside in their natal pride than were Serengeti males. n = number of coalition residencies. **C.** Crater cubs were far more likely to have been fathered by males that were resident in their natal pride. n = number of cubs. **D.** Immigrant coalitions were more likely to gain residence in the Crater when there were fewer resident males on the Crater floor. n = number of immigrating coalitions. **E.** Litters with two Crater parents were less likely to survive long enough to be observed than were litters with an immigrant father. n = number of litters. **F.** Outcrossed litters were no larger than litters with two Crater parents. **G.** Mortality was significantly lower for outcrossed cubs that survived long enough to be observed. n = number of cubs.

perspective, we can now consider its genetic background (figure 10.14). Given the extraordinarily high mortality of the Crater lions in 1963, I assume that the population had already undergone past periods of isolation and inbreeding, thus the survivors of the outbreak were themselves the product of two Crater parents. The surviving nine females mated with three male coalitions that were able to enter the Crater in the following few months and years, giving rise to a large cohort of outcrossed offspring that fueled the rise of the population to its all-time high. But the enlarged population consistently produced large coalitions of nondispersing sons that almost completely blocked further immigration into the Crater for the next three decades. During this long period, generations of inbred Crater lions bred with each other (except for the one male that successfully immigrated in 1993 and teamed up with a Crater male; the pair successfully produced surviving offspring, but their descendants never accounted for more than a sixth of the population: figure 10.14 assumes that the immigrant fathered half the surviving cubs of his coalition), perhaps leaving the overall population susceptible to the die-offs in the 1990s and 2000s.

A new coalition of four males entered the Crater in September 2013—the first-ever successful immigration of a large outside coalition—followed by another singleton paired with a Crater male in 2015 and another outside pair in 2018, representing the largest infusion of "fresh blood" into the Crater since the 1960s. The population subsequently showed signs of escaping its doldrums, exceeding the magic number of eighty lions for the first time in over twenty-five years, as the proportion of "outcrossed" offspring reached 30 to 40 percent of the population. However, the "recovery" was short-lived, and soon returned to fewer than sixty individuals. What happened?

The spike in male immigration resulted from a deliberate conservation effort by Ingela Jansson to reopen a "lion corridor" between the Crater and the much larger Serengeti population. The region between the national park and the Crater floor has a relatively lower prey base except at the height of the rainy season (figure 1.4), the local Maasai routinely retaliate against cattle-killing lions, and young Maasai warriors have often traveled to the area to kill lions for cultural purposes, so itinerant lions traveling from one population to the other would have had to run a serious gauntlet (see chapter 12, particularly figures 12.10 to 12.12). But in 2010, Ingela adopted the "Lion Guardians" model developed by Hazzah et al. (2014) to help Maasai communities improve livestock husbandry and to end the practice of ritual lion killings (see chapter 12). A few years later, Ingela attached the first GPS collars in the NCA, and she was able to track lion movements to and from the Crater.

We had long suspected that "excess subadults" (both male and female) left the Crater floor, only to be speared as cattle-killers or during ritual hunts, but we had no way to know. Ingela's efforts revealed that a proportion of Crater lions do indeed leave, and at least two prides from the Crater have settled to the west of the Crater rim in the past few years, partially accounting for the most recent population decline on the Crater floor. Perhaps these would only have been temporary absences in the past, but now they had found safe haven with no neighboring lions and little conflict with people. Or perhaps the carrying capacity on the Crater floor has somehow dropped to only about sixty individuals, and they would have dispersed anyway only to meet their end at the point of a spear. Perhaps if Ingela's project achieves permanent change, and the area eventually becomes saturated with resident lions, the more outbred animals

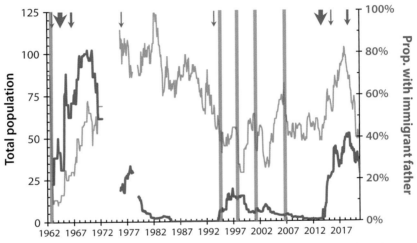

FIGURE 10.14. Parentage of Crater lions. Blue line is the total population, and vertical red lines indicate presumed disease outbreaks, as before. Green arrows represent arrival of immigrant males; widths are scaled according to the number of males entering that year. Green line indicates proportion of the Crater population with an immigrant father.

would be forced to endure higher numbers, and the population on the Crater floor would finally stay at about eighty individuals or so.

But I suspect we'll need another twenty to forty years of data to know for sure.

KEY POINTS

1. The Serengeti lion populations increased in a nonlinear manner in response to (a) increased prey abundance following the elimination of rinderpest; and (b) improved prey accessibility with increased tree density and grass height. Each change coincided with favorable dry season rainfall that increased the lions' access to wildebeest and zebra, and the rapid pace of change resulted from synchronous generation of viable new prides.
2. Greater numbers of nomadic males pass through the Serengeti study area in years with higher wet season rainfall, thereby increasing takeover rates and lowering rates of cub recruitment. The number of nomadic coalitions declined over the course of the study, perhaps because of greater anthropogenic impacts in the rest of the Serengeti ecosystem.
3. During periods of severe drought, the Serengeti and Ngorongoro populations each suffered a substantial die-off caused by co-infection with canine distemper virus and Babesia. During nonsynchronous outbreaks of these and other viral agents, disease inflicted minimal impacts on the Serengeti lions, but the Crater lions showed evidence of multiple die-offs from undiagnosed pathogens.
4. Infanticide and disease outbreaks affect population growth in the Ngorongoro lions, but overall numbers are limited by their food supply. The Crater lions are thinner at higher population sizes, and, excluding disease-outbreak years, population growth rates are highest in years with greater per capita prey availability. Density-dependent growth rates of lions and their primary prey species suggest an estimated carrying capacity of 79.3 to 81.3 lions on the Crater floor.
5. In addition to its geographical separation from adjacent populations, the Crater lions have been further isolated by the resident coalitions that effectively blocked in-migration of males to the Crater floor for decades. The resultant close inbreeding in the Crater population not only appeared to render these lions more susceptible to infectious disease but may also have lowered cub survival.
6. Recent conservation efforts to open a dispersal corridor between the Serengeti and the Crater Highlands have been followed by an influx of immigrant males into the Crater floor, which was in turn followed by a rapid population increase. However, this spike was relatively short-lived, as the same conservation effort has allowed several prides to leave the Crater floor and settle outside.
7. Sociality has two major impacts on population dynamics. First, synchronous pride-level reproduction produced the coarse-grained population shifts following each environmental change in the Serengeti, and, second, the large male coalitions in the Crater kept their population genetically isolated for decades. Solitary species would not show such sudden leaps in population size or become so isolated by the behavior of its own kind.

CHAPTER 11

INTERSPECIFIC INTERACTIONS

Nothing exists for itself alone but only in relation to other forms of life.
—CHARLES DARWIN

We have so far viewed lions in isolation except for their need to obtain enough food to survive and reproduce. But the Serengeti is famous for its rich biological diversity: hundreds of species of birds, scores of ungulate species, and a dozen different carnivores. Although the Ngorongoro Crater may lack giraffe, impala, topi, and wild dogs, the Crater floor sustains a stunning abundance of many of the same species as the Serengeti but within a much smaller space. Top predators are widely presumed to have disproportionate impacts on the lower trophic levels in their ecological food webs, but what impact does the King of Beasts actually have on its domain? Food webs can be studied from many different perspectives, but my primary interest is in addressing whether lions can have such strong effects that they could drive another species to localized extinction. This prospect may seem unlikely given that lions have coevolved with their prey and competitors since time immemorial, but we are coming close to a consideration of the conservation status of an endangered species in a highly modified continent—a continent that is increasingly out of balance.

The boundaries of the Serengeti National Park and surrounding game reserves largely preserve the geographical range of the wildebeest migration. The Serengeti Research Institute was established in the mid-1960s to assist Tanzanian National Parks in conserving the migration (Sinclair et al. 2015), and the Serengeti lion project was initiated shortly thereafter when George Schaller was asked to assess the lions' effects on prey populations (Schaller 1972). Park wardens commonly perceived predators as vermin, and the idea that lions should be actively managed was still so pervasive by 1974 that South African National Parks responded to declining populations of wildebeest and zebra in Kruger Park by culling 355 lions in a single year (Smuts 1975, 1978). In chapter 10, we saw how the lions are limited by their food supply in both the Serengeti and the Crater, but can lions, in turn, actually limit the number of herbivores in either ecosystem?

In addition to the presumed influence of predators on their preferred prey, there has also been concern by conservationists that, as dominant competitors, lions might mean fewer midsized ("meso") carnivores. For example, lions frequently kill cheetah cubs (Laurenson 1994), and after we established a large-scale vaccination program around the Serengeti to protect lions from canine distemper virus (chapter 12), cheetah biologists expressed concerns that by potentially protecting the lion population we might be harming the cheetah population (Chauvenet et al. 2011). Lions can certainly be terrible, horrible, awful, nasty animals that kill members of other species without bothering to eat them, but what are the lions' population-level impacts on the smaller carnivores in their midst?

LIONS AND THEIR PREY

Not all species are vulnerable to lions. Sinclair et al. (2003) showed that adult herbivores weighing more than 1,000 kgs (elephants, hippos, and rhinos) are essentially immune from predation and are effective at protecting their young, whereas species weighing under 100 kgs almost always die from predation (figure 11.1). Larger species like Cape buffalo and giraffe are mostly taken by lions, while spotted hyenas, leopards, cheetahs, wild dogs, caracals, servals, jackals, and wild cats are, respectively, only able to capture prey of smaller and smaller size. But is predation pressure strong enough to limit the population sizes of any of the more common herbivore species?

Serengeti

Rinderpest held the wildebeest and buffalo populations well below their potential from the 1890s until the ring-vaccination program eliminated the virus from the Serengeti in the early 1960s (Sinclair 1979b). Released from the disease, both species grew exponentially until reaching plateaus in the 1970s (see figure 10.1). The wildebeest population has since been limited by the dry season food supply; mortality rates are highest in years with the highest population densities and most deaths result from under-nutrition (Mduma et al. 1999). The Serengeti buffalo also show density-dependent population growth in the parts of the ecosystem that are adequately protected against bushmeat poaching (Dublin et al. 1990; Dublin and Ogutu 2015). The importance of food limitation is well illustrated by the drought of 1993: both species suffered sizable population declines (figure 10.1), and the resultant buffalo die-off played an important role in the CDV/Babesia epizootic the following year (figure 10.8). The recovery of the wildebeest and buffalo from rinderpest fueled sizable growth in the lion population (figure 10.2), both by increasing the lions' food supply in the 1970s and by the wildebeest's effect on woody vegetation in the 1980s (figure 10.3). However, the consequent growth in lion numbers had no measurable impacts on the wildebeest and buffalo populations, thus the two herbivore species appear to be limited almost entirely by their food supply rather than by predation.

Zebra, in contrast, are unaffected by rinderpest, and their numbers have remained essentially unchanged since the 1960s—despite dramatic increases in the number of grazing competitors (figure 10.1) and the overall growth of the lion population (figure 10.2). Migrating zebra show greater sensitivity to lion predation than do wildebeest,

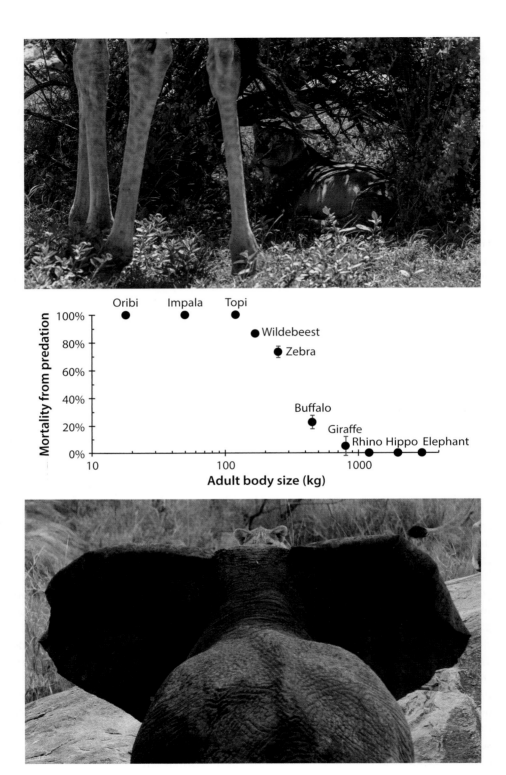

FIGURE 11.1. Effect of body size on predation risks in Serengeti herbivores. Species larger than 1,000 kg are virtually immune to predation; redrawn from Sinclair et al. (2003).

moving more cautiously through high-risk areas (Hopcraft et al. 2014), and lions appear to have a much larger impact on zebra than on wildebeest: while there was only one zebra on average for every 5.67 wildebeest in the Serengeti over the past fifty years, lions fed on one zebra carcass for every 1.55 wildebeest carcasses ($n = 835$ and 1299, respectively). Yet despite this nearly fourfold higher predation pressure on zebras, Grange et al.'s (2004) analysis suggested that the Serengeti zebra population is likely regulated by first-year mortality of foals. Lions take a slightly higher proportion of young zebra than young buffalo, wildebeest, or gazelle (see figure 11.14a), but this disparity is insufficient to account for the much higher estimated mortality rate of zebra foals (61.1%) compared to wildebeest and buffalo calves (25.4% and 33.0%, respectively [Grange et al. 2004]).

Zebras are the lion's only major prey species that live in persistent single-male social groups (Klingel 1969). Replacement-male zebras are infanticidal (Pluháček and Bartoš 2000), thus zebra foals are subjected to an additional source of mortality that does not apply to any of the lions' other prey species. Males suffer higher rates of predation than females in almost every mammalian species (Boukal et al. 2008), and a proportion of zebra foals will inevitably be lost after the deaths of their fathers. However, even if lion predation indirectly increases the mortality rate of zebra foals via sexually selected infanticide, we are still left with the question of why zebra numbers have not dropped in response to long-term increases in the Serengeti lion population. As seen below, the zebra population in the Crater shows clear signs of density dependence with little evidence of being limited by lion predation, so the factors regulating the Serengeti zebra population require further exploration.

Population sizes of the Serengeti wildebeest, zebra, and buffalo populations are all measured across the entire ecosystem, but what about impacts of predation at more local scales? Between 1966 and 2002, Sinclair et al. (2003) regularly monitored a 150 km² area in northern Serengeti where poaching and indiscriminate poisoning removed the majority of lions, leopards, hyenas, and jackals during an eight-year period (1980–1987). In contrast, predator populations remained intact in a 500 km² area in the adjacent Maasai Mara Reserve. During the 1980s, five well-studied herbivore species increased in abundance in the predator-removal area compared to the non-removal area, and all five species declined again after the recovery of the predator populations (figure 11.2). In contrast, giraffe abundance did not increase in the predator-removal area during the 1980s (figure 11.2). These results supported Sinclair et al.'s prediction that species below a threshold body size (in this case, 150 kg) could potentially be predator limited whereas the large-bodied giraffe were not.

But can we attribute these results to lions? Warthog are one of the lions' most important prey species in the Serengeti, so they would have been expected to increase with the disappearance of the local lions, as would topi, as they are one of the top-ten prey species in the Serengeti (see figure 1.6). But oribi and impala are not preferred lion prey, so their population peaks during the predator-removal period must have instead resulted from the elimination of leopards and/or cheetahs. Similarly, Thomson's gazelle are important prey for lions, cheetahs, and leopards, so their population likely benefited from the removal of all three carnivore species. Giraffe did not appear to respond to the loss of predators, but they were in the midst of an ongoing population decline throughout the entire Serengeti ecosystem that was largely caused by

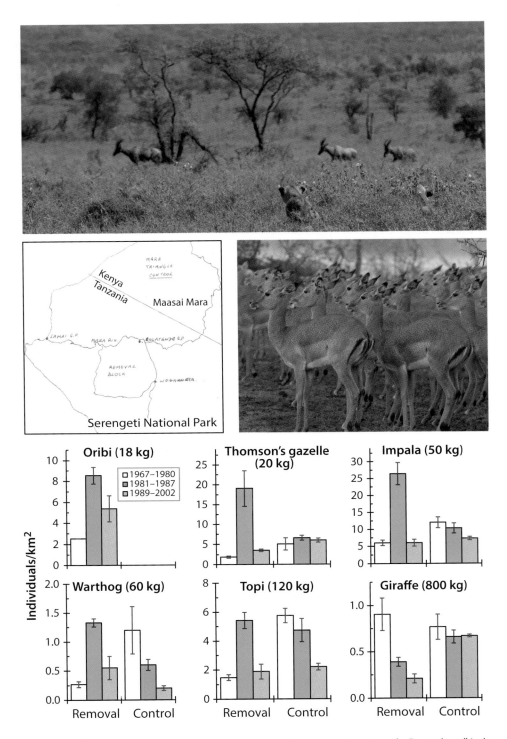

FIGURE 11.2. Effects of a predator "removal experiment" on herbivore population sizes. The "control area" is the Mara Triangle in the Maasai Mara National Reserve; the "removal block" is an area in the northern Serengeti where poachers eradicated the local carnivore populations in the 1980s. Release from predation is associated with significantly higher population size in all species except for giraffe. Redrawn from Sinclair et al. (2003).

increased levels of poaching rather than from lion predation (Strauss et al. 2015).[1] Bushmeat poachers hang snares from trees to target adult giraffes, so giraffe could have been targeted by the same poachers that had eradicated the predators in Sinclair et al.'s study area. Thus, while these data are consistent with the possibility that lions may limit the size of warthog and topi populations, the *lions'* impacts on the other four prey species are less clear.

Although we lack systematic data on warthog and topi numbers within our long-term study area in the Serengeti, forty years of survey data in Kruger National Park suggest that lions limit warthog and tsessebe (the local name for *Damaliscus lunatus*) populations while not having any detectable effect on Kruger's wildebeest and zebra populations (Owen-Smith et al. 2005). So, while the Kruger lion-cull of 1974 had no meaningful impacts on the species it was designed to protect (wildebeest and zebra), it could well have benefited Kruger's warthog and tsessebe.

Ngorongoro Crater

As shown in chapter 10, wildebeest, zebra, and buffalo all show strong signs of density dependence in the Crater, with rapid annual growth at low population sizes versus population declines at high densities (figure 11.3a). The size of each prey population is also correlated with the six-year running average of rainfall, with larger population sizes during wetter time periods (Moehlman et al. 2020). As in the Serengeti in 1993, the 1999–2000 drought in the Crater led to a major die-off in the buffalo (figure 10.12a), which was instrumental to the CDV-Babesia die-off in 2001 (figure 10.8b). After controlling for population density and rainfall, none of the lions' three primary prey species varied with the size of the Crater lion population (figure 11.3b). Thus, although lion numbers depend on per capita prey availability (figure 10.12c), prey populations in the Crater appear to be independent of lion density.

AREN'T PREDATORS AND PREY SUPPOSED TO SHOW POPULATION CYCLES?

Classic ecological theory predicts repeated population cycling between predators and prey (e.g., Volterra 1926; May 1972) as seen in the well-known case of lynx and snowshoe hare. Lynx drive the hare population down to the point where the lynx start to starve; but as the lynx population falls, the hare population recovers, allowing the lynx to recover in turn, and the whole cycle repeats itself every ten years, *ad infinitum*.

---

1 Between 1975–1977 and 2008–2010, the Serengeti giraffe population is estimated to have declined from 10,750 to 3,520, but the cause varied by region (Strauss et al. 2015): poaching primarily removed adults in the western woodlands (an area called Kirawira) and the far northern woodlands (Bologonja, located close to Sinclair et al.'s [2003] study area), whereas giraffe within the well-protected center of the park (Seronera, located within the long-term lion study area) showed a far lower calf–cow ratio, indicating a limited food supply. Over the same time, the Seronera woodlands underwent a threefold decline in the ratio of the palatable *Acacia tortilis* compared to the unpalatable *A. robusta*, a reduction in food supply that resulted from the feeding preferences of giraffe and elephants (Strauss and Packer 2015). Although lion predation attempts may be higher in Seronera than Kirawira (Strauss and Packer 2012), there was no evidence that giraffe were limited by predation in any of the three areas (Strauss et al. 2015).

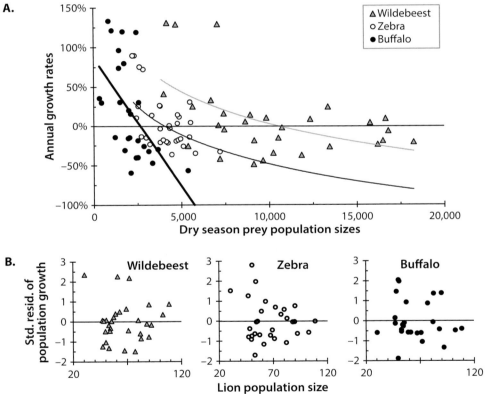

FIGURE 11.3. A. Density dependent population growth of the lions' three major prey species in Ngorongoro Crater Prey. **B.** Standardized residuals of annual population growth versus lion population size each dry season. Residuals control for rainfall and population density of each species. Prey population growth does not vary with the number lions in the Crater.

Population stability, in contrast, is a hallmark of the Serengeti lions (figure 10.2); the large-scale changes in the Crater lion population (figure 10.10) have been driven by inbreeding and disease, and there is little evidence that the lions in either ecosystem have affected the population sizes of their primary prey. Working with theoretical ecologists, we approached the rather surprising stability of the lion's food web from two different perspectives: first, by pretending that lions specialize on a single prey species to a similar extent as lynx on snowshoe hares but recognizing that lions and their prey are both social species. Second, by allowing lions to feed on multiple prey species and assessing whether variations in rainfall might trigger wide swings in the savanna food web by affecting all the lions' prey in the same way.

Stability versus Sociality

Classic theoretical models assume that predators and prey move randomly across featureless landscapes and that prey consumption varies with the density of *individual* predators and *individual* prey. But lions live in prides, their prey live in herds, and sociality fundamentally alters the expected theoretical outcomes. John Fryxell measured the grouping patterns and spatial distribution of all the larger herbivores along a series of ground transects in the Serengeti study area, and he found that all the lions' major prey species exhibit log-linear relationships between *group* density and population density, with slopes significantly less than one (figure 11.4). This means that each prey species forms larger herds in areas of higher population density. Territorial predators only have access to a limited number of herds, and they only capture one animal at a time, so each individual prey's chances of being captured are increasingly diluted at higher population densities, a pattern that greatly reduces predation rates below the levels expected under the classic assumption of random mixing. John then incorporated our long-term data on lion hunting and ranging behavior to develop a set of group-dependent models that fully consider the ecological implications of sociality (Fryxell et al. 2007). Hunting success is only expected to increase asymptotically with foraging group size (as in figure 7.7c), thus hunting groups take fewer prey per capita than do solitary hunters. John's models show that grouping strongly stabilizes population-level interactions between lions and wildebeest: sociality in either wildebeest or lions reduces the impact of predation, and sociality in both species reduces predation even further (figure 11.5a). Sociality thus has as strong an impact on predation as does the annual migration to the short-grass plains where the wildebeest escape from lions for several months each year (Fryxell et al. 1988). Under the observed levels of lion fecundity and mortality, John's models predict a range of conditions under which pride-living lions and herd-living wildebeest could conceivably coexist without undergoing any cycling in their respective populations (figure 11.5b). At higher levels of lion mortality, lion populations would be unable to persist, and, at lower levels of lion mortality, the lion and wildebeest populations would be expected to undergo repeated cycles, as these hypothetically long-lived lions would attain such high population sizes that they would drive down the wildebeest population then suffer the consequences of over-exploitation.

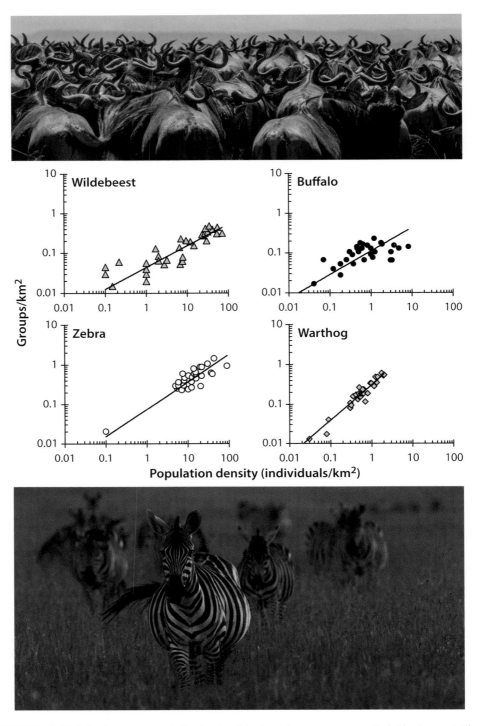

FIGURE 11.4. Herd density versus population density of the lions' four major prey species in the Serengeti. The number of herds increases in a log-linear manner with population density, but the slope is significantly less than 1.0 in all four species. Redrawn from Fryxell et al. (2007).

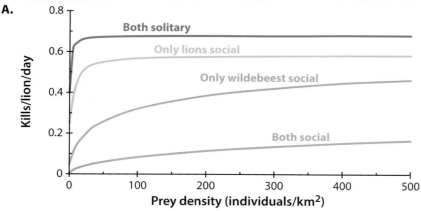

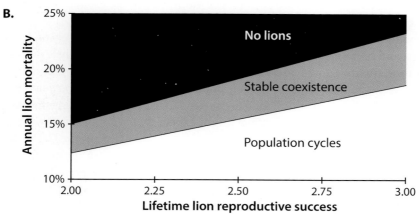

FIGURE 11.5. A. Predicted rates of lion predation on wildebeest, assuming that one or both species are either solitary or social. **B.** Predicted effects of lion reproductive and mortality rates on predator-prey dynamics, assuming that wildebeest and lions both live in groups. Lifetime reproductive success is number of surviving offspring produced over a six-year period. Redrawn from Fryxell et al. (2007).

An obvious shortcoming of the sociality model is that lions—unlike lynx—rely on multiple prey species, consuming significant numbers of wildebeest, zebra, and buffalo in both study areas, not to mention the Serengeti lions' wider diet of warthog, Thomson's gazelle, eland, hartebeest and topi. Thus, lions would be expected to shift their diet depending on the relative availability of each species. However, even in a more complex food web, multiple prey species may respond similarly to the same environmental factor such as a drought, thus generating synchronous population swings throughout the entire food chain (Ripa and Ives 2003). However, if prey species show contrasting responses to a particular perturbation, predators would enjoy a more consistent food supply and, hence, maintain greater stability in population size. In a simple two-prey model, Sinclair et. al. (2013) showed that predator populations would be nearly three times as stable in systems where only one of the two prey species responded to a strong environmental perturbation.

As we've seen, seasonal rainfall patterns in East Africa correlate with the Southern Oscillation Index (figures 10.4 and 10.5), and Sinclair et al. (2013) showed that calf recruitment (as measured by cow–calf ratios) in the Serengeti was significantly associated with SOI in both wildebeest and topi, but recruitment was independent of SOI in zebra, warthog, hartebeest, and waterbuck. Thus, following unfavorable years, lions could potentially shift their diet away from wildebeest/topi to one of these other four species (as seen in Kruger Park (Owen-Smith and Mills [2008]), thereby dampening their own population response to any declines in wildebeest and topi numbers. If so, the lion population would remain relatively constant while briefly driving down the populations of zebra, warthog, and so forth in the years following poor wildebeest and topi recruitment. Unfortunately, we only have relevant long-term population data on wildebeest and zebra, and surveys on these two species were too infrequent to test whether environmentally insensitive species (zebra) declined in the first years following a drop in SOI-sensitive species (wildebeest) (see figure 10.1). Unfortunately, too, annual counts of Kruger herbivore populations cannot be related to lion population size, as no comparable lion study has been conducted in Kruger Park.

What about the Ngorongoro Crater, where the lions feed almost entirely on wildebeest, zebra, and buffalo, and each herbivore species is counted annually? Although cow–calf ratios are not known, recruitment rates likely respond in a similar manner in all three species as they all show greater population growth following years with greater rainfall (Moehlman et al. 2020). Thus, instead of the sort of buffering predicted by Sinclair et al.'s two-species model, we might expect wider population swings in the Crater food web, and this might help explain the rapid decline in the Crater lion population in the early-1980s, which followed a period of below-average rainfall. But unfortunately—again!—there were no herbivore counts in the Crater between 1977 and 1986. Given the subsequent demographic impacts of disease and inbreeding in the Crater lions, the lion population has never again reached a level where it could be strongly limited by prey availability. However, if the lions continue their recovery from the prolonged inbreeding of the 1980s–2000s (figure 10.14), a future dry spell could conceivably trigger a drop in herbivore numbers that is further intensified by heightened lion predation. On the other hand, lions are not the only large predator in this system: spotted hyenas also capture substantial numbers of wildebeest and zebra in the Crater, and, as discussed in the final part of this chapter, hyena numbers are known to decline at higher lion densities. So, any impacts of a thriving, outbred lion population on the herbivores in the Crater food web might potentially be ameliorated by a concomitant decline in the Crater hyena population.[2]

THE SAVANNA LANDSCAPE OF FEAR

Lions may or may not have significant population-level effects on their major prey species in the Serengeti and Ngorongoro, but lions certainly have measurable impacts on the way that individual prey animals behave and distribute themselves across the savanna landscapes. Such *indirect* effects of predation have even proven important in the classic case of lynx and snowshoe hares. Although lynx and hare are both solitary (and hence more vulnerable to population cycles), hares are not only directly removed from the prey population by successful hunts, but survivors of unsuccessful chases can be sufficiently stressed to suffer from reduced fertility (Krebs et al. 2018). Thus, the indirect and direct effects of predation combine to cause the hare population to decline at higher lynx densities. So even if, for example, group living helped to stabilize lion-wildebeest dynamics via the removal of only a single individual from each herd, whole herds of escaping wildebeest could potentially suffer group-level costs from physiological stress or from the avoidance of high-quality feeding sites in a landscape of fear. We couldn't possibly evaluate the physiological profiles or reproductive performance of an individual wildebeest or zebra after escaping from a lion attack.[3] But the fear of predation can also alter the locations where prey animals spend their time (Brown et al. 1999; Laundré et al. 2001, Atkins et al. 2019)—and ambush predators such as lions are likely to inspire the greatest fear of all: Is a lion hiding under that bush? Somewhere near this waterhole? Maybe, maybe not—better play it safe. Thus, we sought to develop a methodology for monitoring how wildebeest, zebra, buffalo, and so on all distribute themselves across the landscape, avoiding dangerous areas and seeking refuge in other areas.

Ali Swanson set out a grid of over two hundred camera traps in a 1,000 km^2 portion of our Serengeti study area (figure 11.6; Swanson et al. 2015). On any given day, each camera trap only captured a minuscule proportion of the animals in its vicinity, but we were able to maintain the grid continuously for the final five years of the lion project and accumulated such a large volume of data that Ali and Margaret Kosmala teamed up to establish "Snapshot Serengeti," a citizen-science project where online volunteers counted and classified the animals in millions of images. Each photo was presented at random to ten different people viewing the images online in dozens of countries around

---

[2] A final point about possible lion impacts on herbivore numbers: if a prey population is primarily regulated by parasitic infection rather than by predation, lions could actually *increase* prey population size by selectively removing highly infectious individuals and thereby "keeping the herds healthy and alert" (Packer et al. 2003).

[3] Examining the bone marrow of over 750 wildebeest, Sinclair and Arcese (1995) found that individuals killed by predators were in better physical condition than those dying from other causes, but victims of predation were in worse condition than living animals, suggesting that wildebeest may often enter dangerous areas as they lose condition, thus predation works jointly with starvation to limit prey populations.

FIGURE 11.6. Locations of the "SnapshotSerengeti" camera-trap grid. Modified from Swanson et al. (2015).

the world, and their consensus classifications showed 97 percent agreement with the "expert classifications" made by ecologists such as Ali, David Bygott, and myself, who had spent considerable time in the Serengeti (Swanson, Kosmala, et al. 2016).

Anderson et al. (2016) analyzed the first three years of Snapshot data to measure the lions' spatial associations with eight of the most common large herbivores in the Serengeti. Lions were photographed more often at the same camera-trap sites as all three of their most important prey species (buffalo, wildebeest, and zebra) as well as one nonpreferred species (impala); lions were photographed less often than expected at sites with topi (figure 11.7). The positive associations between lions and their preferred prey presumably reflected the lions' positioning themselves in areas where wildebeest, zebra, or buffalo might eventually appear, whereas topi presumably took greater care to avoid areas occupied by lions. Impala are the most abundant browsers in the Serengeti woodlands, yet they only account for 24 out of 2,700 carcasses (0.9%) in the woodland lions' diet. Thus, the positive association between lions and impalas likely reflects the lions' daytime habit of sleeping in the shade of an impala-friendly tree that had been fitted with a camera trap. "Associations" between predator and prey were measured over a three-year period, so any impala and lion that were photographed at the same tree would have arrived several hours or even days apart.

Although a camera grid appears to be broadly useful for examining predator-prey relationships, we not only had to overcome the bias of lions sometimes sleeping in the shade of the only tree in a broad area of open grassland, but we also had to deal with the sparsity of the lion imagery compared to our detailed knowledge of lion ranging behavior (Swanson, Forester, et al. 2016). Thus, Meredith Palmer used the long-term data to develop two "lion layers" for a GIS analysis of the Snapshot imagery. First, the lions show persistent preferences for specific landscape features (chapter 9), thus a prey animal could conceivably use these cues to minimize its risk of *encountering* a lion. Second, prey might avoid areas where lions make most of their kills (chapter 9), so Meredith made *predation-risk* maps for each prey species.

Among the many possible ways to think about a landscape of fear (see Moll et al. 2017 for a review), we focused on how prey animals might respond to the tradeoff between feeding success and predation risks. An animal that forsakes feeding opportunities by spending time in a safer area is suffering an important *indirect* effect of predation; an animal that forages in a larger group in a dangerous area is similarly reducing its individual food intake rate while diluting its personal chances of being captured in a lion attack. Meredith used monthly satellite imagery to estimate plant productivity via the Normalized Difference Vegetation Index (NDVI), as Anderson et al. (2016) had found that each herbivore species was most often found at camera-trap sites with preferred levels of NDVI (e.g., high for bulk-feeding buffalo, moderate for wildebeest and short grass for Thomson's gazelle). Thus, she could test how each prey species balanced feeding with the risks of encountering—or being captured by— lions. These risks vary significantly with the phase of the moon (being highest at the new moon and lowest at the full moon, figure 1.17), so she could treat the lunar cycle as a naturalistic experiment.

Looking solely at nighttime photos, Meredith found that buffalo, wildebeest, zebra, and gazelle all alter their behavior across the lunar cycle and that these changes are largely consistent with a landscape of fear (figure 11.8). For example, during the wet

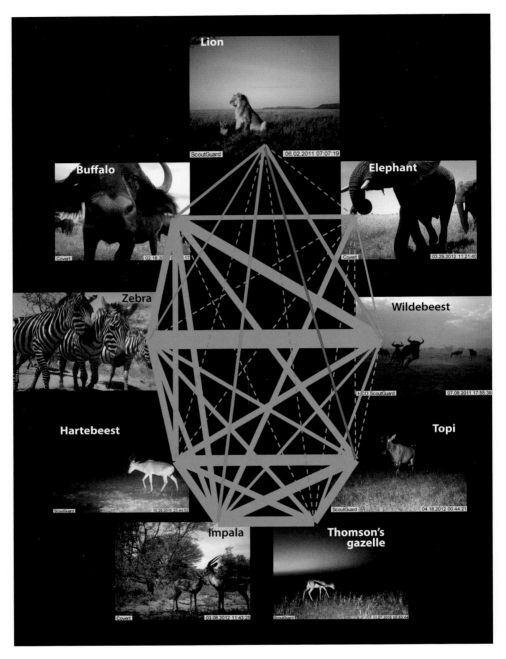

FIGURE 11.7. Associations among species in the SnapshotSerengeti camera grid. Green lines indicate positive associations between lions and their prey; red line indicates a negative association between lions and topi. Blue lines indicate positive associations among herbivore species. Dotted lines show nonsignificant associations; width of each solid line is proportional to the coefficient of each significant association. Redrawn from Anderson et al. (2016).

season, wildebeest spend more time overall in low lion-encounter-risk areas and most often enter high encounter-risk areas on the brightest nights near the full moon. Zebra form larger aggregations in high predation-risk areas in the darkest nights of the wet season. During the dry season, zebra form larger aggregations in high-food-abundance areas during the darkest nights, presumably reducing their individual predation risks while remaining in essential grazing areas when food is most limited. Finally, buffalo spend more time in high-food-abundance areas during the brightest nights of the wet season, presumably so they can continue feeding during the safest nights of the lunar cycle.

LIONS VERSUS THE OTHER LARGE CARNIVORES

Lions face little serious feeding competition from other carnivores. Wild dogs, cheetahs, and leopards readily abandon their captured prey carcasses to even a lone scavenging lion; mobs of spotted hyenas can oust a singleton female lion from her kill, but groups of lions gain far more meat from hyenas than vice versa (Kruuk 1972; Schaller 1972). In terms of prey selection, lions mostly feed on buffalo, zebra, and wildebeest, whereas wild dogs, cheetahs, and leopards all specialize on prey the size of Thomson's gazelle and impala. Spotted hyenas almost never catch buffalo, and, when hunting zebra and wildebeest, hyenas take more juveniles than do lions (see figure 11.14a). Thus, none of these four carnivore species are likely to limit the size of the lions' most important prey populations, which, as we've seen, are largely food limited. Yet lions readily kill wild dogs, cheetahs, leopards, and spotted hyenas (Palomares and Caro 1999; also see figure 11.12). Lions do not kill these animals for food, they just choke and shake their victims then leave the bodies behind. All four species pose clear threats to lion cubs (Palomares and Caro 1999; also see figure 4.8), and thus lions would benefit from eliminating as many cub predators as possible. But do lions have measurable population-level impacts on any of these other carnivore species as, say, wolves have on coyotes (Berger and Gese 2007)? And has interspecific competition played a significant role in the evolution of lion sociality?

Lions versus Wild Dogs and Cheetahs

Wild dogs and cheetahs are largely restricted to the plains portion of our Serengeti study area. The plains lion population tripled in size between the mid-1960s and late 1990s, but this increase had contrasting impacts on wild dogs and cheetahs: whereas the wild dogs disappeared from the area, the cheetah population remained unchanged (figure 11.9a and 11.9b).[4] In South Africa, lions, wild dogs and cheetahs have all been reintroduced to a number of small, fenced reserves that range in size from about 200 to 1,000 km².

---

4 The inverse relationship between lion and wild dog numbers was first noted by Jeannette Hanby and David Bygott (1979), but a later hypothesis suggested that wild dogs disappeared from the Serengeti because of stress induced by a vaccination program intended to protect the species against rabies (reviewed by Jackson et al. [2018]). However, Jackson et al's analysis demonstrated that handling does not "reactivate latent viruses" in the Serengeti wild dogs and that the earlier effort caused no measurable harm to vaccinated packs.

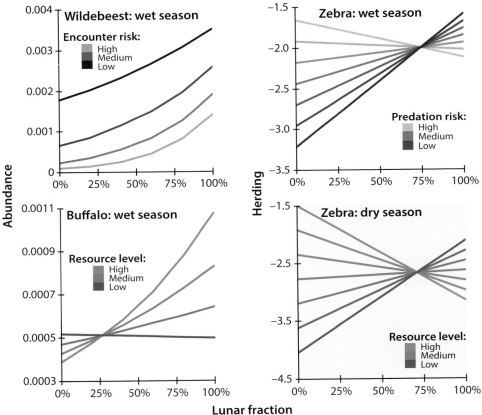

FIGURE 11.8. Abundance estimates and herding tendencies of three herbivore species versus lunar fraction at different levels of predation-related risks (top) or resource availability (bottom). Wet season is indicated by white background; dry season by tan. See text for details. Redrawn from Palmer et al. (2017).

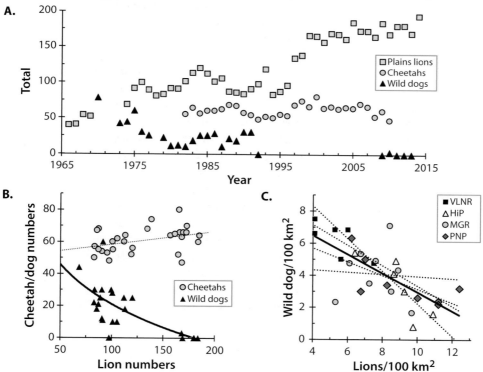

FIGURE 11.9. Coexistence between lions, wild dogs, and cheetahs. **A.** Population sizes of lions, cheetahs, and wild dogs on the Serengeti plains. **B.** Wild dog and cheetah numbers versus lion population size on the Serengeti plains. Wild dog numbers were significantly lower in years with more lions; cheetah numbers remained unchanged. **C.** Wild dog numbers in four fenced South African reserves were consistently lower when lions were more abundant. Redrawn from Swanson et al. (2014).

Whereas lions thrive in these restored ecosystems (see chapter 12), wild dog numbers decline in circumstances when lions are most abundant (figure 11.9c), while reintroduced cheetah populations are unaffected by lion numbers (Swanson et al. 2014).

Why are wild dog populations so much more vulnerable to lions? The answer seems to be the dogs' extraordinarily broad landscape of fear. Radio-tracking data shows a striking pattern in the ranging patterns of wild dogs in relation to lions: wild dogs occupied low-lion-density areas during the 1980s, but these spaces were completely filled by the early 2000s, by which time the dogs remained well outside the national park (figure 11.10)—and have remained outside the lion study area even after the reintroduction of several new packs to the Serengeti by conservationists and the continued growth of the wild dog population to the east of the National Park (Jackson et al. 2019). Dividing the data from the 1980s by season further emphasizes how wild dogs remain in areas where they are least likely to encounter lions (figure 11.10), whereas cheetahs occupy similar areas as lions (figure 11.11a). Research in Botswana likewise shows that cheetah often remain close to lions (figure 11.11b). Though the Serengeti cheetahs occupy the same areas as lions, they are careful to remain separated in time: cheetahs don't show up at the same camera trap as a lion until at least twelve hours later (figure 11.11c). Cheetahs are the fastest land mammal, so even though lions frequently kill cheetah cubs, cheetahs show a number of behavioral responses that allow them to live close to lions. For example, when female cheetahs feed alone, they consume their prey as rapidly as possible, whereas females with cubs prioritize vigilance over speed of eating (Hilborn et al. 2018). Thus, cheetahs manage to maintain their numbers by playing a never-ending game of "chicken" with the lions and, presumably, relying on their great speed to escape, whereas wild dogs keep their distance—even to the point of surrendering large tracts of open habitat to the lions. Interestingly, wild dogs in South Africa's Hluhluwe iMfolozi Park seek out landscape features that help them to avoid detection by lions; thus ecosystems with greater landscape heterogeneity provide greater opportunity for lion–wild dog coexistence than the relatively homogenous Serengeti plains (Davies, Tambling, et al. 2021).

Lions versus Leopards

Leopards are woodlands animals, so we almost never encountered leopards in the Crater or the Serengeti plains. We have seen a woodlands pride kill an adult female leopard (figure 11.12), and George Schaller (1972) famously saw a leopard kill two lion cubs, but we know almost nothing about leopard ecology in the Serengeti, nor can we estimate the impacts that lions and leopards have on each other, as we lack population and telemetry data on the Serengeti leopards. By far the best data on lion-leopard coexistence come from studies by Guy Balme and his colleagues in South Africa. Camera-trap data show broad similarities in habitat use by the two species in ten different reserves in the bushveld of Limpopo and KwaZulu-Natal, although there are a few notable differences: leopards prefer more rugged topography and thicker brush, while lions avoid the thickest brush and prefer areas with larger prey (figure 11.13a). Given their similar patterns of overall habitat use, leopards, like cheetahs, would be expected to show behavioral adaptations to minimize risks from lions. Leopards are

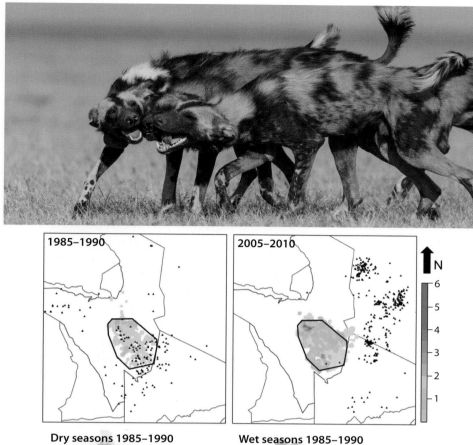

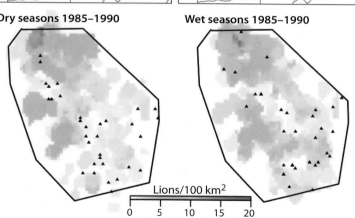

FIGURE 11.10. *Top maps:* Locations of radio-collared wild dogs in the Serengeti ecosystem versus local lion densities in the long-term Serengeti study area. Triangles show wild dog sightings; colors indicate lions per 100 km². Wild dogs resided within the lion study area when lion densities were low in the late 1980s but remained outside the study area by the 2000s. *Bottom maps:* Dry and wet season locations of radio-collared wild dogs versus seasonal lion densities within the long-term study area. The wild dogs significantly avoided higher-use lion areas. Redrawn from Swanson et al. (2014).

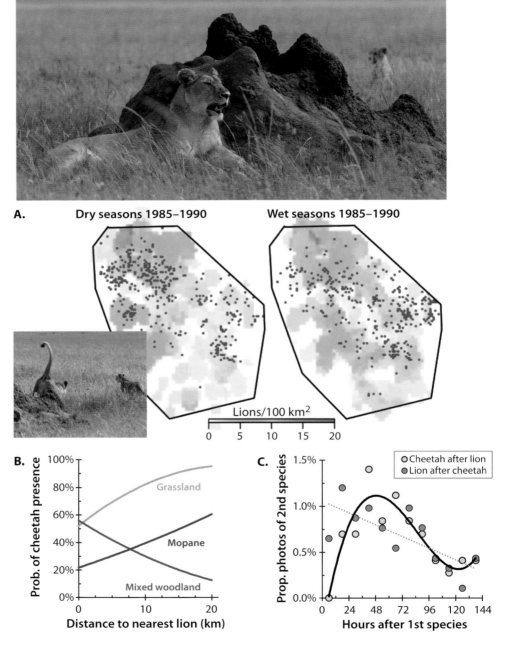

FIGURE 11.11. A. Locations of radio-collared cheetahs in the late 1980s versus local lion densities in the Serengeti study area. Cheetahs utilized similar areas as the lions. Redrawn from Swanson et al. (2014). **B.** Relationship between the risk of cheetahs encountering lions as determined by the distance to the nearest lion in the Okavango ecosystem. Cheetahs were closest to lions in mixed woodlands and furthest in grasslands and mopane woodlands. Redrawn from Broekhuis et al. (2013). **C.** Time lags between photographs of lions and cheetahs at the same Serengeti camera-trap site. Whereas lions were consistently more likely to appear at a camera site in the first few days after a cheetah, cheetahs never came to the same camera within twelve hours after a lion. Redrawn from Swanson, Forester, et al. (2016).

well known for carrying carcasses up trees, and Guy has found that leopards typically hoist prey carcasses that are about the same size as themselves, as these are too large to consume in a single sitting, but leopards lose nearly all their heavier carcasses to scavengers (figure 11.13b). Spotted hyenas steal a far higher proportion of carcasses from leopards (83% cases of interspecific scavenging on leopard kills) than do lions (15% of interspecific scavenging), and hoisting appears to be most effective against hyenas: once a hoisted carcass has been detected, hyenas only steal 19.5 percent of carcasses while tree-climbing lions steal 62.5 percent (Balme, Miller, et al. 2017).

Can lions suppress leopard populations? There is no clear relationship between lion and leopard numbers across the ten reserves studied by Miller et al. (2018), though any overall pattern may be obscured by a subset of reserves only containing enough prey to support a limited number of leopards and lions whereas other reserves can support higher numbers of both. The best test would be to measure changes within the same park as in figure 11.9, and data from a long-term study in Sabi Sands reveals that leopard and lion numbers are positively correlated at low population densities, but leopard numbers decline when the lion population exceeds about eighty-five individuals in the 625 km^2 reserve (figure 11.13c). Interestingly, camera-trap studies in India suggest that leopard numbers show a similar sensitivity to tiger densities, with both species increasing at lower densities but leopards declining above a threshold density of tigers (Kumar et al. 2019).

Lions versus Spotted Hyenas

Spotted hyenas have the greatest dietary overlap with lions (Hayward 2006). They are the only species that can supplant lions from prey carcasses (Cooper 1991; Höner et al. 2002), and they are also the only carnivore that has ever been seen to kill an adult lion (Loveridge et al. 2010). On the other hand, lions frequently kill hyenas (figure 11.14; Kruuk 1972; Watts and Holekamp 2009),[5] and scavenging lions obtain as much or more meat from hyenas as vice versa (Watts and Holekamp 2008). The two species seem to be able to coexist because of contrasting feeding strategies. First, hyenas are able to digest the skin and bones of remains left behind by foraging lions (Kruuk 1972), and, second, hyenas capture a higher proportion of juvenile prey than do lions (figure 11.14a).

Our Snapshot Serengeti data emphasize the extent to which lions and hyenas associate with each other: each species is most likely to appear at a camera-trap site in the first twenty-four hours following the arrival of the other (figure 11.14b). Périquet et al. (2014) reported a positive correlation between lion and hyena densities across twenty-six ecosystems, although this pattern was largely driven by a single data point from the Kalahari Desert, where large herbivores, lions, and hyenas were all extremely scarce. Höner et al. (2005; 2012) resumed monitoring the spotted hyenas in Ngorongoro Crater thirty years after Hans Kruuk's landmark study in the 1960s. The Crater hyena population was at its highest recorded level in 1965–1967 (385 adults), during

---

[5] Lions were the primary source of mortality for hyenas in the Ngorongoro Crater (Kruuk 1972) and Maasai Mara (Watts and Holekamp 2009).

FIGURE 11.12. Group of three female lions kill an adult female leopard in the Serengeti. Photos by S. Mwampeta.

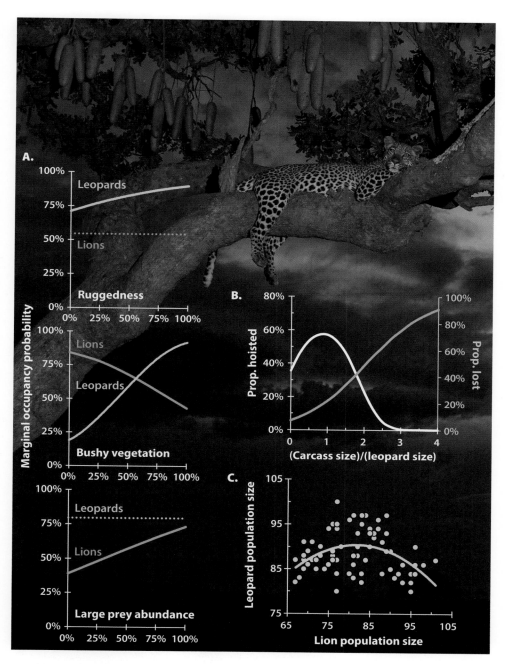

FIGURE 11.13. A. Marginal occupancy probability of leopards and lions in ten South African reserves as a function of topographic ruggedness, bushy vegetation, and relative abundance of large prey. Solid lines represent significant relationships; dotted lines are nonsignificant. Redrawn from Miller et al. (2018). **B.** Caching behavior of leopards in Phinda Reserve, South Africa. Leopards mostly hoist carcasses of similar body weight as themselves and lose most of the larger carcasses to other predators. Redrawn from Balme, Miller, et al. (2017). **C.** Leopard population size versus lion numbers in Sabi Sands Reserve. Leopard numbers show an inverse-U-shaped relationship with lion density. Redrawn from Balme, Pitman, et al. (2017).

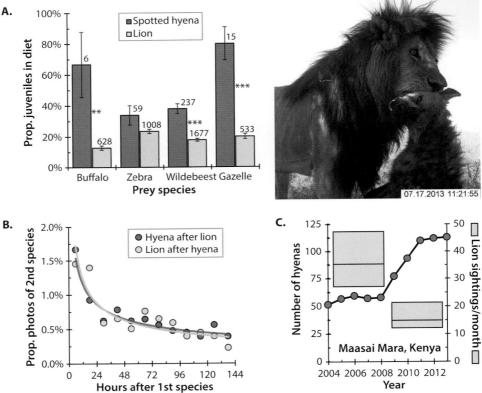

FIGURE 11.14. Lions and hyenas. **A.** Age composition of each prey species in the diet of spotted hyenas and lions. Hyenas in the Ngorongoro Crater feed on a significantly higher proportion of juveniles than do lions. Hyena data are from Höner et al. (2002). **B.** Time lags between photographs of lions and hyenas in the same Serengeti camera-trap site. Each species was significantly more likely to appear at a camera site in the first hours and days following the presence of the other species. Redrawn from Swanson et al. (2016). **C.** Population size of spotted hyenas in Maasai Mara versus the average number of lion sightings per month by the hyena research team over two five-year periods. Redrawn from Green et al. (2018).

which time lion numbers averaged only about 30 individuals, as these were the first few years after the lion die-off in 1963 (see chapter 10). When Höner et al. resumed hyena monitoring in 1995, the Crater lion population had recently declined from about 80 individuals to less than 50, and lion numbers remained relatively constant for the next thirteen years, during which time the hyena population grew from 117 to a peak of 368 adults in 2011 (O. P. Höner, personal communication). Hyenas only have one to two cubs per litter and an interbirth interval of nearly eighteen months (Holekamp et al. 1996), so it is possible that the steady increase in hyena numbers through the late 1990s was at least in part due to the Crater lion population's persistent fall below their fifteen-year-long peak starting in the early 1990s.[6]

Maasai pastoralists regularly graze livestock in Kay Holekamp's long-term hyena study area in Maasai Mara, and, beginning in 2009, increased human-lion conflict caused a substantial drop in the local lion population, as the number of lion sightings recorded by her field team dropped by half compared to the previous five years (Green et al. 2018), and the fall in lion sightings was associated with a doubling in the hyena population (figure 11.14c). Thus, data from both the Crater and Maasai Mara are consistent with lions being capable of driving down hyena numbers, especially at high lion densities (Watts and Holekamp 2008). Note that large-scale changes in lion numbers were either caused by disease (Ngorongoro) or human-lion conflict (Maasai Mara) rather than by competition with spotted hyenas.

LIONS, HYENAS, AND POSSIBLE "TROPHIC CASCADES"

Whereas the Crater lions prey almost exclusively upon buffalo, wildebeest, and zebra, the Crater hyenas show a striking preference for juvenile Thomson's gazelle (Kruuk 1972, Höner et al. 2002). With such a simple dietary divergence between the two predators, any impacts of lion numbers on hyena abundance might potentially alter population growth in the Crater gazelle (e.g., Polis and Holt 1992). Thomson's gazelle is one of the most abundant prey species in the Crater yet their population growth shows much weaker evidence of density dependence than the lions' three primary prey species (compare figure 11.15a with figure 11.3), suggesting that gazelle numbers might be more strongly affected by top-down processes such as predation rather than by food limitation alone. Thus, it is noteworthy that Thomson's gazelle numbers in the Crater were highest in years when lions were most common (figure 11.15b).

The lion-hyena-gazelle linkage in Ngorongoro Crater is remarkably similar to the indirect effects of wolves on pronghorn antelope: by reducing coyote numbers, wolves reduce the abundance of the primary predator on pronghorn fawns, thereby allowing pronghorn numbers to increase (Burger et al. 2008). How general might the lion-hyena/wolf-coyote pattern be in savanna ecosystems? In contrast to the Crater, the lions' diet is much less specialized in the Serengeti (including a much higher proportion of gazelle, see figure 1.6), thus the Serengeti food web is far more complex. However, annual population counts are entirely lacking for all the Serengeti herbivores—and many of

---

6 Ferreira and Funston (2020) found that hyena numbers fell when/where lion numbers rose in six regions of Kruger National Park, further suggesting that lions limit hyena populations.

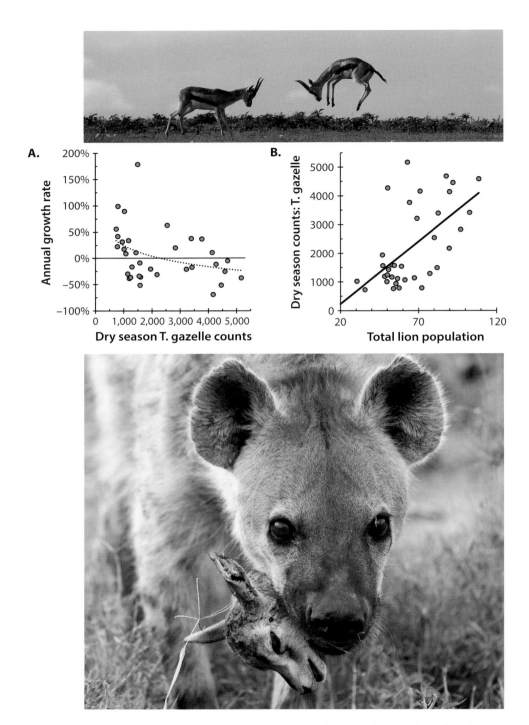

FIGURE 11.15. Possible "trophic cascade" in Ngorongoro Crater. **A.** Population size of Thomson's gazelle in the Crater is only weakly density dependent. **B.** Thomson's gazelle numbers in the Crater show a strong positive correlation with lion numbers.

these species are also subject to significant levels of bushmeat poaching (e.g., Dublin et al. 1990; Sinclair and Beyers 2021; Strauss et al. 2015). Even in the best-studied sites in South Africa, hyena counts are virtually nonexistent, so it is impossible to say whether similar trophic cascades might exist elsewhere or whether a food web as complex as in the Serengeti is more or less resistant to year-to-year changes in predator abundance than the relatively simple web in the Crater.

LIONS VERSUS HYENAS AND THE EVOLUTION OF LION SOCIALITY

Various authors have commented on the importance of sociality in enabling the coexistence of lions and hyenas. A single hyena could never supplant a solitary lion from its kill, but "mobs" of hyenas can usurp carcasses from feeding lions (see Cooper 1991, Lehmann et al. 2017), and the consequent pressure on lions to defend their kills has been suggested to contribute to the evolution of lion sociality (Cooper 1991). We have certainly seen cases where lone female lions lose their kills to hyenas, as in this case:

> **Ngorongoro Crater 17-Sep-1980 14:20.** I find the Lake pride female, Laiti, alone with her four 4-mo old cubs. She has just caught a pregnant zebra mare. The body cavity is open, and she has pulled out the near-term fetus, but little else has been eaten by the time I arrive. Mother and cubs feed intermittently over the course of the afternoon. By sunset, Laiti's belly size has increased from 3.0 to 2.75; her cubs' bellies have enlarged from 2.5 to 1.75. At 19:00, about 20 hyenas show up at the kill. Laiti starts to defend the carcass, but her cubs are all nervous of the hyenas, and they start to retreat toward a cluster of tall reeds in the Mandusi swamp. Laiti abandons the kill and remains close behind her four cubs as they disappear into dense cover.

With four small cubs, Laiti was in a particularly difficult situation, as was SBG when she caught the zebra far from the safety of her own territory (chapter 8). But how do lone females ordinarily cope with hungry hyenas? And are the losses high enough to favor group living?

Our most comprehensive data on this topic again come from the ninety-six-hour watches in the mid-1980s, as we focused so closely on the foraging success of lone females. Of the seventy-four carcasses obtained by females during that study, about a third were acquired by singletons. Compared to females in groups, lone female lions scavenged a higher proportion of their meals from hyenas, and singletons also lost a higher proportion of their carcasses to hyenas (figure 11.16). In terms of gains and losses to hyenas, *scavenging* singletons obtained carcasses averaging nearly 30 kg of edible biomass, whereas *hunting* singletons typically managed to consume all of their own prey carcasses without losing any meat to hyenas (figure 11.17a). Because bone-crushing, skin-eating hyenas are better able to utilize the remains from mostly eaten carcasses, I separated relatively intact carcasses from those that had been reduced to less than a quarter of the muscle tissue before first being acquired by the lions (figure 11.16). While in proximity to hyenas, singleton females fed on eleven relatively intact carcasses, eating two carcasses down to skin and bones, abandoning another four carcasses before reducing them to skin and bones, losing one carcass to another lone female lion, and being chased away by the hyenas from four relatively intact carcasses. Singletons that were chased away by hyenas had already eaten an average 12.1 kg per carcass and lost a remaining 50 kg to the hyenas, whereas singletons that finished or abandoned their carcasses had already consumed a median of 25.5 kg and left a median of approximately 20 kg uneaten by the time of their departure (figure 11.17b).

Thus, being displaced by hyenas certainly meant that singleton female lions lost measurable amounts of food, but, on the other hand, feeding with a second female would have meant a 50 percent reduction in food intake at almost all the singleton's other carcasses, as she would have had to share the meat equally with her partner. Assuming that a pair of females could have eaten their fill from all four of the carcasses that singletons lost to hyenas (and that they also would have won the encounter with the lone outside female), the average payoff per carcass for each member of the pair would have been 14.9 kg as opposed to the observed level of 19.7 kg per carcass for singletons (figure 11.17c). Thus, habitually dividing the food with a second lion would have cost substantially more than losing meat to hyenas from a small proportion of carcasses. Of course, this exercise ignores all the other advantages of group living—including the ability to maintain a territory farther away from the hyenas' preferred hunting grounds (Kruuk 1972)—but it does demonstrate that lions are unlikely to have developed sociality *solely* in response to feeding competition with hyenas.

THE LION'S PLACE IN THE WEB

While lions have measurable effects on the abundance and/or distribution of certain species, many gaps remain in our understanding of the Serengeti and Ngorongoro food webs. Lions appear to have the largest population-level impacts on warthog, topi, and wild dogs. Buffalo, wildebeest, and zebra all modify their behavior in areas with higher risks of lion predation; however, despite enormous efforts by the Serengeti Ecological Monitoring Program, only a few years' data are available on population sizes of the dominant herbivores and no meaningful numbers have been collected for most of the lions' other prey species in the Serengeti. Annual population counts are available for the most important lion prey in Ngorongoro, but the Crater lions have been held so far below their carrying capacity by inbreeding and disease for the past twenty-plus years that they have had little chance to exert full pressure on their primary prey populations. Yet it is noteworthy that Thomson's gazelle may benefit from a trophic cascade in the Crater via the lions' impacts on hyena numbers. The only other holistic long-term study of savanna ecology comes from Kruger Park, but the Kruger research program lacks longitudinal data on any of its carnivore populations. Regardless of the component interactions, however, food webs in all three ecosystems appear to be reasonably stable, despite considerable variation in their constituent parts. But the biggest challenges are no doubt yet to come, as climate change is expected to produce greater variation in rainfall and continued human population growth will place ever-greater demands for land and bushmeat.

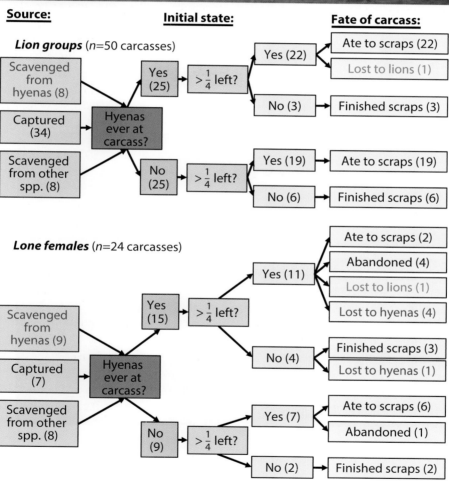

FIGURE 11.16. Origins and fates of all carcasses obtained during the ninety-six-hour follows. *Top:* All foraging groups containing more than one adult lion. *Bottom:* Foraging by lone female lions. Lone females scavenged a significantly higher proportion of carcasses from hyenas and lost a higher proportion of relatively intact carcasses to hyenas. Note: "scraps" refers to "skin and bones."

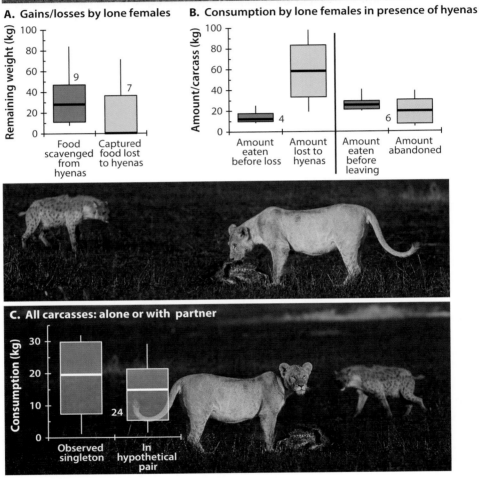

FIGURE 11.17. Spotted hyenas versus lone female lions during the ninety-six-hour follows. **A.** Amount of edible biomass scavenged from each other by hyenas and lone female lions. **B.** Amount of food eaten by lone females before losing or abandoning carcasses to hyenas (left), and amount of food remaining at each carcass when abandoned (right). **C.** Observed feeding success of singleton females across all carcasses versus per capita food intake of each female in a hypothetical pair that never lost any food to hyenas (see text). Sample sizes refer to the number of carcasses.

Before leaving this topic, I want to make two final points. First, despite the obvious oversimplification of John Fryxell's single-prey/single-predator model, lion sociality inevitably contributes to the stability of the savanna food web—if only for the fact that a pride of lions almost always consumes all the meat and offal of its captured prey, whereas singleton females often abandon partially eaten carcasses (figures 11.16 and 11.17b). Thus, if four female lions only ever foraged solitarily, they would inevitably kill more prey animals than a fission-fusion pride of four females that typically hunted together and coalesced at larger kills. Second, whereas leopards and spotted hyenas both thrive at lower lion densities, their numbers appear to suffer at higher lion densities. Thus, a world that is utterly dominated by lions could conceivably be a different world, and, as we will see in chapter 12, it might really be possible to have "too many lions": many of the small, fenced reserves in South Africa have resorted to the use of contraception and culling their lions in order to keep their parks from becoming "predator pits."

KEY POINTS

1. Lions apparently limit warthog and topi populations in the northern Serengeti, but wildebeest and buffalo in the Serengeti and Ngorongoro Crater are both limited by their dry season food supply rather than by predation. Whereas the Crater zebra population also appears to be food limited, the forces regulating the Serengeti zebra are unclear, although lion predation is unlikely to be the primary factor.
2. While classic ecological models predict instability in predator-prey dynamics, the inefficiencies in prey consumption introduced by group foraging and herd living can help account for the relative constancy of the Serengeti lion population and the fact that their most common prey is food limited rather than subject to predator-induced population cycling.
3. Balancing food intake with predation risk requires prey species to vary their behavior across a landscape of fear according to the time of year and the phase of the moon, as food is limited in the dry season and hunting success is highest during the darkest nights of the lunar cycle. Wildebeest, zebra, and buffalo all modify their nocturnal behavior in ways that reflect this fundamental trade-off.
4. Despite their tendency to kill cheetah cubs, lions do not have significant impacts on cheetah population sizes. But lions do have strong negative effects on African wild dog populations, and high lion densities may also reduce hyena and leopard numbers. Lions may also enhance the Thomson gazelle population in Ngorongoro Crater by reducing the number of spotted hyenas.
5. Spotted hyenas are the lions' primary feeding competitors, and foraging groups of lions almost always gain or maintain prey carcasses during encounters with hyenas, whereas lone female lions often lose carcasses to hyenas. However, a lone female would lose even more meat by sharing the contested carcasses with a second female.

CHAPTER 12

LION CONSERVATION

> The lion shall be sentenced to public humiliation and degradation, since it is clear that the king of animals, nature and everything else is man.
> —DMITRI PRIGOV

Why worry about the future of the lion? Beyond their exalted status in the human imagination and their inherent impacts on African savannas, lions are convenient indicators of ecosystem health. If the lions are thriving, most of their prey species must be, too, as is the vegetation that supports the herbivores and the physical environment that sustains the plants. But lions have become increasingly threatened over the past fifty years. When I first arrived in the Serengeti in 1978, I largely viewed the lion as the *Drosophila* of big cats—easy to observe and reassuringly abundant compared to such endangered species as tigers and pandas. But the lion's status has changed. The total population of wild lions across all of Africa may be as low as twenty thousand to thirty thousand individuals (Bauer et al. 2015), and anthropogenic pressures have grown even in one of the best protected parks in all of Africa: lions living in the middle of the Serengeti are vulnerable to infectious diseases carried by domestic dogs in the surrounding villages, livestock are being herded ever closer to the park boundary, trophy hunters have overharvested lions in the adjoining game reserves, and the sheer force of Tanzania's rapidly growing human population (from 12 million in 1966 to over 60 million in 2020) weighs heavily on the future. Across the country as a whole, habitat conversion is likely to continue or even accelerate, and a weakening political will to conserve the nation's wildlife estate led the Tanzanian government to open 7,070 km^2 of protected area to farming and pastoralism in 2019.

By the late 1990s, we expanded our research efforts to address a range of conservation topics, and we subsequently initiated a series of conservation interventions in communities surrounding the greater Serengeti ecosystem and other parts of Tanzania as well as in several other African countries. Over the past twenty years, the African lion has attracted a growing cohort of young conservationists who might one day turn the tide, but it may only be a matter of time before the lion is reduced to similar circumstances as the tiger or the panda.

INFECTIOUS DISEASE

Canine Distemper Virus and Ring Vaccinations

Sarah Cleaveland was studying viral diseases of domestic dogs in villages near the northwest boundary of the park when the fatal CDV outbreak struck the Serengeti lions in 1993–1994. CDV had been endemic in these villages since at least 1992, and, because domestic dogs are at least ten times as abundant as any other carnivore in northern Tanzania, we assumed that village dogs served as the reservoir species for the disease—especially as none of the dogs had ever been vaccinated (Roelke-Parker et al. 1996). Genetic data revealed a close link between CDV in the domestic dogs and wild carnivores (Carpenter et al. 1998), and the temporal patterning of infection across the ecosystem suggested that the virus had emanated from the villages northwest of the park then spread to nearby lions, hyenas, and so forth before moving through the entire Serengeti/Mara ecosystem over the following year and spreading further to village dogs to the east and southwest (figure 12.1).

In 2002, Sarah, Andy Dobson, and I received NSF funding to explore disease dynamics in the Serengeti carnivores. The centerpiece of our grant was a ring-vaccination program for domestic dogs in the surrounding villages. Like measles and rinderpest, canine distemper is a morbillivirus: infection only lasts for a few days, and survivors enjoy lifelong immunity. Assuming that dogs are the primary "reservoir" for distemper, we sought to control the disease in wildlife by focusing entirely on village dogs in a 10 km band around the park. Working with veterinary officers in the neighboring districts, Sarah and her team vaccinated about 32,000 domestic dogs each year, and the project secured additional funding from Lincoln Park Zoo to extend the vaccinations for an additional seven years. Sarah had piloted a smaller-scale vaccination program from 1997 to 2002, which had little impact on the spread of CDV in either the dogs or the lions, but once she had established the larger-scale program, the incidence of the disease in domestic dogs was greatly reduced, although the lions still experienced minor outbreaks over the next ten years (figure 12.2), and a pack of wild dogs was struck by the disease in 2007 (Goller et al. 2010). Thus, even though the vaccination program appeared to have reduced the prevalence of the disease during subsequent lion outbreaks, the virus was not eliminated from either domestic or wild carnivore populations.

The persistence of CDV suggests the existence of a complex maintenance community of host species instead of a simple "spillover" of the virus from domestic dogs into wildlife. We knew that spotted hyenas, bat-eared foxes, and two species of jackal became infected with CDV in the 1994 outbreak (Roelke-Parker et al. 1996), but we had no idea how often the virus was transmitted from one host species to another. We therefore asked Meggan Craft to develop a series of mathematical models to assess the potential importance of interspecific disease transmission. At one extreme, the virus might only have been transmitted from one lion to the next, so Meggan used a network-modeling approach to simulate contact rates within and between lion prides (figure 12.3a) and only allowed the disease to spread via lion-to-lion transmission. Owing to a lack of empirical data, Meggan varied the transmission rate across all possible values, and identified the rates that reached the observed prevalence of 95 percent of the study population while mimicking the speed with which the virus could have spread from our study prides at the center of the Serengeti to the lions in the Maasai Mara, approximately 100 km in about eight months (figure 12.3b). Her models suggested that the observed prevalence of the disease could only have reached 95 percent via a transmission rate between prides that was *higher* than 20 to 30 percent (figure 12.3c). However, the slow advance of the outbreak from the middle of

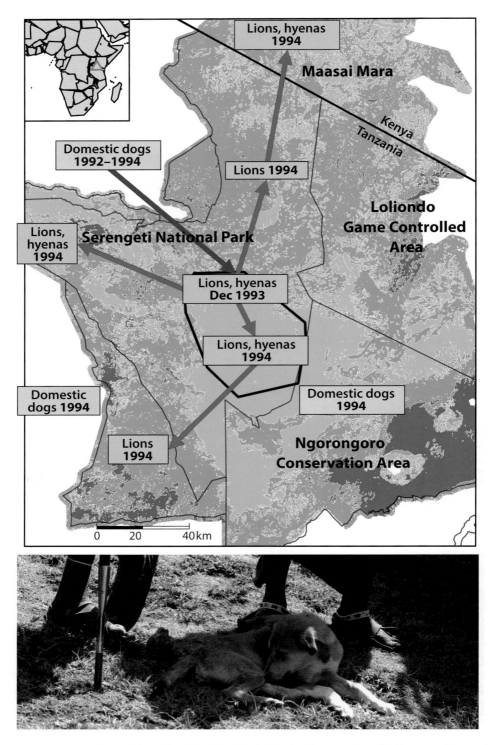

FIGURE 12.1. Approximate timing of the 1993–1994 outbreak of canine distemper virus. The disease was first noted in domestic dogs in villages located to the northwest of the Serengeti National Park then passed to wildlife and to villages in and around the ecosystem. Modified from Cleaveland et al. (2007). Photo by CP.

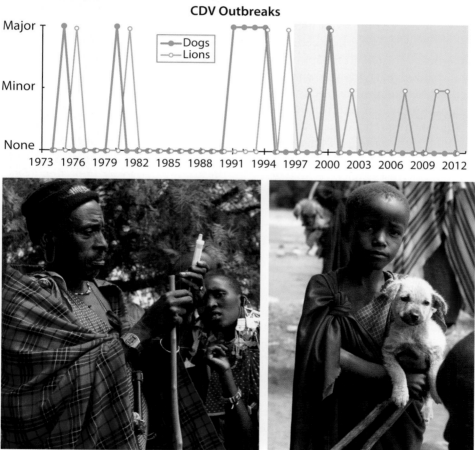

FIGURE 12.2. Timing of CDV outbreaks in domestic dogs and Serengeti lions between 1973 and 2012. Light blue shading indicates timing of pilot vaccination program in the villages to the northwest of Serengeti National Park; darker blue shading indicates the full-scale ring-vaccination effort. "Major" outbreaks were associated with seroprevalence greater than 50 percent; otherwise, outbreaks were considered "minor." Redrawn from Viana et al. (2015). Photographs show Maasai veterinary assistants and dog owners during a Vaccination Day in the Loliondo Game Controlled Area. Photos by CP.

the Serengeti to Maasai Mara (approximately 12 km/month) was only plausible if the transmission rate had been *below* 10 percent (figure 12.3d). It was therefore unlikely that such a slow-moving outbreak could have resulted in such a high prevalence of infected lions (figure 12.3d), if the disease were indeed spread *solely* by lion-to-lion transmission (Craft et al. 2009, 2010).

Different carnivore species often aggregate at carcasses—we've already seen how lions interact with hyenas (figures 11.14–11.17), and jackals are also frequent scavengers at lion kills (figure 12.4a). Hyenas are the most likely vectors for disease spread between domestic dogs and wildlife as they range over broad areas, even leaving the park and entering villages in the same region where the 1994 CDV outbreak likely originated (figure 12.4b). Village surveys to the west of the park revealed that 21 percent of households reported *weekly* contacts between their family dogs and spotted hyenas, while 43 percent of households to the east of the park reported *daily* interactions with hyenas (Craft et al. 2016). Jackals are abundant and evenly distributed across the Serengeti ecosystem, and jackals often approach to within 5 m of feeding lions (figure 12.4a). Thus, Meggan developed a series of stochastic "susceptible-infected-recovered" (SIR) models to test the potential role of interspecific disease transmission in the CDV outbreak. Here, she tracked the likely date that each Serengeti study pride had been infected with CDV, as the virus hopscotched around the study area and adjacent prides were sometimes infected after gaps of almost four months (figure 12.5a). The SIR model again showed that lions were unlikely to sustain an epidemic by themselves, whereas a modest level of transmission between lions, hyenas, and/or jackals could have produced a widespread spatial outbreak that followed an irregular pattern whereby adjacent lion prides would be infected at irregular intervals (figure 12.5b).

Thus, both of Meggan's models were consistent with the Serengeti lions being part of a multihost disease network, as are the findings from a recent whole genome analysis of viruses collected from domestic dogs, bat-eared foxes, lions, and hyenas during the 1993–1994 outbreak (figure12.6). Weckworth, Davis, Dubovi, et al.'s (2020) study suggests that the virus first moved from domestic dogs to wild carnivores in the early months of 1993 then moved from hyenas to lions on multiple occasions over the following year (figure 12.6a): half of the sampled hyenas in the long-term study area died before the first case in the lions, and the hyena and lion strains often differed by only one to two mutations (figure 12.6b). Classic disease models predict that disease outbreaks can only persist if the total population of susceptible plus infected individuals exceeds a "critical community size" (CCS), and the CCS is expected to be smaller for a highly infectious disease like CDV that spreads via aerosols than for canine rabies, where an infected individual may only transmit the virus to one or two other victims before death (Anderson and May 1991). Our Serengeti disease program was more successful at controlling rabies (Hampson et al. 2009; Lembo et al. 2010), but animal bites by rabid dogs only generated short-lived chains of rabies transmission in Serengeti carnivores, leaving the dogs as the clear maintenance population (Lembo et al. 2008). In contrast, once CDV had entered the Serengeti, the disease spread between lions, jackals, hyenas, and bat-eared foxes for over a year, and the combined population of these species in addition to smaller carnivores like mongooses, genets, and civets was likely sufficient to exceed the CCS despite the efforts of our dog vaccination program against CDV.

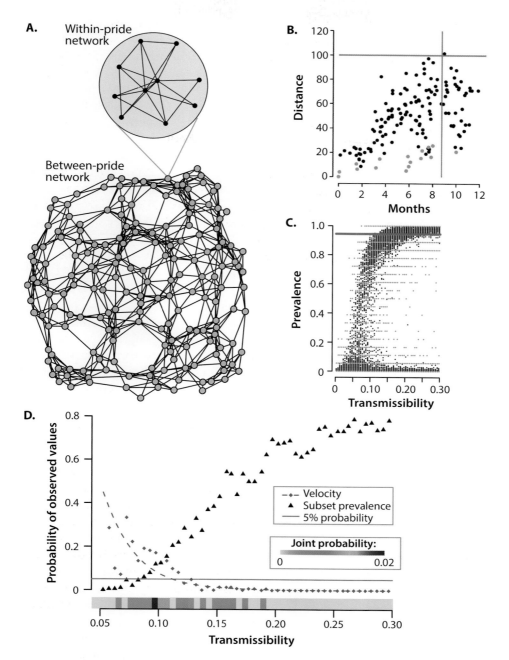

FIGURE 12.3. Network analysis of disease transmission in the Serengeti lions. **A.** Schematic representation of a contact network within and between lion prides. **B.** Representative example of a simulated epidemic that swept from the long-term study area through the entire Serengeti-Mara lion population. Points indicate the time and distance from first infection to each newly infected pride (brown, study prides; black, other prides). Red lines represent the duration (thirty-five weeks) and distance traveled (100 km) in the 1994 CDV epidemic. **C.** Simulated prevalence over a range of transmissibility values in the entire population of 180 prides (black) and the subset of 18 long-term study prides (green). Each point represents the results of a single simulation. The red line represents the observed prevalence in the 1994 outbreak, as estimated from the 18 study prides. **D.** Proportion of simulations that generated the velocity and prevalence of the 1994 outbreak. Diamonds indicate the fraction of simulations that took at least thirty-five weeks to reach 100 km. Triangles show the fraction of simulations that infected at least 17 of 18 prides in the simulated study population. The red line indicates the range of transmissibility where each pattern was observed in at least 5 percent of runs. The joint probability is the fraction of simulations that exhibited both the observed velocity and prevalence. Redrawn from Craft et al. (2009 and 2010).

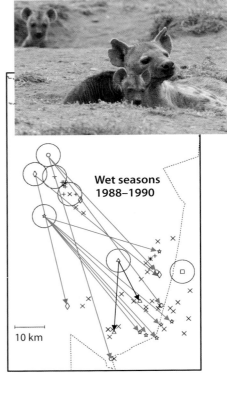
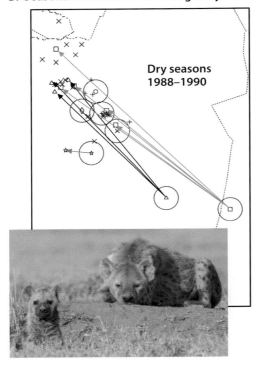

FIGURE 12.4. **A.** Time spent by jackals and spotted hyenas at each distance from feeding lions in the Serengeti study area ($n = 100$ carcasses). If present, jackals typically remained within 15 m of the lions; hyenas remained 25 to 50 m away. **B.** Commuting patterns of spotted hyenas in seven different clans in the southeastern Serengeti; redrawn from Hofer and East (1993a). Symbols indicate separate clans; large circles highlight "central denning locations"; lines highlight some of the longer distances traveled by radio-collared members of each clan.

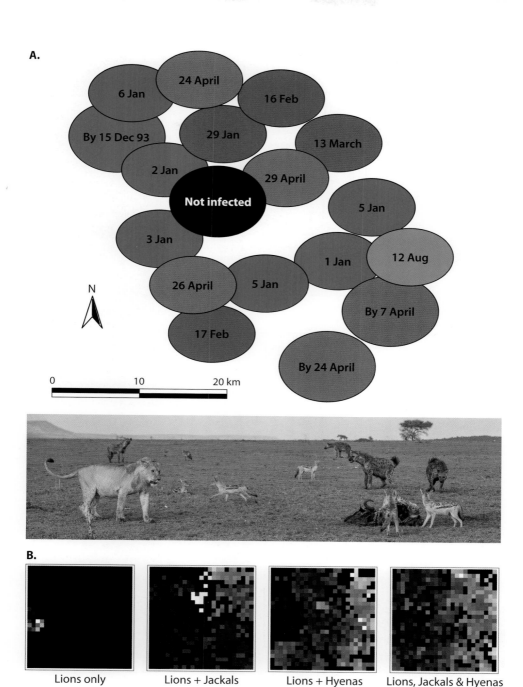

FIGURE 12.5. Timing and location of CDV infections in the 1994 outbreak. **A.** Timing of initial infection in the eighteen Serengeti study prides. Colors transition from red to green at a rate commensurate with elapsed time after the first reported cases of the disease; seven prides were infected in the first month of the outbreak, while the seventeenth pride was not infected until August 1994. **B.** Sources of infection in a multihost simulation model. The color of each grid cell represents the source of infection in a landscape of adjacent lion prides (blue indicates transmission from one lion pride to another; green represents jackals infecting lions; brown, hyenas infecting lions); colors fade from early (dark) to late (light) infection; uninfected cells are black. Lion-only outbreaks typically die out quickly; three-species outbreaks eventually infect the most prides, and the spatio-temporal pattern most closely resembles the observed outbreak. Both redrawn from Craft et al. (2008).

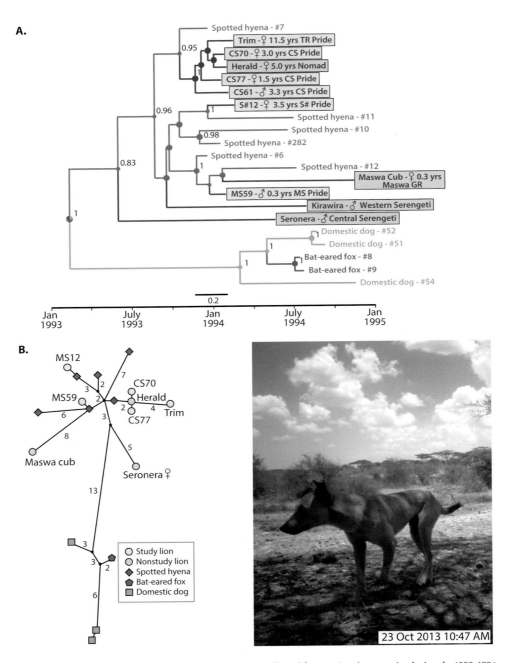

FIGURE 12.6. Evolutionary relationships of CDV genomes collected from various host species during the 1993–1994 outbreak. **A.** Reconstruction of host species transitions. Branch and node color indicate the most probable host state. Larger circles at each node indicate a host state probability age equal to or greater than 0.70, except for the root, where a pie chart shows the probability of each host. Nodes are labeled with an estimated node age greater than or equal to 0.80. The name, sex, and age are given for each lion from the long-term study area (highlighted in ochre boxes); sex and age (where known) are listed for lions that originated from outside the Serengeti study area (yellow boxes). **B.** Mutational steps separating near whole-genome CDV sequences. Branch lengths are annotated with the number of mutational steps separating nodes. Two bat-eared foxes share an identical sequence, while all other symbols represent a single sequence. Thirteen nucleotide substitutions separate all lion/hyena CDV sequences from domestic dog/bat-eared fox CDV sequences. Both figures redrawn from Weckworth, Davis, Dubovi, et al. (2020).

CDV only inflicted significant mortality when the lions were co-infected with high levels of Babesia (figure 10.9), and these conditions have arisen too infrequently to justify a large-scale dog vaccination program focused solely on this one pathogen, as the conservation benefits to wildlife are so limited. A more sensible approach would be to vaccinate the lions themselves during severe droughts when coincident infections of CDV and Babesia are most likely to arise. However, by successfully controlling *rabies*, our vaccination program conferred substantial public health benefits to the rural communities surrounding the Serengeti—and proved the feasibility of attempting an even larger-scale rabies eradication program across Tanzania (Lembo et al. 2010). Prior to our studies, the control of canine rabies was considered to be impossible in rural Africa because of abundant wildlife and the presumed inaccessibility of village dogs. But the occasional spillover into wild carnivores only results in short-lived chains of rabies infection, and sufficient numbers of village dogs can be vaccinated so as to achieve necessary levels of vaccination coverage for ultimately eradicating the deadly disease (Cleaveland and Hampson 2017). In this context, simultaneously vaccinating against CDV reduces overall mortality of vaccinated dogs, thereby helping to maintain "herd immunity" against rabies.

Bovine Tuberculosis: Disease Impacts versus Genetic Health

Mycobacterium bovis or bovine tuberculosis (bTB) was introduced to Africa via cattle imported from Europe in the nineteenth century. We documented a few isolated cases of bTB in the Serengeti in 1984–1987 (figure 12.7a), but the prevalence of the disease remained low over the next fifteen years with no measurable impact on the study population (Cleaveland et al. 2005). But in 1995 a lion in South Africa's Kruger National Park was discovered to be suffering from bTB, and a surveillance program in 1998–2000 revealed that nearly half the sampled lions were infected with the disease (figure 12.7b). Deaths from bTB were sufficiently common that the Kruger veterinarians raised the prospect of an imminent crash in the lion population (Keet et al. 2010). However, park-wide surveys showed no change in population size between 1976–1978 and 2005–2006, so the question arose as to whether the impact of bTB was serious enough to justify an extensive disease-control effort. However, the Kruger lions are not censused on the basis of individual markings but are instead counted at call-up stations—thus, the park veterinarians countered, the bTB outbreak could have started so recently that healthy lions from uninfected prides might be replacing the animals that had died in the disease hotspots. bTB was still spreading from south to north, so immigration from the north might be masking any losses in the southern prides; the full impact of the disease wouldn't be apparent until the entire population had been affected.

bTB infection is followed by a prolonged latency period before progressing to obvious disease symptoms and death. The bacterium is primarily spread to lions by consumption of infected buffalo carcasses (Keet et al. 1997) and, possibly, by lion-to-lion transmission. The eventual population impacts therefore depend on the risk of exposure to the disease, the transition rate of an exposed lion to the infectious state, how many other lions might be infected by each infectious individual, and the mortality rate of disease-shedding individuals. These parameters would be virtually impossible

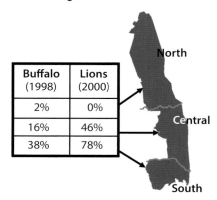
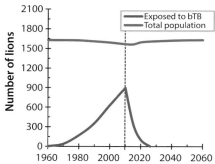
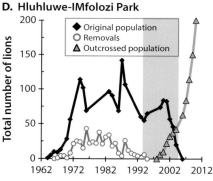
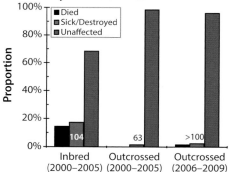

FIGURE 12.7. Bovine tuberculosis in African lions. **A.** A bTB-infected male from the Serengeti showing the dilated eyes typical of the few cases observed in our long-term study area. **B.** Observed prevalence of bTB in Cape buffalo and lions in each region of Kruger National Park. **C.** Estimated spread and impacts of bTB on the Kruger lion population. Simulation model assumes that the disease entered Kruger in the 1960s and is successfully eliminated from buffalo in 2010. Modified from Kosmala et al. 2016. **D.** Population changes in the lions of Hluhluwe-iMfolozi Park. Pale red bar indicates time period of maximum health impacts from bTB. Modified from Packer et al. 2013 with additional data from Trinkel et al. 2011. **E.** Health impacts of bTB on inbred versus outcrossed lions in HiP. Data from Trinkel et al. (2010 and 2011).

to measure directly, but Margaret Kosmala developed an Approximate Bayesian Computation (ABC) framework that estimated transmission and mortality rates from the lions' demographic patterns (mostly relying on age-specific reproduction and survival of the Serengeti lions), the disease prevalence estimates in Kruger from 1998–2000, and the changes in lion population size in each part of Kruger between the 1970s and 2000s (Kosmala et al. 2016). Feeding these estimates into an agent-based model on lion population dynamics, the likely long-term impact of bTB on the Kruger lion population turns out to be almost imperceptibly small.

According to the ABC model, most lions that are exposed to bTB would be expected to remain in a latent state for several years where they neither suffer from the disease nor transmit it to other lions. Given the typical mortality rate, only about 5 percent of these exposed individuals would be expected to survive long enough to develop symptoms and become contagious; the rest would have already died from other causes. Upon reaching the infectious state, the model predicts an 82 percent annual mortality rate, the equivalent of a further ten to fifteen months of life. Thus, because so few lions ever become contagious, and the contagious phase is so brief, lion-to-lion transmission would barely contribute to the overall spread of the disease. While this may seem like a lot to have inferred from just a few simple measures of disease prevalence and population size, it is noteworthy that Margaret's model estimates the transition rate from exposed to infectious to be about one lion in twenty compared to one in ten in untreated humans (Bates 1984)—rates that would be expected to vary between species but would be likely to remain within the same order of magnitude. Given these estimated parameter values, the ABC model predicts that the lion population would only be expected to experience a 3 percent decline by the time the disease completed its spread throughout the park, and, if a disease-control program were to completely eliminate bTB from the Kruger buffalo, the lion population would take about seven to eight years to fully recover (figure 12.7c).[1]

Although bTB only had minor population impacts on the Serengeti and Kruger lions, the disease led to the near extinction of the lions in South Africa's 960 km² Hluhluwe-iMfolozi Park (HiP). However, like the Crater lions, the HiP lions had a history of genetic isolation and close inbreeding. The HiP lions descended from seven founders in 1965; the population grew so rapidly that park management removed dozens of lions in the 1970s–1980s in order to hold the population below about seventy-five individuals (figure 12.7d). By 1990, the population's reproductive performance had declined to the point where lion removals were no longer required, and the lions started showing obvious ill health from bTB in 1992. Between 1999 and 2001, the park introduced sixteen disease-free lions in an attempt to restore the population's genetic viability (Trinkel et al. 2008). In the following five years, the translocated lions and their offspring showed significantly better health status than the "pure-bred" HiP lions (figure 12.7e). By 2006, the descendants of the translocated individuals had completely replaced the purebred HiP lions, and the population regained the same degree of vigor as in the 1960s (figure 12.7d). So, in the same way that an influx of immigrant

---

[1] "Completely eliminating" bTB from Kruger would be no small thing as it would require the culling of thousands of infected prey animals.

males may have helped to revitalize the Crater lion population in the 2010s (see figure 10.14), the "genetic rescue" in HiP overcame the population's extreme susceptibility to infectious disease. However, even though the purebred HiP lineage has disappeared, the descendant population still shows lower levels of genetic variation than in open systems like Kruger or the Serengeti (Miller, Druce, et al. 2019). Thus, although genetic rescue overcame the short-term impacts of bTB, repeated translocations would be necessary to assure continued population health, as will also be true for the many small, fenced reserves now found across much of Africa (see below).

HUMAN-LION CONFLICT

Rainfall divides human livelihoods into two broad categories in rural Tanzania: dry regions are only suitable for raising livestock; agriculture is only possible where rainfall exceeds 50 mm per month (figure 12.8). This basic dichotomy leads to strikingly different conflicts between lions and humans. In pastoralist societies, livestock is the primary source of food, income, and cultural wealth. Pastoralists accompany their herds during the daytime, remain within earshot at night, and defend their animals with spears, as they have presumably done since the earliest days of domestication. In contrast, subsistence farmers remain unarmed while commuting to their fields and tending their crops. Agriculturalists possess nothing of direct interest to carnivores, but their crops are often magnets to herbivores, presenting farmers with the dilemma of protecting their sole source of income versus becoming potential prey for man-eating lions. In either situation, lions pose a profound threat to human well-being, and retaliation by local communities presents a fundamental challenge to lion conservation, which not only results in hundreds of lions being killed (see Somerville 2020 for an exhaustive review) but also converts adjacent areas into "attractive sinks" for lions living well within the boundaries of the protected areas (Woodroffe and Frank 2006; Loveridge et al. 2010).

Pastoralists versus Lions

After completing his lion research in Ngorongoro Crater, Bernard Kissui shifted his focus to the nearby Tarangire National Park. Like the Serengeti, Tarangire is home to a large-scale migration, but Tarangire's wildebeest, zebra, and buffalo all leave the national park each wet season, and the park lions and hyenas follow the herds to Maasai pastoralist areas where they kill cattle and "shoats" (sheep and goats). As with wild prey, lions take the largest livestock species, while hyenas and leopards specialize on the smaller shoats (figure 12.9a). Cattle hold great cultural and economic value compared to shoats, killing a lion confers tangible cultural benefits (see below), and lions are far easier to spear than hyenas and leopards (Kissui 2008). Thus, the Maasai are far more likely to kill lions in retaliation for livestock depredation (figure 12.9b), killing nearly one lion for every livestock lost to lions (figure 12.9c). In fact, retaliatory lion killings were sufficiently common around the national park that they likely

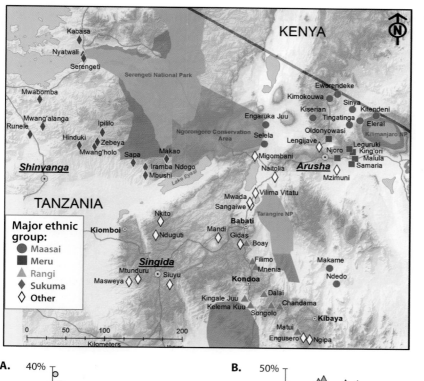
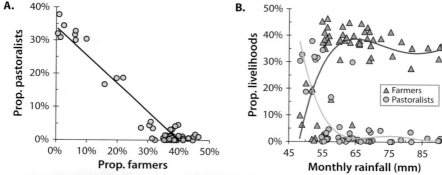

FIGURE 12.8. Fifty-six villages surveyed by Savannas Forever–Tanzania in 2006–2008. Map shows major ethnic group in each village. **A.** Proportion of families in each village that engaged in farming or pastoralism. **B.** Relationship between local rainfall and livelihood strategy in each village. Pastoralism is largely restricted to villages receiving less than 55 mm of rain each month; agriculture is only feasible in villages that receive at least 55 mm of rain per month.

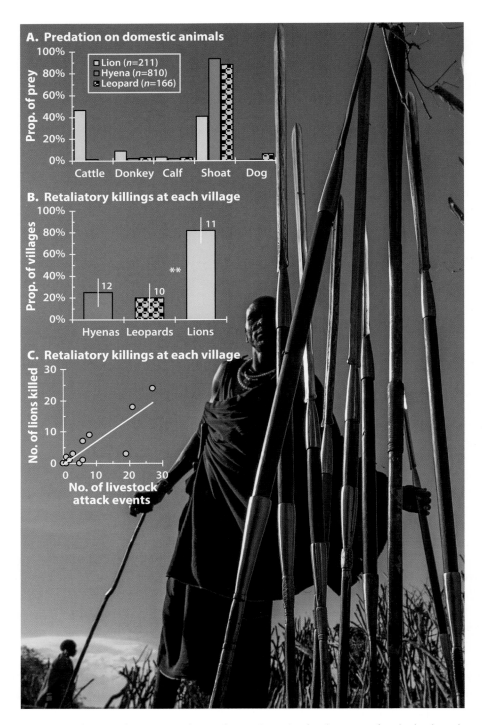

FIGURE 12.9. Predation on domestic animals around Tarangire National Park. **A.** Type of stock taken by each carnivore species. n = number of attack incidents by each species. **B.** Proportion of villages suffering predation by each species that killed the animals in retaliation. n = number of villages. **C.** Rate of retaliatory killings on lions in response to livestock attack events. Each point indicates a separate village. All redrawn from Kissui 2008. *Photo:* Maasai leaving his spear at a community gathering.

contributed to an overall decline in the Tarangire population between 2002 and 2010 (Packer, Brink, et al. 2011).[2]

Although the lions living in our Serengeti and Crater study areas are far better protected from the conflicts experienced by the Tarangire lions, both of our study populations come into contact with pastoralist herders in the Ngorongoro Conservation Area. By the late 1950s, the Maasai had been evicted from many of the newly created national parks in southern Kenya (Amboseli and Nairobi) and northern Tanzania (Lake Manyara and Tarangire), and Maasai leaders complained of being punished for having successfully coexisted with wildlife. In response, the British authorities designated the Ngorongoro Conservation Area as a multiple land-use area where wildlife has the same conservation status as within a national park (Fosbrooke 1972) but where Maasai pastoralists are also allowed to live with their livestock (figure 12.10). However, the human population in the NCA has grown about tenfold over the past half-century, whereas livestock numbers have remained relatively constant (figure 12.10a). The World Bank estimates that pastoralists require a per capita minimum of three to four "Tropical Livestock Units" to stay above the poverty line (de Haan 2016), and the Maasai in the NCA fell below this threshold by the late 1980s (figure 12.10a), resulting in severe poverty and magnifying the adverse impacts of lion predation on local livelihoods.

The southeastern boundary of our Serengeti study area runs along the boundary between the national park and the NCA, and a proportion of Serengeti lions follow the migration into the NCA each wet season (figure 12.11). In addition, Maasai herders from the NCA sometimes seek water and ungrazed pastures inside the national park during the dry season. Between 2000 and 2014, we found five radio-collared lions from the Serengeti and one lion from the Crater floor that had been killed by Maasai (red dots in figure 12.11 show the location of the carcasses). We only collared a small proportion of the Serengeti study population, and uncollared Crater lions are virtually impossible to monitor on their temporary forays outside the Crater floor, so these six killings probably indicate a far larger number of lion mortalities over the past two decades.

In response to the first known killing, I asked Dennis Ikanda to study Maasai-lion conflict in the NCA. Starting in 2001, he surveyed forty-three homesteads each month for three years and found that lions were more likely to attack herds that were relatively unprotected, either because only a small number of herders had been tending a large herd (figure 12.10b) or the herd was being guarded by children (who are typically armed with sticks) rather than by adults armed with spears (figure 12.10c). Although we considered most lion killings in the NCA to have been retaliation for livestock depredation, a number of young Maasai spear lions as part of a cultural rite of passage, known as *ala-mayo*, after entering the *morani* (warrior) caste. Retaliatory killings are permitted by conservation authorities, but unprovoked attacks on lions had been banned by law for decades. Nevertheless, Dennis was able to infer that *ala-mayo* was still practiced in the NCA.

---

2 Livestock killing by spotted hyenas and leopards can intensify negative attitudes toward lions via "contagious conflict" whereby local people develop a generalized hostility toward all large carnivore species after losing their livestock (Dickman et al. 2014).

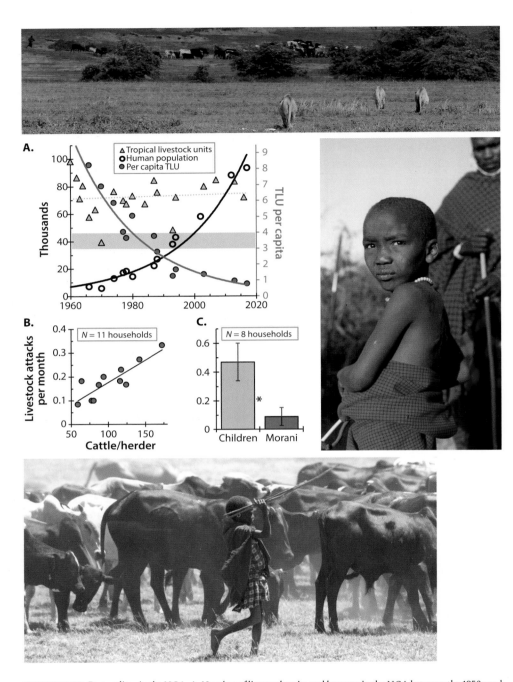

FIGURE 12.10. Pastoralism in the NCA. **A.** Number of livestock units and humans in the NCA between the 1950s and 2016. "Tropical Livestock Units" are defined as 1 cow = 0.6 TLU, and 1 sheep/goat = 0.1 TLU. The total number of TLUs in the NCA remained relatively constant over the past sixty years, while the human population grew exponentially. Solid lines indicate significant trends; horizontal red bar indicates the minimum number of TLUs per capita necessary to remain above the poverty line. **B.** Frequency of lion attacks on livestock versus the number of cattle tended by each herder. **C.** Frequency of lion attacks when livestock were tended by children or by *morani* (warriors). All graphs redrawn from Ikanda and Packer (2008). *Top photo:* Three lions approaching a herd of Maasai cattle on the floor of Ngorongoro Crater. *Middle photo:* Maasai child scarred by a lion bite while attempting to protect his livestock when he was seven years old. *Bottom photo:* Unassisted child herder in the NCA. Photos by CP.

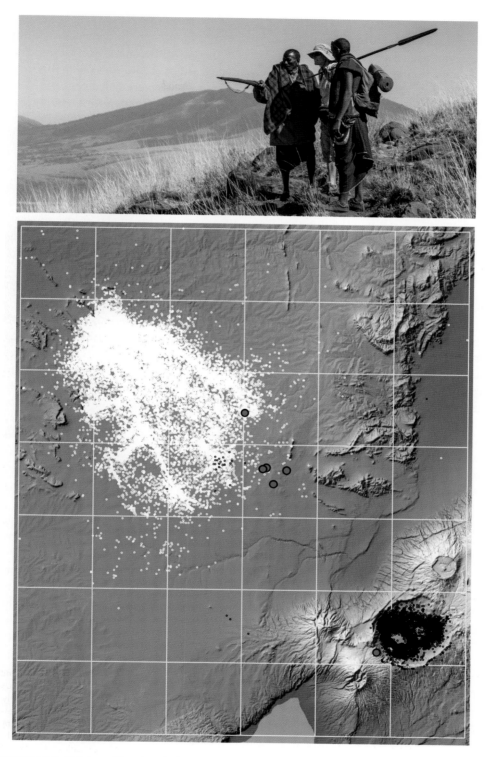

FIGURE 12.11. Sightings of lions born in the Serengeti study area (white dots) and Ngorongoro Crater (black dots), and locations where lions were killed by Maasai (large red dots). Gridlines are 20 × 20 kms.

Separating the data by habitat (woodlands versus plains) and season (wet versus dry), the ratio of lions killed compared to cattle losses was highest on the open plains during the wet season (figure 12.12a). In other parts of the NCA, the rate of lion killings was roughly one lion for every cow, which was broadly similar to Kissui's (2008) findings in Tarangire. But in the wet season on the plains, six lions were killed per cow. Most of these excess killings occurred just outside an area called Angata Kiti, a valley that runs east-west between two high ridges near the northern boundary of the Ngorongoro Conservation Area (visible directly to the right of the four clustered lion-killing sites in figure 12.11). Nomadic lions originating from the Serengeti National Park frequently follow the seasonal migration of wildebeest as they pass from the park through Angata Kiti to the easternmost plains of the NCA. But Angata Kiti also drew young men into the NCA from other parts of Maasailand. The Serengeti lions had not killed any cattle in Angata Kiti, nor had the itinerant young men from outside the NCA lost any livestock. The *morani* had specifically come to engage in *ala-mayo*, and the timing of the lion killings emphasizes their cultural dimension: most lions were killed in the years following the initiation of new age-cohorts of *morani* (figure 12.12b).

As Goldman et al. (2013) have noted, motivations for lion killings often overlap: retaliatory killings not only help the *morani* eliminate active cattle killers but also reaffirm the *morani*'s protective role in Maasai society; nonretaliatory killings help individual *morani* gain prestige and select new leaders, as well as prevent young lions from becoming habituated to people and livestock. But regardless of its cultural significance, *ala-mayo* was still illegal, thus complicating our investigations into the problem. When Dennis discussed our findings with Maasai elders in the NCA, they expressed concern over the continued practice of *ala-mayo* but said there was little they could do about any *morani* that entered the NCA from elsewhere. The elders were fully aware of the vulnerability of larger herds and the risks of relying on child labor (figures 12.10b and 12.10c), but times were changing, they said, and many of their young men had left the NCA in search of paying jobs. There was a shortage of skilled help, so they forced children to stay at home to herd livestock rather than leave home to attend school (Packer 2015).

In 2010, Ingela Jansson started a community-engagement program in the NCA that was modeled after Leela Hazzah and Stephanie Dohlrenry's successful Lion Guardians project near Kenya's Amboseli National Park (Hazzah et al. 2014). By hiring *morani* to act as an early warning system and providing them with radio-telemetry equipment, the young men become advocates for lions: livestock losses are reduced, and conservation conflicts are mediated. Direct community engagement is essential, as not only are local people expected to bear the brunt of living with dangerous animals (Dickman et al. 2011), but negative attitudes toward wildlife can persist for decades after threats have been eliminated (Dickman 2010). In establishing a community-based conflict-mitigation program in the NCA, Ingela hired a dozen Lion Scouts from local settlements, who not only helped to improve livestock husbandry, but their informants' surveillance in and around Angata Kiti resulted in dozens of lion-hunting *morani* (many of whom had indeed come from outside the NCA) being arrested by the NCA authorities, greatly reducing the appeal of *ala-mayo*.

As described in chapter 10, the overarching goal of Ingela's KOPE lion project has been to reduce the level of human-lion conflict inside the NCA to the point where

A. Habitat/season of killings

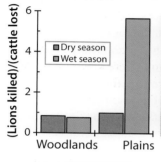

B. Yrs of killings & age cohorts

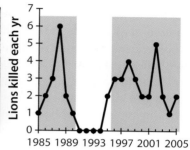

FIGURE 12.12. Ritualistic lion killings in the Ngorongoro Conservation Area. **A.** Seasonal differences in the ratio of lions killed per livestock-attack event in each part of the NCA. **B.** Number of lions killed by Maasai in the NCA each year from 1985 to 2005. Shaded areas indicate years when *moranis* were active. The cohort initiated in 1983 was elevated to elder status in 1991; the 1997 cohort had not been elevated to elder status by January 2007. Both redrawn from Ikanda and Packer (2008). *Top photo: Moranis* practice throwing spears prior to a lion hunt in the NCA (photo by CP). *Inset photograph:* 5-year-old Serengeti female, MH35, after being fatally speared in the NCA. *Bottom:* NCA *morani*.

lions from the Serengeti can travel unimpeded to the Ngorongoro Crater and achieve a level of gene flow that reduces the impacts of inbreeding in the Crater population. But any successes of these conflict-mitigation projects should be viewed with caution. For example, Mkonyi et al. (2017a; 2017b) confirmed the results from Kissui's earlier research around Tarangire and further showed that fortified *bomas* (corrals) greatly reduced the incidence of livestock depredation. However, their fortified bomas were preferentially located near the boundary of Tarangire National Park, which, like the Lion Guardians program outside Amboseli, can potentially encourage increased livestock utilization along the boundaries of the protected areas. Grass is increasingly becoming a limited resource throughout Africa, so any action that improves the access of livestock to the pastures in or near the boundaries of the protected areas will inevitably reduce the amount of pasture available for grazing wildlife. Veldhuis et al. (Veldhuis, Ritchie, et al. 2019) found that the "boundary degradation" caused by excessive livestock grazing has squeezed wildlife into the core of the Serengeti/Mara ecosystem (see bottom photo in figure 12.22)—effectively reducing the size of the protected areas by about 20 percent. Thus, without a holistic approach to livestock husbandry that limits grazing access to pastures in and around the protected areas and reduces the stocking rates of neighboring cattle herds, successful conflict-mitigation efforts will risk diminishing the land available to the lions' primary prey species even further.

Subsistence Farmers versus Lions

Between 1990 and 2008, nearly a thousand Tanzanians were attacked by lions. The attacks were unprovoked, two-thirds of the victims died, and many were eaten. The number of cases reached a peak between 1999 and 2004 (averaging nearly ninety attacks per year), and most were located in the agricultural areas between Tanzania's largest city, Dar es Salaam, and the border with Mozambique (figure 12.13). In 2006, the Tanzanian government asked for our help in determining the causes of the outbreak and finding ways to reduce the loss of life. Working with staff from the Wildlife Division of the Ministry of Natural Resources and Tourism, Dennis Ikanda and Hadas Kushnir visited the homes of hundreds of survivors and families of victims. Here are a few of their stories:

Lindi District 21-Jun-1995. At 10:00 p.m., Mohammed Mandingi, aged 45 yrs, was sleeping on the side of the road near the market. Neighbors heard him saying, "Don't take me; where are you taking me?" Since he was considered crazy, they just thought he was talking to himself. The next morning, they found blood and traced it to where he had slept. They found his remains and put them under a nearby tree, and a game scout waited in the tree. At around 18:00, the lion came back to the remains and was lying near the body when the ranger made a noise. The lion looked up, and the ranger shot and killed it with one bullet.

Lindi District 19-Dec-1995. Children had seen a lion and gone inside a house, but 15 yr old Tabia Rashid Manditi was worried her mom would get mad at her for being late, so at 10:00 p.m. she decided to walk home. The lion grabbed her, and she screamed, but her mother came too late. Tabia was 20 m from her home and 10 m from the house she had left.

Lindi District 5-Jun-2003. At 18:25, 4-yr-old Hamisi Mkilwa's grandmother was bathing him in front of the hut on their *shamba* (farm plot). When she lifted him out of the basin, a lion grabbed him from her hands and ran away with the child. The family called for help and searched all over, but they did not find the remains until later.

Rufiji District 15-Feb-2004. At 11:00 p.m., Hamisi Mkugwa, 70, was sleeping in the hut on his *shamba* when a lion came in through the roof and attacked him. His wife tried to fight off the lion. The lion released him, but it was too late.

Rufiji District 26-Jun-06. At 11:00 p.m., Nyamwegio, 50, and her grandchild Subra, 4, were sleeping in a bed on the ground floor of the hut in their *shamba*. Lions broke through the door, waking them up. The grandmother hid under the bed while the grandchild wrapped herself in a mat. An older boy, who was sleeping in the second story of the hut, came down and started hitting the lion with a large stick. Hearing the noises, neighbors came; one man had a gun, and they killed the lion that was inside the house. A second lion that was outside the hut escaped. Grandmother and granddaughter both survived their injuries.

We surveyed wildlife abundance in sixteen districts that had reported lion attacks in the previous ten years and found that lion attacks were most common in districts with the lowest abundance of the lions' preferred prey (e.g., wildebeest, zebra, buffalo, etc.) (figure 12.14a). This pattern was unsurprising as man-eating has long been presumed to be driven by a lack of food. However, the number of attacks was highest in districts with the greatest abundance of a prey species that we didn't normally associate with lions: the bush pig (*Potamochoerus larvatus*) (figure 12.14b). Bush pigs are native to Africa, they are major crop pests, and they are only active at night. Subsistence farmers cannot afford to lose their crops, so they sleep in simple shelters (*dungus*) in the middle of their *shambas* to prevent crop damage by bush pigs (see figure 12.15). Lions in these areas primarily prey upon bush pigs and hence follow the pigs into agricultural fields where they encounter vulnerable humans; hence, lion attacks are most common at harvest time (figure 12.14c).

In the Serengeti, lions are least successful catching wildlife during the brightest nights of the lunar cycle, thus they are hungriest around the full moon (figure 1.17a). Lion attacks on humans are most common between sunset and moonrise during the first week *after* the full moon (figure 12.14d). Lions mostly attack people in the darkest hours of the night, and humans are most active from sunset to about 10:00 p.m. During the week before the full moon, the moon rises *before* sunset, providing people a certain amount protection against lion attack during those moonlit evenings. But the moon rises an hour later each night, and, starting the night after the full moon, the moon rises *after* sunset, leaving any people who happen to be outside in the darkness vulnerable to lions that may not have fed over the previous week. Thus, the full moon accurately indicates that the risks of lion predation will increase dramatically in the

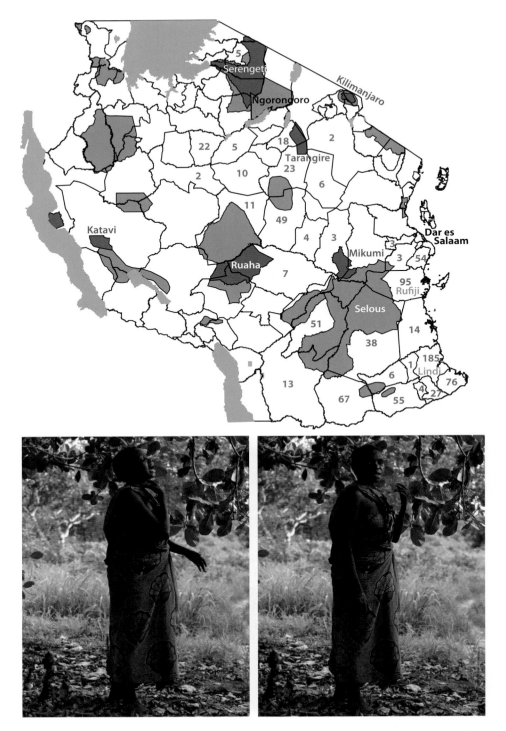

FIGURE 12.13. Number of reported lion attacks on humans in each district of Tanzania between 1990 and 2008. National Parks are highlighted in dark green; Game Reserves in light green. Redrawn and updated from Packer, Ikanda, et al. (2005). *Photos:* Woman in Lindi District describes how her child's throat was crushed by a female lion about 30 m away from their hut; the mother chased off the lion, but her child died a few hours later. Photos by CP.

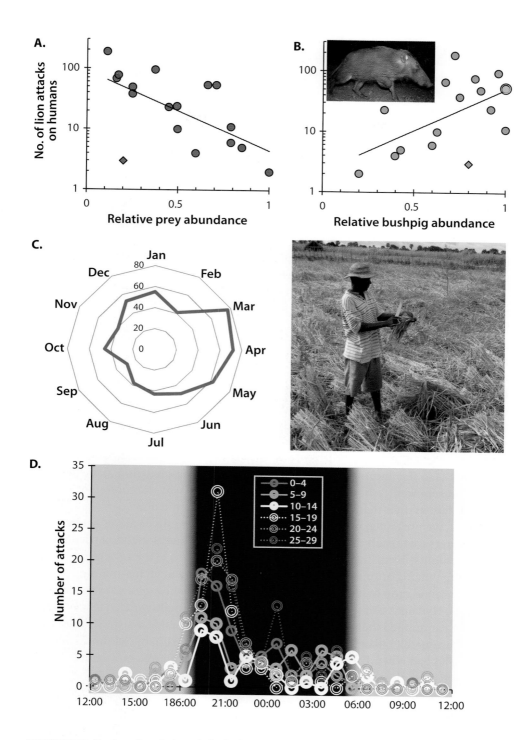

FIGURE 12.14. Number of attacks in each district in Tanzania with a recent history of lion incidents versus **A.** relative abundance of midsize prey; and **B.** relative abundance of bushpigs. The diamonds in **A.** and **B.** indicate Kilosa District where the Tanzanian Government moved people out of a high-lion-risk area in 1992. Redrawn from Packer, Ikanda, et al. 2005. **C.** Monthly number of lion attacks. Attacks are significantly more common during harvest time (March–May). **D.** Time of day of lion attacks versus moon phase. Attacks are far more common in the first week after the night of the full moon. Redrawn from Packer, Swanson, et al. (2011). Photos by CP.

following evenings, perhaps helping to explain the many myths and superstitions about the full moon (Packer, Swanson, et al. 2011).

What caused the number of lion attacks to increase in the late 1990s? Hadas mapped the attack locations in the two worst-hit districts, Rufiji and Lindi (see figure 12.13). The growing human population had recently converted large areas of wildlife habitat to small-scale agricultural fields, thus providing new opportunities for bush pigs and increasing the amount of time that people spent in vulnerable locations (Kushnir et al. 2014). How did villagers perceive the risks of being taken by lions? Hadas and her field assistants interviewed dozens of survivors and their families and found that people generally recognized where and when the risks were highest. People's perceptions of risks of lion attack were positively correlated with actual risks across different locations, time of day, as well as age group, gender, and activity (figure 12.15a). However, as people in Western societies fear terrorist attacks more than the risks of dying from influenza, villagers in Lindi and Rufiji greatly exaggerated the risk of lion attack compared to the risks of drought or disease (figure 12.15b).[3]

These are among the poorest districts in Tanzania. Few people have guns, and the wildlife authorities are often slow to respond to reports of animal attacks. Pig populations are virtually impossible to exterminate, even in wealthy countries in Europe and North America, but an effective crop-protection strategy would eliminate the need to sleep in a makeshift shelter that is vulnerable to lion attack. We therefore tested various crop-protection techniques in two villages in Rufiji District, Kipo, and Ndundu Nyikanza. Between April 2009 and September 2010, we monitored a series of 25 × 25 m plots that were (a) enclosed by physical barriers (either wooden fencing or a 1.5-meter-deep trench); (b) surrounded by irritants (metal noisemakers or flashing red lights); or (c) left unguarded as control plots. Trenches and fences were equally successful, whereas the bush pigs caused as much damage in the plots with noisemakers and lights as in the unprotected controls (figure 12.15c). The extent of damage in the control/noisemaker/flashing-lights plots (75%) explained why the subsistent farmers in Rufiji slept in their fields despite the associated risks from lions.

In 2011, we presented our findings to village representatives along with public-safety posters that recommended (a) avoiding walking alone at night, particularly on moonless nights; (b) raising the height of sleeping platforms in the agricultural fields; (c) erecting fencing around ablution blocks and homes; and (d) maintaining fences/trenches around crop fields (figure 12.15). After seeing the test plots, several farmers in Ndundu Nyikanza had spontaneously built their own simple wooden fences (figure 12.15) and maintained them for seven years without any outside assistance or discussion. Unfortunately, all of the fences were pulled down in 2016 by Sukuma[4] livestock herders who had recently settled in the area and insisted on free movement of their cattle—even into the residents' long-established rice and maize fields.

---

3 By the time Dennis and Hadas interviewed the survivors and victims' families, people clearly recognized that the attacks had involved actual lions. But many villagers had initially believed the culprits to have been "spirit lions" that were fabricated by sorcery (Packer 2015). This belief was so strong during a similar man-eating outbreak in neighboring Mozambique that twenty-four people were lynched by angry mobs in 2002–2003 (Israel 2009).

4 Tanzania's largest ethnic group, the Sukuma are Bantu-speaking agropastoralists who have recently expanded from their traditional stronghold near Lake Victoria to most parts of the country as a result of rapid population growth, overgrazing, and declining soil quality (Salerno et al. 2017).

The number of lion attacks in southern Tanzania dropped dramatically between 2004 and 2005 and remained low until 2008 (the last year that the Wildlife Division released their data) (figure 12.16). The decline pre-dated our conservation efforts and arose from the realization by local people that lions habitually return to the remains of their half-eaten prey:

BBC 15-Jun-2004. Dead wife bait traps killer lion . . .

When Selemani Ngongwechile found his wife's half-eaten body, he calmly poisoned it, knowing the lion would return for the rest of its "meal." After killing Somoe Abdallah near her home and eating her upper body, the lion might have gone for a drink, [the local police commander, Simon Dau,] said. This is when Mr. Ngongwechile found her remains hidden in a bush. But instead of panicking, he put the poison in her corpse and waited inside his house for the lion to return . . . "Nowadays there are very few animals in the area for the lions to hunt. So instead of starving they decide to kill humans rather than keep on looking for antelopes or deers [sic] which are scarce," Commander Dau said.

Packets of rat poison and insecticide are cheap and widely available (see top photo in figure 12.16), and media coverage of Ngongwechile inspired other villagers to lace the remains of half-eaten lion kills and butchered sheep/goat carcasses, and the man-eaters were soon extirpated from southern Tanzania (Packer 2015).

Man-eating carnivores not only return to the carcasses of their half-eaten victims, but they also tend to be repeat offenders, and, in the case of lions, individual man-eaters are often part of a larger outbreak. For example, in the two villages in Rufiji District where we conducted the pig-exclusion study, nine people were attacked in Kipo over twenty-eight months in 2001–2004, and another nine people were attacked in Ndundu over eighteen months in 2002–2004. In Simana, one of the hardest-hit villages in Lindi, eleven people were attacked over a twenty-month period in 2003–2004 (see bottom photos in figure 12.16). Although attacks stopped in Rufiji District after the death of a male that had been named "Osama" by the villagers of Ngorongo, the forty-seven attacks attributed to that one lion (see photos in figure 12.17) actually involved multiple individuals, including several adult females, and Osama would not have been fully grown when the string of attacks first began (Packer 2015). Likewise, man-eating lions killed dozens of railway workers in Tsavo, Kenya, in the late 1890s, and additional victims were attacked even after the first pair of males had famously been shot by Colonel John Henry Patterson in 1898 (Peterhans and Gnoske 2001). Thus, given an initial attack, additional people will be at risk either from the same or neighboring prides. Combining our Tanzanian surveys with similar data on hundreds of tiger and leopard attacks in India and Nepal, Nick Fountain-Jones used an epidemiological tool, SaTScan (Kulldorff 1997), to characterize discrete clusters of attacks by the three big-cat species. In comparing the timing and location of attacks in each geographic area, Nick approached the problem in the same way as public health agencies investigate spatiotemporal clusters of infectious disease and cancer, and the results showed clear outbreaks by each species (figure 12.17). Lion outbreaks involve far more victims, persist for longer periods, and cover larger geographic areas than do

FIGURE 12.15. A. Perceived risks of lion attack were positively correlated with actual risks across different locations, time of day, as well as age groups, gender, and activities of victims. Black line shows expected trend if perceptions perfectly mirrored the actual risks. **B.** Actual risks of drought or disease (blue bars) versus comparative perceived risks from lion attacks (orange bars); horizontal line indicates actual risk of lion attack. **C.** Bushpig damage in 25 × 25 m test plots in Rufiji District in 2009–2010. Irritants were either flashing lights or noisemakers; barriers were low wooden fences or trenches. **A.** and **B.** redrawn from Kushnir and Packer (2019). *Top photo:* Agricultural shelter (*dungu*) in Rufiji District. *Bottom left photo:* Villager shows us the fence he built after seeing the successful exclusion plots. *Bottom right:* Swahili poster distributed to villagers at the conclusion of the pig-exclusion study. Photos by CP.

leopard or tiger outbreaks, presumably because lions often learn the behavior from their pridemates, whereas a solitary leopard or tiger only acquires the behavior on its own (Packer et al. 2019)[5].

The outbreak of man-eating in Tsavo closely followed the Great Rinderpest epizootic that struck East Africa in the 1890s and removed most of the lions' natural prey (Peterhans and Gnoske 2001), and the outbreak in southern Tanzania in the 1990s coincided with the loss of prey following widespread habitat conversion to subsistence farming (figures 12.14a and 12.14b). Though the latter outbreak had largely ended by the time we started our research in Rufiji and Lindi, continuing habitat loss may someday lead to renewed waves of lion attacks, and the SaTScan analysis could potentially help wildlife managers predict the approximate spatial area and timeframe of successive attacks.

SPORT HUNTING

Over the years, we collared about three dozen nomadic and resident males in the Serengeti. As their name implies, nomads wander widely, and our VHF collars could only be detected from about 5 to 10 km away, so we had little idea of their ultimate fates. Some may have gained residence in prides well outside our long-term study area; some might have been killed by other males; some might have died in a poacher's snare. However, each transmitter had a label with our address, and, in 1992, an American sport hunter reported that he had shot one of the collared nomads over 100 km outside the lion study area in the northern part of the Loliondo Game Controlled Area. Even resident males can move long distances when prey is scarce, and we tracked the collar of one resident that had been shot in the Ikorongo Game Reserve, a mere 500 m outside the park boundary.

For most trophy species, sport hunting is unlikely to harm the population as a whole. Hunters only shoot males, and male ungulates, for example, rarely associate with their young, so a father's death would not be expected to alter the survival of his offspring. But in an infanticidal species like the lion, shooting males could increase the frequency of male takeovers and, hence, conceivably raise cub mortality to the point where the entire population spirals downward. To test these ideas, Karyl Whitman spearheaded the development of a virtual lion population model that we called "SimSimba." Each individual in the simulated population reproduces and dies according to the precise demographic parameters of the Serengeti lions, and the death of a resident male has the same impact on takeovers, infanticide, dispersal, and so on in each simulated pride as in a Serengeti pride (Whitman et al. 2004). Karyl had conducted her fieldwork in the Maswa Game Reserve located along the southwest boundary of Serengeti National Park, and she had seen how the hunting blocks had been so badly overhunted that few adult males remained in the area, so the hunting companies filled their quotas by shooting males as young as two to three years of age.

---

5 Tigers and leopards also differed from lions in the timing of their attacks on humans. In contrast to lions, the majority of tiger and leopard attacks took place during the day, and their nighttime attacks showed no obvious relationship with moon phase.

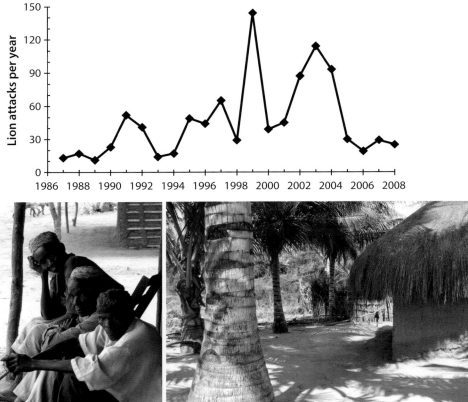

FIGURE 12.16. Number of lion attacks reported to the Wildlife Division of the Tanzanian Ministry of Natural Resources and Tourism each year from 1986 to 2008. *Top photo:* Rat poison salesman in Dar es Salaam. *Bottom left:* Village elders in Simana, Lindi District, where eleven people were attacked by lions in separate incidents over a twenty-month period in 2003–2004. *Bottom right:* One of the lion-attack sites in Simana village. Top photo by CP; bottom photos by Hadas Kushnir.

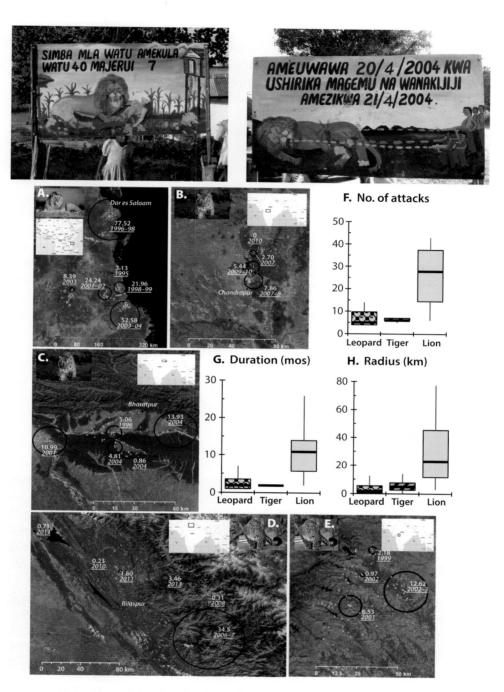

FIGURE 12.17. Location of **A.** lion, **B.** and **C.** tiger and **D.** and **E.** leopard attacks in Tanzania, India, and Nepal. Note the different spatial scales in each map. Bar graphs show significant differences between "outbreaks" by each species. All modified from Packer et al. (2019). *Photos:* Signs in Ngorongo Village celebrating the killing of the last individual involved in an outbreak in Rufiji District, Tanzania. The left-hand sign reads: "The man-eating lion has eaten 40 people and injured 7." The second sign reads: "Killed on 20/4/2004 by a group of villagers; buried 21/4/2004." Photos by Hadas Kushnir.

Karyl therefore used SimSimba to vary trophy offtakes not only by the *number* of males shot each year but also by the minimum *age* that a male could be included in the annual harvest. As suspected, there was little risk of overharvesting in a noninfanticidal version of SimSimba that prevented incoming males from harming their predecessors' dependent offspring. In this hypothetical situation, removals only mattered if hunters were to eliminate so many sperm-bearing males that the females failed to get pregnant (figure 12.18a). But if Karyl ran SimSimba with the full impacts of infanticide, the simulated population was far more sensitive to overharvesting, and the impacts depended on the ages of the harvested males (figure 12.18b). By shooting out all the older males and including too many four-year-olds in their quota, hunters would drive down the population by over half; but with an age minimum of at least six years, the overall population would remain stable, no matter how many of the older males were shot each year (Whitman et al. 2004; Packer et al. 2009). These results can be intuited as follows: males typically sire their first cubs at about four years of age (figure 1.21a) then need another two years to raise them successfully (figure 3.2). Thus, if hunters only shot males that were at least six years of age,[6] the females would typically start their next breeding cycle at about the same time that they would have bred with the prior residents, thereby allowing their reproductive performance with successive sets of replacement males to match females in unhunted populations.

We parameterized SimSimba with as much detail as possible, but it was still just a model. Real lions were being shot across large parts of Africa and the lion was becoming an increasing focus of conservation concern. Although hunting associations in Mozambique, Tanzania, Zambia, and Zimbabwe had officially embraced our recommendations for an age minimum, there were questions about corruption and compliance (see Lindsey et al. 2013), and even some of the more ethical hunting companies doubted the feasibility of estimating a lion's age before deciding whether it was old enough to shoot (Packer 2015). Was there any way we could derive a sustainable hunting quota if age determinations proved unenforceable and/or impracticable? Sport hunters continued to post photographs of their lion trophies on the internet for several years after we had published SimSimba, so we knew that males as young as two to three years of age were still being shot (inset to figure 12.18). Thus, a good test of SimSimba would be to determine whether standard practices had actually driven down any of the hunted lion populations and to use past offtakes to estimate the levels of offtake where hunting harvests had been too high.

No reliable population estimates were available for lions in any of Tanzania's game reserves, but we were able to inspect the lion offtake patterns from all 170-plus hunting blocks that had been active between 1995 and 2008. The results were clear: blocks with the highest initial lion offtake showed the steepest subsequent declines in lion harvest, whereas harvests remained relatively constant at lower offtake levels, sug-

---

6 Our original paper showed a strong correlation between a male's age and the amount of dark pigmentation in the otherwise pink flesh of his nose (Whitman et al. 2004) (compare the pink noses of the young males in figures 1.15 and 5.6 with the black noses of the males in figures 1.20 and 1.21), and a similar relationship has been found in multiple lion populations across Africa (Miller et al. 2016). The advantage of nose coloration over alternative methods is that nose coloration can be assessed before taking a shot. A major disadvantage, however, is that the tip of the male's nose may often be obscured by blood, as most lions are shot at baits.

gesting that lions had been overhunted in about half of the 44 hunting blocks in the Selous Game Reserve as well as in most of the blocks in the rest of the country (figures 12.18c and 12.18d). If age assessments were truly infeasible, our analysis suggested an alternative strategy: annual harvests should be limited to one lion per 1,000 km² in the Selous, whereas quotas should only be one lion per 2,000 km² in other parts of Tanzania.

No further data are available from Tanzania, but intensive trophy hunting in Zimbabwe had measurable impacts on lion population size, and survival rates of cubs and adults both increased when hunting offtakes were reduced (Loveridge et al. 2016). Similarly, lion populations in Zambia rebounded after a three-year moratorium on lion hunting (Mweetwa et al. 2018), and, rather than returning to business as usual after the moratorium, Creel et al. (2016) recommended that the Zambian wildlife authorities set an age minimum of seven years in combination with an annual quota of less than one lion per 2,000 km². While it is not clear whether any governmental agency in Africa has successfully implemented an age-minimum policy, Begg et al. (2018) demonstrated the feasibility of a strictly enforced six-year age minimum in Mozambique's Niassa Reserve, which is administered by an independent conservation group.

We started investigating the impacts of trophy hunting during the heyday of sport hunting for African lions. In Tanzania, sport hunters shot about two hundred lions each year from 1990 to 2010, but offtakes dropped to fifty lions in 2011 and fell further to only eighteen lions in 2018, with overconsumption likely accounting for much of the decline. In 2015, the United States Fisheries and Wildlife Service classified lions from eastern and southern Africa as "threatened" on the US Endangered Species Act and required exporting countries to prove that hunting had positive impacts on lion conservation, thus virtually ending the importation of lion trophies to the United States from Tanzania, which represents about half the hunting market.

While Tanzania's National Parks plus the NCA protected 58,000 km² of lion habitat, lions were once hunted in about 146,000 km² of Tanzania's hunting blocks, and lions generated 15 percent of the hunting industry's total revenues (Lindsey et al. 2012). Like Loliondo, Ikorongo, and Maswa around the Serengeti, many of the hunting blocks adjoin the perimeter of a national park and thus potentially serve as buffers between high-density wildlife areas and human-dominated regions. But by 2019, nearly two-thirds of Tanzania's hunting blocks had been abandoned: wildlife numbers had fallen to the point that the blocks were no longer profitable.[7] The empty blocks now present the profound challenge of somehow conserving vast tracts of land across the country, with similar potential losses in Mozambique, Zambia, and Zimbabwe. Without any sort of conservation status, the abandoned hunting blocks will risk losing their natural vegetation to agriculture and any remaining wildlife to livestock.

---

7 In the two first-ever auctions held by Tanzanian Wildlife Authority, only seven of twenty-six hunting blocks were successfully auctioned in June 2019 and two of twenty-four in November 2019 (*Citizen Newspaper* [Tanzania] January 12, 2021), while only another ten of thirty were auctioned in January 2021. Also in 2019, the Tanzanian government excised 30,893 km² from the Selous Game Reserve to form the country's largest national park, with the remaining 20,155 km² of the Selous still subdivided into hunting blocks, though approximately 1,000 km² of this combined area will be flooded by a new hydroelectric dam within the ecosystem (Baldus 2021).

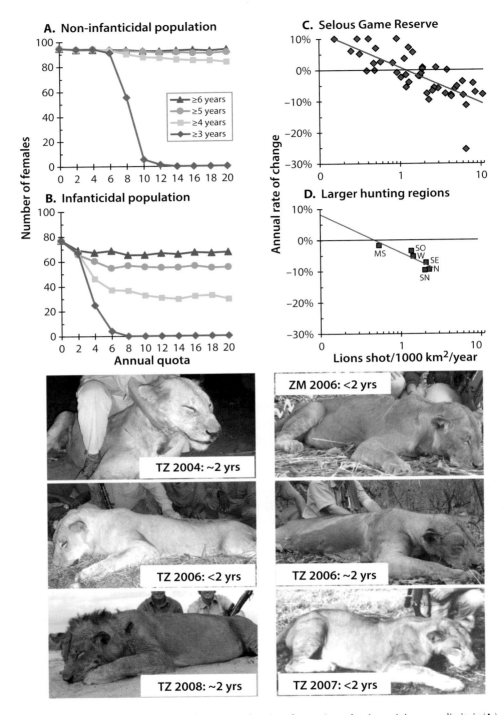

FIGURE 12.18. Simulated effects of trophy hunting as a function of quota size and various minimum age limits in (**A.**) hypothetical populations where males are not infanticidal and (**B.**) infanticidal populations. Impacts are measured in terms of the number of adult females in the population after thirty years; note that infanticidal populations are smaller and more vulnerable to overhunting. Redrawn from Whitman et al. 2004. Proportional change in lion harvests versus average harvest in 1996–1999 in (**C.**) the Selous Game Reserve ($n = 44$ blocks) and (**D.**) the six ecosystems outside of Selous (MS, Maasai Steppe, twenty-four blocks; N, northwestern Tanzania, four blocks; SE, Serengeti, eight blocks; SN, new blocks outside Selous, sixteen blocks; SO, old blocks outside Selous, seven blocks; W, western Tanzania, fifty-four blocks). In both cases, higher initial harvests were followed by declining offtakes. Redrawn from Packer, Brink, et al. (2011).

BUSHMEAT HUNTING

A few months after I first arrived in the Serengeti, an elderly female lion appeared in the middle of our long-term study area. A poacher's snare had cut a deep groove all the way around her neck and throat. She had tugged so hard on the anchored wire that she had crushed her esophagus. She was emaciated, unable to eat, unable to drink, so we asked the park rangers to euthanize her. Every few years another snared lion would enter the study area, and, although we never knew the precise origins of any of these animals, we could be certain they had come from somewhere out west. People living along the western boundary of the Serengeti ecosystem primarily rely on agriculture for their livelihoods, but they also consume considerable quantities of bushmeat.[8] The threat of bushmeat hunting in the early days of the park helped inspire Bernard Grzimek to write "Serengeti Shall Not Die" in 1959, and a severe economic crisis in the 1980s prevented anti-poaching units from patrolling the western boundaries, resulting in major declines in animal numbers over much of the ecosystem (see chapter 10). Renewed anti-poaching activities largely curtailed widespread poaching in the middle of the national park, but significant numbers of animals continued to be taken, and the communities along the western boundary still consumed significant amounts of bushmeat (Loibooki et al. 2002).

From 2007 to 2010, Dennis Rentsch and his field assistants studied bushmeat consumption by conducting monthly "dietary recall" surveys in 132 households in eight villages just outside the Ikorongo and Grumeti Game Reserves on the northwestern side of the Serengeti. Bushmeat hunters mostly originate from villages located immediately adjacent to the protected areas (figure 12.19a), and the average household reported consuming 0.79 to 6.68 kg of bushmeat each week, with quantities varying across the year. About half of their intake consisted of migratory wildebeest, which closely corresponded to the wildebeest's overall abundance relative to other herbivore species in this part of the ecosystem (figure 12.19b). Dennis's comprehensive surveys covered the northern half of the western boundary, and short-term surveys in six villages just outside the Maswa Game Reserve showed similar household diets, so he was able to calculate the number of wildebeest required to feed the 600,000 people that lived immediately adjacent to the entire boundary.

Assuming that a butchered wildebeest carcass provides 26.4 kg of meat, we estimated that a total annual offtake of 98,000 to 141,000 wildebeest would be needed to meet the demand (Rentsch and Packer 2015). Previously, Mduma et al. (1999) had predicted that the Serengeti wildebeest population could only sustain a harvest of 80,000 animals a year. However, their estimate assumed equal offtakes of males and females; our higher estimates could still be sustainable if poachers mostly harvested males. Given that local communities are growing by about 3.5 percent per year, bushmeat is also eaten in villages located farther away from the park, *and* the park is being squeezed by overgrazing pastoralists (Veldhuis, Ritchie, et al. 2019), it is hard to imagine how the Serengeti wildebeest population can possibly remain at the same size as

---

8 Lions are mostly by-catch of snares set by bushmeat hunters. Snaring was so rare *inside* our long-term study areas that it had no meaningful impacts on the study populations, but snares are a major source of lion mortality in Zambia (Becker et al. 2013), Mozambique (Begg et al. 2018), and other parts of Africa (Lindsey et al. 2017).

seen the past forty years. The importance of the wildebeest to the Serengeti lions, of course, can hardly be overstated.

THE FUTURE

Across the entire continent, savanna habitat shrank by about 75 percent between 1960 and 2010 (Riggio et al. 2013), and, by 2015, lions had likely vanished from as many as sixteen African countries (Bauer et al. 2015). Add the impacts of infectious disease, human-lion conflict, poorly managed sport hunting, and illegal bushmeat consumption, as well as an emerging demand for lion parts in traditional Chinese medicine (Everatt et al. 2019),[9] and there seems little hope for the future of the African lion. But it's not all doom and gloom. South Africa, for example, is home to more wild lions in the twenty-first century than at the beginning of the twentieth, and lions in the Serengeti and Ngorongoro Crater still thrive. So why have some populations persisted while others collapsed or disappeared, and, beyond the most obvious successes and failures, how can we even assess the extent to which a population is being adequately protected?

Using an ecological model developed by Andrew Loveridge and Susan Canney (2009) that predicts lion carrying capacities from prey biomass, rainfall, and soil nutrients, we were able to estimate the conservation effectiveness of thirty-eight protected areas in eleven countries by measuring how closely the observed population matched the expected value in each park (Packer et al. 2013). Three patterns stood out: First, *fencing*: no matter the size of the reserve, fenced populations always matched or exceeded the expected carrying capacity (figure 12.20a). If lions are enclosed within perimeter fences, they (and their habitat) are isolated from any potential conflicts with rural communities (Hunter et al. 2007), and their population sizes can reach full potential. Lions in many of these reserves are so successful, in fact, that managers often have to treat the females with contraceptives to limit the size of the population (McEvoy, Miller, et al. 2019). Second, *funding*: lion populations in most unfenced parks appeared to be far below their predicted densities, and the extent of this shortfall depended on the park's management budget (figure 12.20b). A world-renowned national park like the Serengeti generates so much revenue from tourist fees that park managers can mount effective anti-poaching campaigns (Hilborn et al. 2006) and initiate meaningful community-conservation projects. However, even within Tanzania, the rarely visited Katavi National Park, for example, is chronically underfunded and lacks the necessary resources to protect its wildlife and engage with local communities. Third, *people*: for a given level of management funding, lions fare worst in unfenced reserves that are surrounded by the highest densities of humans (figure 12.20b). Impacts from poaching and human-wildlife conflict inevitably increase with the number of people in the vicinity of a park

---

9 The international market for lion bones arose from the "canned hunting" industry of South Africa, where the postcranial bones of captive lions shot by tourist hunters were exported to Asia as a substitute for tiger bone wine (Williams et al. 2017). Other than Everatt et al.'s (2019) study site in Zimbabwe, the demand for bones has largely been met by the thousands of captive lions in South Africa, and bone collection from wild lions has mostly been opportunistic rather than targeted (Coals et al. 2020). We occasionally found lion carcasses with the paws removed, as lion claws have both cultural and monetary value, and Maasai would also remove the tail as part of *ala-mayo*, but we otherwise saw no evidence that lions were killed for their body parts in the long-term study areas.

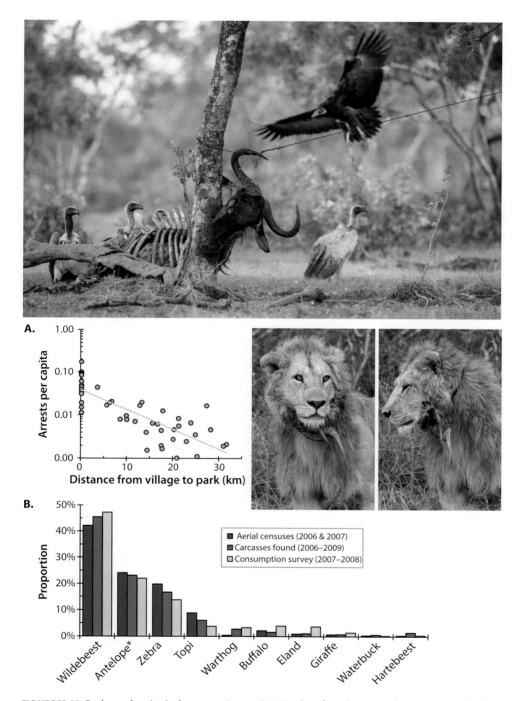

FIGURE 12.19. Bushmeat-hunting in the western Serengeti. **A.** Number of poachers arrested per capita in each village during 2004–2009 declined with distance from the Grumeti and Ikorongo Game Reserves. **B.** Relative proportions of wildlife species consumed in the eight primary study villages closely matched their relative abundance as estimated from aerial counts in Grumeti and Ikorongo Game Reserves and the relative proportions of carcasses found by anti-poaching patrols inside the Serengeti National Park. "Antelope*" = unidentified herbivore species. Redrawn from Rentsch and Packer (2015). *Photo:* Serengeti male that became resident in the long-term study area after being snared elsewhere in the Park.

(e.g., Metzger et al. 2010), and rural populations are growing faster than the national averages in most parts of Africa (Wittemyer et al. 2008; Joppa et al. 2009).

So, what are the solutions? Should every park in Africa be fenced? Where will the money come from to build those fences or to protect the lions inside the reserves that cannot be fenced? What will happen if, as predicted, the human population in Africa triples between 2020 and 2100? These are the key challenges for lion conservation for the foreseeable future, and they will require interventions on a previously unimagined scale.

Fencing

Ideally, people living closest to dangerous wildlife could be offered sufficient incentives to voluntarily move a safe distance from lion habitat (Sinclair and Beyers 2021). But past efforts to relocate indigenous people have led to lasting bitterness and controversy (Adams and Mulligan 2003). Given the massive scale of the problem, the most realistic solution will be to erect physical barriers that safely separate people from wildlife. Following the construction of wildlife-proof perimeter fences, lions were successfully restored to over forty different reserves in South Africa (Miller et al. 2013), as well as to Rwanda's Akagera National Park and Malawi's Majete and Liwonde National Parks. Benin and Kenya employ perimeter fences to separate lions from high-human-occupancy areas adjoining Pendjari and Lake Nakuru National Parks, respectively, and a fence has long separated Nairobi National Park from adjacent urban neighborhoods. South Africa and Kenya have two of the most advanced economies on the continent, and human population densities in Benin, Malawi, and Rwanda all exceed $100/km^2$. Assuring the safety of local people is clearly essential for conserving dangerous animals in these countries, and, given Africa's rapid human population growth, demands for protective wildlife barriers will inevitably intensify. However, care must be taken to maintain the lions' genetic diversity in these small reserves. Many hold too few prides and male coalitions to generate the normal mortality from male takeovers and intergroup conflicts, causing these fenced lion populations to outstrip their food supply and requiring frequent management interventions such as euthanasia and contraception (Miller et al. 2013; McEvoy et al. 2019).

Poorly positioned wildlife fences can also cause more harm than good to lion conservation (Hayward and Kerley 2009). For example, veterinary fencing caused massive animal die-offs of Botswana's migratory herbivores in the 1970s (Owens and Owens 1984), and disease-control fences continued to reduce wildlife populations for decades thereafter (Mbaiwa and Mbaiwa 2006). But Botswana's fences were erected to prevent disease transmission from wild ungulates to cattle, and they were erected over the strenuous objections of conservationists. Animal movements are not just blocked by physical barriers: high levels of human activity can create hard boundaries that are essentially impermeable to wildlife (Tucker et al. 2018). So how can we determine where "mitigation" fences could be most strategically located to minimize the detrimental impacts on wildlife while simultaneously maximizing human safety? Here I will consider elephants in tandem with lions, as no physical barrier will be lion-proof unless it is also elephant-proof.

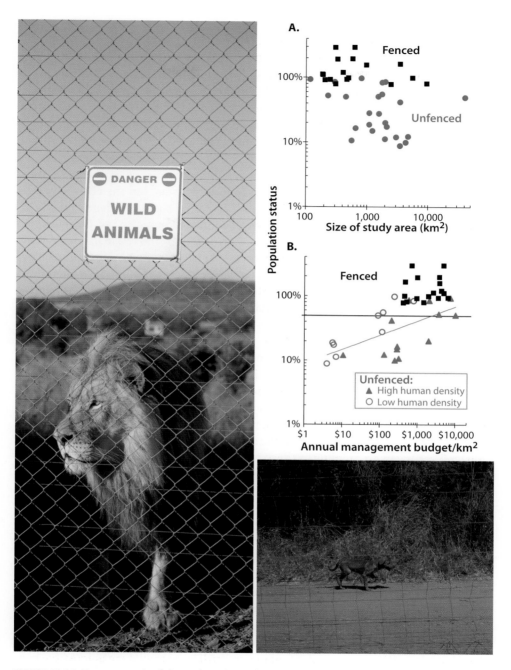

FIGURE 12.20. Percentage ratio of observed population density to predicted carrying capacity for African lions in fenced (black squares) and unfenced (red symbols) reserves according to (**A.**) the size of the survey area and (**B.**) the management budget per square kilometer of survey area. The red regression line is for all unfenced reserves; parks located next to low human population densities have significantly higher lion densities for a given management budget. The horizontal line indicates 50 percent of the potential carrying capacity. Redrawn from Packer et al. (2013). *Photos:* Lion in the fenced Dinokeng Reserve, South Africa; domestic dog outside the fence in Akagera Park, Rwanda. Both photos by CP.

Enrico DiMinin identified areas across Africa where remaining populations of lions and elephants are located within 10 km of high densities of humans, cattle, *and* crops (figure 12.21). Because human activities in these areas already block wildlife movements and otherwise disrupt ecosystem services (e.g., the unfenced western boundary of the Serengeti ecosystem [top photo in figure 12.22]), any physical barriers would mostly act to protect humans from wildlife and vice versa. Enrico's analysis showed that while only 8 percent of the cumulative perimeter of the lions' and elephants' remaining ranges adjoin the highest-risk areas, they contain over half of Africa's remaining lions and a quarter of its elephants (DiMinin et al. 2021). The 9,482 km of fence lines proposed in figure 12.21 would cost about $90 million to build, and the total maintenance costs would be on the order of approximately $5 million per year, but the costs would be spread across fourteen countries. Further, by calculating an "Equivalent Annual Annuity" of fence construction and maintenance in terms of money saved from lowered livestock losses and reduced crop damage, Enrico found that these barriers would actually pay for themselves.[10] Given the estimated $1 *billion* per year required to secure lion habitat across Africa (see below)—this would provide a good conservation return-on-investment and allow funding for on-the-ground conservation efforts to be focused on parts of lion range where they would be needed most.

Funding

Without a sizeable increase in conservation funding, lion numbers will continue to fall. In the first decade of this century, management budgets in several African countries were only about $10/km^2/year, and, unsurprisingly, the unfenced lion populations were far below their potential (figure 12.20b). But at the highest funding levels, unfenced parks can protect lions as successfully as the fenced reserves, and a simple regression line run through these data suggests a necessary annual budget of $2,000/km^2 to maintain an unfenced lion population at least 50 percent of its potential carrying capacity (figure 12.20b). Lindsey et al. (2018) expanded this approach with data from 115 fenced and unfenced protected areas across Africa, and their far more comprehensive statistical model suggested a minimum annual budget of $1,271 per km^2 to maintain lion populations at least 50 percent their potential. Taking the approximately one million km^2 of remaining lion habitat into consideration, they estimated a total funding shortfall of about a billion dollars a year.

Most Western countries finance their parks through taxes paid by their entire citizenry: Americans pay for the management of Yellowstone National Park whether or not they ever visit in person. But Africa is the poorest continent on earth, and most nations lack the tax base to follow a similar funding model. Thus, wildlife parks in Africa have long been expected to pay for themselves through user fees—either entrance fees for photo-tourists or trophy fees for sport hunters. But very few tourist destinations have the international appeal of a Serengeti or Ngorongoro Crater, and gate revenues from

---

10 Enrico's financial analysis did not attempt to incorporate the hidden economic impacts from the *threat* of wildlife attack, which can be substantial (Barua et al. 2013; Thondhlana et al. 2020). Additionally, our interviews on lion attacks in southern Tanzania revealed deep emotional scars in family members and survivors that could never be erased by any sum of money.

the most popular parks in each country must be redistributed to support the less visited parks. In areas where photo-tourism would never raise the necessary revenues, sport hunting was originally seen as a solution for protecting the "million miles of bloody Africa," that Ernest Hemingway famously described in 1935 in *Green Hills of Africa*. But, as we've seen, the hunting blocks are fast disappearing because they are no longer economical,[11] and interest/participation in sport hunting has declined rapidly throughout the world over the past twenty years (Chardonnet 2019), and, of course, all forms of international tourism collapsed during the COVID-19 pandemic (Lindsey et al. 2020).

But, again, it's not all doom and gloom. In the past twenty years, a number of conservation organizations like African Parks, Frankfurt Zoological Society, Wildlife Conservation Society, and World Wildlife Fund (WWF), as well as wealthy eco-philanthropists like Dan Friedkin, Greg Carr, and Paul Tudor Jones, have made the necessary financial commitments to cover funding shortfalls in large tracts of land (see Pringle 2017; Baghai et al. 2018). These efforts have restored failing parks and reserves in Angola, Benin, Central African Republic, Chad, Congo, Democratic Republic of Congo, Malawi, Mozambique, Rwanda, Tanzania, Zambia, and Zimbabwe (see Packer and Polasky 2018). Meanwhile, a growing conservancy movement has successfully partnered with local pastoralists to secure wildlife areas in northern Namibia (Jones 2010) and around Kenya's Maasai Mara National Reserve (Blackburn et al. 2016). Going forward, lessons from these successes may both inspire an even greater philanthropic footprint and provide blueprints for more cost-effective methods for conserving African lions.

People

Almost every conservation issue in this chapter has resulted from the enormous growth in human populations over the past half century, whether we consider the villages around the Serengeti, all of Tanzania, or the entire continent of Africa, and we can only expect these problems to intensify as populations continue to expand. Lions still roam over an area of approximately one million km^2, but about four million km^2 of Africa's land area will need to be converted to agriculture by the year 2060 in order to feed the coming wave of humanity, greatly exacerbating extinction risks of large mammals across the continent (Tilman et al. 2017). Half of global population growth between 2012 and 2050 is projected to be in Africa, and sub-Saharan Africa is predicted to be the only region with a total fertility rate above replacement by 2050 (3.2 children versus 2.1 at replacement) (UNDESA 2013). At current rates, the population of sub-Saharan Africa will reach 2.2 billion people by 2050, but if the region could move toward replacement-level fertility by 2050, its population would grow to only 1.8 billion, and crop demand would only be two-thirds of the baseline projection (Searchinger et al. 2018).

Birth rates decline in response to lower infant mortality, greater educational opportunities for girls, and higher economic status for women, all of which are included in the seventeen Sustainable Development Goals (SDGs) of the United Nations and

---

11 Even at its peak, sport hunting only generated about $15 million per year in Tanzania (Packer 2015), while annual funding of $150 to 300 million would have been required to effectively protect approximately 150,000 km^2 of hunting blocks.

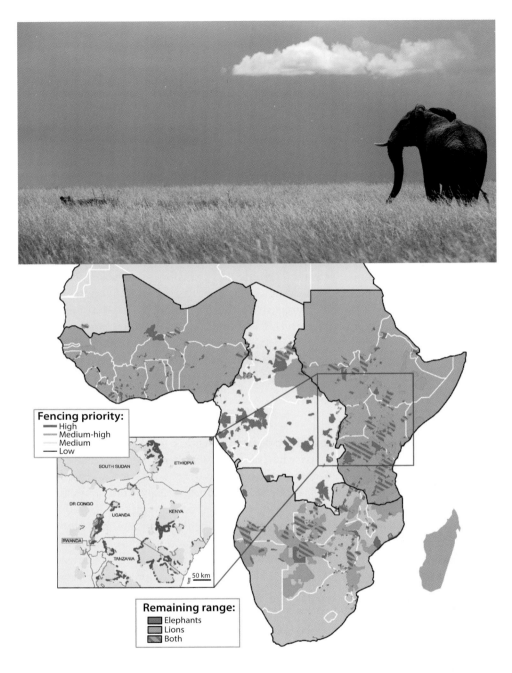

FIGURE 12.21. Remaining ranges of African elephants and lions. **Inset.** Fencing priorities in East Africa. "Extended ranges" (indicated by darker gray) include all protected areas where elephants or lions are found plus species ranges located in nonprotected areas. High fencing priority: proximity to high human density, cattle density, and cropland; moderate priority: any two high-human activities; low priority: one high-human activity. From DiMinin et al. (2021).

FIGURE 12.22. Human impacts in the greater Serengeti ecosystem. *Top:* Illegal cattle trail crossing the northwestern boundary of Serengeti National Park. *Bottom:* Dust storm caused by livestock grazing in Ngorongoro Conservation Area.

the World Health Organization, so the underlying solution to wildlife conservation will ultimately rely on improvements in human well-being—not just making people safe from dangerous animals but also promoting human prosperity.[12] In 2006, Susan James and I established an organization in Tanzania that we called Savannas Forever–Tanzania (SFTZ). Working with Monique Borgerhoff Mulder, Deborah Levin, Joe Ritter, and Kari Hartwig, we developed survey tools that could assist international development agencies such as the World Bank, USAID, and the United Nations' World Food Program to assess the effectiveness of their aid programs (Packer 2015). Although our survey data contributed to several scientific publications (e.g., Aichele et al. 2014; Lawson et al. 2014; Rentsch & Packer 2015; Salerno et al. 2016), SFTZ's primary mission has focused on helping the aid agencies improve the lives of rural villagers across a meaningful geographical scale.

This challenge is particularly important with agriculture, as farming is the most important economic activity in rural Africa, and increased demand for crops is met by expanding the amount of land converted to cropland in less developed nations (Polasky et al., in review). Crop yields in the wealthiest nations have grown dramatically since the advent of the Green Revolution, but these improvements have not yet reached Africa. If development projects were to help close the so-called "yield gaps" in farm production by 80 percent, the demand for new land clearing between 2010 and 2100 could be reduced from 4.3 million km^2 to 1.95 million km^2 (Tilman et al. 2017). Such improvements could be achieved by planting legumes and cover crops, using improved seeds, and implementing drip irrigation and crop rotation—and thereby minimizing many of the negative impacts of intensive agriculture, as a number of strategies are able to maintain or even improve yields while lowering application of nitrogen fertilizers ("integrated soil-crop management" in China [Cui et al. 2018]; narrow-strip intercropping [Li et al. 2020]; and precision agriculture [Sela et al. 2016]). Malawi, Rwanda, Zambia, Ghana, Mali, and Senegal have already increased crop yields by 20 to 80 percent, so it is not unreasonable to suggest that similar gains could be made in countries like Tanzania and Mozamibique, though this would require massive improvements in infrastructure to transport crops to market—improvements that would only be feasible with assistance from the international community.

Similarly, livestock programs currently being piloted in the conservancies around Maasai Mara are increasing *meat* production by encouraging pastoralists to transition their cattle away from functioning primarily as a cultural asset to instead serving as a source of cash revenue. Several of the Mara Conservancies have recently introduced faster-growing cattle breeds, ensured that only the best herders tend the stock, provided improved veterinary services, reinforced the *bomas*, and transported cattle to urban areas for slaughter—while simultaneously aiming to lower the stocking rate by 50 percent and thereby reducing grazing competition with wildlife. Individual Maasai landowners negotiate annual rents with local tourist operators and receive a secondary payment each month from livestock sales, while the conservancy keeps 40 percent of the livestock receipts to cover the costs of husbandry, transportation, and pasture restoration. The program is too new to know whether it is the magic bullet

---

12 With rapid attainment of the SDGs, the population of sub-Saharan Africa is estimated to reach only 1.6 billion by the year 2100 (Vollset et al. 2020).

for solving the problem of human-wildlife conflict in the pastoralist areas and closing an additional yield gap in food production.

Increased production can increase demand as abundant supplies lead to new uses and lower prices (Jevons 1865). Meat consumption rises with economic development, and wealthier countries devote a significant proportion of their cropland for livestock forage, thus cheaper meat may amplify consumption and accelerate the rate of habitat conversion to cropland, unless producers are incentivized to grow foods for direct human consumption rather than as cattle feed (Polasky et al. in review). However, Polasky et al.'s projections of future cropland requirements suggest that accelerated economic development in the world's poorest nations could close yield gaps and hasten the demographic transition to smaller family size to an extent that would far outweigh any increase in per capita food consumption (Polasky et al. in review). Thus, doing what's best for the people of Africa would also be best for nature.

Climate Change

While there's hope that human population growth can be stabilized and more food could potentially be generated from less land—at least in the near term—there are still the impending effects of climate change to consider. Climate models predict more rapid increases in land temperature across Africa than in the rest of the world, a heavier and more variable rainfall regime in equatorial Africa—with more frequent swings between drought and flood—and a general drying of the climate in the southern third of the continent (Niang et al. 2014). These trends will be expected to reduce crop productivity, increase the vulnerability of livestock to disease, exacerbate food insecurity, increase infant mortality, and reduce educational opportunities. A climate refugee crisis has already emerged in the Sahel, and widespread political instability could well expand to the point where any of today's predictions about food production and population growth will largely be irrelevant.

Over the course of writing this book, I have often felt embarrassed to be devoting so much effort to the behavior, ecology, and conservation of African lions when our entire planet faces such an uncertain future—an uncertain future for all our children and grandchildren and anyone who might someday read these pages. But if we do manage to avoid the most apocalyptic scenarios, we should at least try to maintain space for all the rest of nature, though it won't be easy. From their position atop the savanna food web, lions are vulnerable to a bewildering variety of hazards. Lions connect to the daisy chain of disease transmission between multiple carnivore species, and they are susceptible to the pathogens carried by infected prey. Lions require so much space that many of Africa's protected areas are too small to sustain populations with a healthy level of genetic variability. The lions' reliance on large prey makes them especially dangerous to the livestock species that are most valued by pastoralists, their sociality makes outbreaks of man-eating lions particularly persistent and wide-ranging, and their infanticidal behavior makes them unusually vulnerable to sport hunting. Being temperature sensitive, male lions will likely grow smaller and blonder manes as a result of climate change (West and Packer 2002), and many of the lions' water-dependent prey species, like wildebeest, zebra, and buffalo, will likely be replaced by smaller species (Veldhuis, Kihwele et al. 2019).

Yet if lions weren't so large, dangerous, and downright *interesting*, they wouldn't hold the same place in the human imagination. Man may be the "king of animals, nature and everything else," but the lion deserves its flagship status, its special place in the pantheon of species that mankind will respect, and perhaps even manage to protect, for generations to come.

KEY POINTS

1. Canine Distemper Virus (CDV) outbreaks regularly strike the Serengeti and Ngorongoro but only inflict significant mortality when the lions are co-infected with high levels of blood parasites. A large-scale vaccination program in domestic dogs failed to protect lions from CDV even though the multivalent vaccine successfully controlled canine rabies. Despite having minor impacts in large populations, bovine tuberculosis threatened an inbred lion population in South Africa until genetic diversity was restored.
2. Human-wildlife conflict is a major source of lion mortality in pastoralist areas surrounding the national parks. Lions are killed in direct proportion to the number of livestock taken in villages around Tarangire National Park, and lions are killed both in retaliation and for ritual purposes in the Ngorongoro Conservation Area. Recent programs are showing signs of success in improving livestock husbandry, reducing retaliatory killings, and engaging local communities in conservation efforts.
3. Lion attacks on humans have been widespread in Tanzanian districts with scarce prey except for bushpigs, a nocturnal crop pest. Subsistence farmers typically sleep in their fields to protect against bushpig damage, despite recognizing their vulnerability to lions. Low-cost fencing successfully protects crops and was subsequently adopted by farmers in one field-test village. Outbreaks of man-eating are more widespread and persistent in lions than in leopards or tigers, presumably because lions learn from pridemates.
4. Simulation models predict that overhunting of adult males can reduce lion populations via social instability and associated infanticide unless offtakes are restricted to males old enough to have reared a cohort of young. Harvesting of younger males was common until the early 2010s, and empirical data from across Africa and from individual hunting blocks in Tanzania confirmed that overhunting was widespread and had population-level consequences. An age minimum has since been adopted in several African countries.
5. Bushmeat consumption and illegal livestock grazing threaten to lower prey populations, and competition for land will intensify if Africa's population triples by the end of this century. However, perimeter fences protect lions in high-human-density areas and can even pay for themselves by reducing livestock losses and crop damage by elephants. Accelerating economic development in sub-Saharan Africa will be vital for improving crop yields and hastening the demographic transition to smaller family sizes.

AFTERWORD

> Do not be daunted by the enormity of the world's grief. Do justly now, love mercy now, walk humbly now. You are not obligated to complete the work, but neither are you free to abandon it.
> —THE TALMUD

It has been an immense privilege to have spent so many years with so many lions. In looking back over my decades of research, I feel even more privileged today than during my time living in the Serengeti. But this good fortune has always been coupled with a profound sense of responsibility. Lions were never considered a conservation priority until early this century, so I started out feeling free to tackle questions without any immediate practical application. But by studying a species as iconic as the lion, I always wanted to make sure we got the story right—which sometimes meant waiting years before we had gathered a sufficiently large sample to be confident in our findings. We took eleven years to observe enough crèches to measure nursing preferences of mothers and cubs, and we hauled our lion dummies all the way to Tsavo to find enough previously untested animals to finish off the mane-choice experiments.

But basic research can often have unforeseen practical value. When we first started collecting blood samples, the goal was to perform paternity testing. Even though we had initially only sought insights into the lions' mating system and kinship patterns, our archived blood samples later proved essential for deducing the timing and severity of past disease outbreaks. We had mostly focused on the impacts of male takeovers on cub survival to understand how females coped with the constant threat of losing their cubs to a new set of husbands, but these same data were instrumental in estimating the consequences of excessive sport hunting and the potential population impacts of bovine tuberculosis.

But in 1994, the Serengeti lion die-off demanded immediate practical action, and I was lucky: after the first few lion fatalities, Steve O'Brien funded the investigation with resources from his Laboratory for Viral Carcinogenesis at the National Cancer Institute. Melody Roelke, an American vet with extensive experience working with Florida panthers, happened to be based in the Serengeti at the time, she knew how to implement rigorous disease-sampling surveys, and she sent necropsy samples to Linda Munson at the University of Tennessee, who soon recognized the telltale signs of distemper. Max Appel's serological tests at the Cornell Veterinary Lab quickly revealed the extent of the current outbreak and the existence of prior CDV outbreaks, and Sarah Cleaveland's studies of village dogs confirmed that domestic dogs were the source of the virus. Sarah's overarching goal was rabies control, so she was the perfect collaborator for NSF's new program on the Ecology of Infectious Diseases. The fatal Crater outbreak struck shortly thereafter, enabling Linda Munson and Karen Terio to investigate the importance of co-infection in CDV's devastating impacts on lion mortality following buffalo-killing droughts.

The stars don't often align like that, especially for some random field biologist who had never thought that lions—the global icon of power and strength—could ever be felled by anything but another member of its own kind. Lions might still be their own worst enemy—once your pridemate got sick, there was no social distancing, no mask mandate, so you were destined to get it, too—but the source of infection to each new pride linked back to jackals and hyenas and, ultimately, domestic dogs living somewhere outside the national park. We found the funds to start vaccinating thousands and thousands of village dogs, and even though we didn't fully protect the Serengeti lions from CDV, we likely saved hundreds of rural villagers who would have otherwise died of rabies.

When we were asked to investigate man-eating, several international organizations were calling for a ban on lion trophy hunting, and the Tanzanian government wanted to highlight the *hundreds* of lion-attack victims over the previous few years as evidence of too many lions rather than too few. The man-eating problem had been overlooked for far too long; newspaper accounts were coming out on a weekly basis. The stars lined up again insofar as the government provided case records and field staff, and a German aid worker, Rolf Baldus, had already compiled enough government data to help guide our investigation. Dennis Ikanda and Bernard Kissui both knew lions thanks to their work in Ngorongoro and Tarangire, and they both had experience interviewing local people in their research on human-lion conflict. After we compiled the survey data and uncovered the links between lions and nocturnal crop pests, we found funding to make posters and to test alternative crop-protection strategies. But lion poisonings had solved the man-eating outbreak by then, outside interest soon waned, and I must admit that my sympathies were always with the local people rather than with the lions.

It wasn't until traveling south to the villages in southern Tanzania and personally meeting with mothers who had lost their children to lions, families whose brothers and fathers had been eaten by lions that had come into their fields and villages looking for prey, that I became convinced of the overwhelming importance of perimeter fencing. I had worked in South Africa by then, and human-lion conflict was never an issue in that part of the continent. In fact, lion habitat had been restored throughout South Africa precisely because the newly restored reserves were always surrounded by wildlife-proof fences, and neighboring communities were always assured of their personal safety. What greater responsibility than to protect the lives of people from lions? I keenly felt the moral injunction: *Primum non nocere*. First, do no harm.

I have always been drawn to the challenge of solving a difficult problem. Watch and listen, decide what to measure, compile the data, decide whether any sort of clarity has been achieved, then repeat. The anecdotes, illustrations, and statistical analyses in the past twelve chapters reflect our attempts to advance numerous lines of investigation, and while I believe we have achieved varying degrees of resolution, the final word is unlikely to have been written on any one of them. The oldest cliché in science is "More research is needed." But among all these questions, a few have most obviously defied any easy answers, and I hope that future generations will someday answer them, though many may never be resolved.

Here's my top-ten list:

1. How is paternity distributed among the largest male coalitions? Does the variance continue to increase with coalition size (as suggested by the genetic analysis), thus discouraging unrelated males from teaming up to form quartets or larger? Or do individual males attain similar reproductive success over multiyear timescales (as implied by the consortship analysis), and, if so, are unrelated partners cognitively incapable of maintaining a social unit larger than three individuals?
2. How cooperative are lions during group hunts? Our statistical analysis of a coarse physical measure (distance traveled each hunt) suggested three contrasting strategies (refrain, pursue, and conform), but we relied on direct observation from a distant vehicle during the day. GPS-tracking devices can collect fine-scale measurements of individual-hunting tactics day and night, though it would still be difficult to relate such data to the movements of individual prey animals without the use of a lifelike, highly mobile, and easily repaired mechanical warthog/wildebeest/zebra/buffalo.[1]
3. By lagging behind during an approach toward outsiders, are individual lions enabling their opponents to more accurately assess the size of their advancing group, or are they facilitating the eventual encirclement of their opponent? Or something else? Cheap, easily repaired, lifelike robot lions, please—both to stage mock incursions with and without laggards and to remain stationary until being surrounded.
4. Can landscape complexity and local-scale population density be measured in ways that could rigorously test whether geography predicts typical pride size within and across highly diverse ecosystems? Typical pride size varies from one part of Africa to another; average pride size in two of the Serengeti woodland "neighborhoods" was 50 percent larger than in a third. Can the size and shape of individual lion neighborhoods be predicted by the hydrology and topography of each landscape?
5. What are the proximal mechanisms that facilitate/maintain lion sociality? Do lions have more oxytocin receptors than leopards or tigers? Lions perform better than other big cats on cognitive tests (Borrego and Gaines 2016)—how important is their greater "intelligence" in maintaining social bonds, or do they receive a stronger positive reinforcement from physical contact with or proximity to other lions?
6. What will happen to the Crater lion population should it attain a similar level of genetic diversity as the Serengeti lions? Will population numbers stabilize at around eighty individuals? Will future generations of large coalitions once again

---

1 Natalia Borrego and I set out life-sized cardboard cut-outs of warthogs, wildebeest, and buffalo in several South African reserves. We pulled a cut-out on a cart, but the lions only approached until they viewed the cut-out obliquely. We tried a life-size rubber impala, but it looked like a square block from the rear. Perhaps future tests could use lifelike robots capable of lowering and raising their heads to "graze" and occasionally look alert.

effectively seal off the Crater floor from in-migration by outside males despite the restoration of a more contiguous population of resident lions between the Serengeti and Ngorongoro?

7. What are the primary determinants of ecological stability and resilience in lion-dominated ecosystems? Does stability result primarily from the sociality of predators and their prey? How important is food web complexity? How essential is asynchrony of prey responses to environmental perturbation? Will Africa's protected areas maintain their resilience in the face of impending changes in rainfall and temperature?[2]

8. During the peak years of lion hunting in Tanzania, sport hunters generated less than 10 percent of the funding necessary to conserve the country's hunting blocks (Packer 2015). Now with the widespread abandonment of hunting blocks in Tanzania, Zambia, Zimbabwe, and Mozambique, what else can be done to rescue these vast tracts of wildlife habitat and possibly even restore them in the future?

9. Can economic development strategies be designed to help African countries achieve the Sustainable Development Goals before their growing food demands cause irreversible losses in the continent's biodiversity? Can yield gaps be closed and human population growth be slowed in time to prevent catastrophic impacts from habitat conversion, bushmeat consumption, and grazing competition between livestock and the lions' prey?

10. Who will cover the costs of lion conservation in the decades ahead? Will the international community develop a workable system of financial incentives (e.g., biodiversity "credits" analogous to carbon credits) that cover costs to local communities of living with a globally valued species? Will African range states someday support their own reserves through tax revenues rather than entrance/trophy fees? After all, India has successfully conserved its lions despite widespread poverty and a population of over a billion (Banerjee et al. 2013).

I hope that even the most subtle aspects of lion behavior will help current and future generations of lion biologists gain a better idea of what it is they are trying to understand and conserve. Lion conservation is hard—and it's especially hard because lions are a social species. We rely on tourist revenues to fund conservation of Africa's parks, but tour companies tend to build their lodges in the same lion hotspots that are essential for the evolution of group territoriality: picturesque river confluences with abundant water and plentiful game. Entire prides feed on the same tick-infested buffalo carcasses at the height of drought, sneeze on each other once a single pride member has come down with distemper. Once lions become man-eaters, the behavior persists until the entire pride has been eliminated, and cattle-killing may follow a similar pattern. The small South African reserves might even become predator pits because management practices inadvertently encourage the formation of small prides—thus losing the advantages of sociality on the stability of predator-prey dynamics.

---

2 A recent examination of the necessary conditions for stable coexistence between lions and each of their eight most common prey species (and that incorporated more comprehensive parameter estimates than our prior analysis) suggested that the Serengeti lions are likely *forced* to include a broad range of prey, as they would be unable to survive on an exclusive diet of any single species (Fryxell et al. 2022). Thus, the lion is inevitably part of a complex food web.

No lion in the Serengeti or Ngorongoro chooses to live alone, but stuff happens, and an unlucky female may be forced to depart from her extended family, while another may end up the sole survivor of its pride or coalition. But outside the well-protected national parks and reserves, I fear that the intricacies of lion life, the large prides with their large coalitions, their crèches and their joint defense of a shared territory, may rarely if ever reach full expression. Given the lions' understandable skittishness around well-armed humans, direct observations in the high-conflict areas are difficult to impossible, so we simply don't know much about the social lives of lions in homogenized landscapes now devoid of the prey base that supported the densities necessary to favor group living. In the Crater and Serengeti, the solitary female cuts a pathetic figure, furtive, vulnerable, only marginally connected to any one spot on the map, condemned to almost never raise her cubs to maturity. But without the pressures of neighboring prides breathing down their necks, lone females do often manage to persevere in the smaller South African reserves (McEvoy 2019) and in the highly disturbed pastoralist areas in northern Kenya (Bhalla 2017).

Will the lion endure into the next century? Almost certainly—at the very least, lions thrive so well in captivity that their reproduction must be curtailed to keep numbers at manageable levels. But will the species still display the full range of its social repertoire?

ACKNOWLEDGMENTS

Our understanding of the lions could only have been possible with the dedicated efforts of the field team of the "Serengeti Lion Project," who drove out almost every day to look for lions and record relevant details of their behavior, diet, and reproductive status. For the first eighteen years, lions could only be located opportunistically, meaning that we all had to drive around the Serengeti study area or the floor of the Ngorongoro Crater, eyes peeled, hoping to spy a patch of yellow fur in the tall grass or a feline silhouette on top of a pile of rocks—and data collection was often sporadic and frustrating, as our early field assistants, Jamie Douthit, John Fanshawe, Monique Borgerhoff Mulder, and Sarah Cairns, could readily attest. But after attaching our first radio collars in 1984, we recruited a succession of graduate students (Dave Scheel, Jon Grinnell, Peyton West, Grant Hopcraft, Bernard Kissui, Anna Mosser, Dennis Ikanda, Meggan Craft, Ali Swanson, Natalia Borrego, Meredith Palmer) and post-docs (Karen McComb, Rob Heinsohn) to investigate specific questions about lion behavior or ecology. Beginning in 1990, we employed full-time field assistants to perform the basic lion-monitoring duties: Cyprian Malima, Kiyungi Kiyungi, Sarah Legge, Pamela Bell, Audie Hazenberg, Maria Finnigan, Iain Taylor, Henry Brink, Kirsten Skinner, Ingela Jansson, Candida Mwingira, Patrik Jigsved, John Stewart, George Lohay, Daniel Rosengren, Stan Mwampeta, and Norbert Isaac, who ensured that every pride was observed several times each week. I cannot thank them enough, as every post-doc, graduate student, field assistant, collaborator, and predecessor—regardless of their role—made essential contributions to the Serengeti Lion Project.

I also wish to thank the Tanzanian Commission for Science and Technology and the Tanzanian Wildlife Research Institute for permission to conduct research in the Serengeti and Ngorongoro Crater, and I am extremely grateful to Tanzania National Parks and the Ngorongoro Conservation Area Authority for their support while working within their jurisdictions. The Wildlife Division of the Tanzanian Ministry for Natural Resources and Tourism generously provided data from the hunting blocks and the man-eating areas. I am indebted, too, to the two anonymous referees who carefully waded their way through my first draft—their queries and suggestions greatly improved the final product—and to Dana Henricks for her thorough editing.

Our work was initially funded by a two-year grant from the H. F. Guggenheim Foundation, followed by several grants from various organizations until our first three-year grant from the National Science Foundation. NSF then funded the project for over three decades via their programs in Animal Behavior, Population Biology, Biocomplexity, Ecology of Infectious Disease, and Opportunities for Understanding through Synthesis (OPUS). But most important to the continuity of the lion project was their program in Long-Term Research in Environmental Biology (LTREB), which arose largely thanks to the efforts of Mark Courtney. It may be rare to hear anyone express genuine warmth for a program run by the federal government, but I cannot thank them enough.

Second only to NSF is the National Geographic Society, which came through with timely grants in the 1980s, '90s, '00s, and '10s that helped strengthen the long-term lion study and expanded our efforts to sustain the SnapshotSerengeti camera grid. In a similar vein, funds from Panthera helped with our work on human-lion conflict; donations from Paradise Wildlife Park helped initiate the dog vaccination program following the CDV outbreak; Lincoln Park Zoo continued the vaccinations after our disease grant from NSF had expired; and USAID funded the village livelihood surveys by Savannas Forever–Tanzania.

Frankfurt Zoological Society's country representative, Markus Borner, not only provided flight time during our aerial radio-tracking efforts for thirty years, but Markus also did more than anyone to stimulate my interest in the larger questions about wildlife conservation in East Africa. He was also a near-constant source of good cheer. Which brings me to Barbie Allen, who not only provided enormous practical help in making life bearable during times when economic hardships in Tanzania made it difficult to find fuel and food, but her boundless generosity and razor wit also made life much more fun. Markus and Barbie were a big part of the reason that I *enjoyed* coming back to Africa for so many years.

Finally, I must also give belated thanks to the J. S. Guggenheim Foundation for their Fellowship way back in 1990 when I had promised to write a book like this during my sabbatical but instead wrote *Into Africa* (1994). I apologize for having taken so long, but I really did need all that extra time.

APPENDIX

TABLE A.1. *Weights and biomass in kilograms for the ten most common prey species in the Serengeti and Ngorongoro Crater*

Species	Sex	Live wt.	Prop. offal	Dress wt.	Prop. meat	1 Intact	2 With guts	3 No guts	4 3/4	5 1/2	6 1/4
Cape buffalo	M	751	33%	381	80%	429.9	367.9	305.9	191.2	114.7	38.2
Cape buffalo	F	447	30%	226	81%	269.6	226.5	183.4	114.6	68.8	22.9
Eland	M	501	25%	301	84%	327.7	289.7	251.8	157.4	94.4	31.5
Eland	F	300	23%	180	84%	202.5	177.0	151.6	94.7	56.8	18.9
Giraffe	M	1097	29%	548	83%	685.3	570.2	455.1	284.4	170.7	56.9
Giraffe	F	717	27%	358	83%	462.4	380.0	297.5	186.0	111.6	37.2
Grant's gazelle	M	62	24%	36	83%	40.9	35.6	30.2	18.9	11.3	3.8
Grant's gazelle	F	42	24%	24	83%	28.3	24.3	20.3	12.7	7.6	2.5
Hartebeest	M	143	29%	82	83%	87.4	77.6	67.7	42.3	25.4	8.5
Hartebeest	F	126	27%	73	83%	80.1	70.5	60.9	38.1	22.8	7.6
Thomson's gazelle	M	20	28%	11	83%	12.8	11.0	9.1	5.7	3.4	1.1
Thomson's gazelle	F	16	25%	9	82%	10.5	9.0	7.5	4.7	2.8	0.9
Topi	M	130	31%	71	84%	79.1	69.4	59.8	37.3	22.4	7.5
Topi	F	109	29%	56	84%	67.8	57.5	47.2	29.5	17.7	5.9
Warthog	M	87	32%	48	85%	51.7	46.4	41.1	25.7	15.4	5.1
Warthog	F	53	32%	34	86%	31.5	30.2	28.9	18.0	10.8	3.6
Wildebeest	M	201	37%	102	81%	108.0	95.3	82.6	51.6	31.0	10.3
Wildebeest	F	163	32%	82	82%	96.2	81.9	67.6	42.2	25.3	8.4
Zebra	M	248	32%	136	84%	146.7	130.6	114.5	71.6	42.9	14.3
Zebra	F	219	32%	120	84%	129.7	115.4	101.2	63.3	38.0	12.7

Source: Data are mostly taken from Sachs (1967) and Ledger (1968).
Note: Lions consume the offal (heart, liver, spleen, etc.) before eating the muscle tissue, so we estimated the amount of biomass remaining at a carcass according to the six right-hand columns, ranging from "1" for an intact carcass of a freshly killed prey animal to "6" for less than a quarter of the muscle tissue remaining. Calves/fawns/foals of each species were assumed to weigh about an eighth of their mothers' weights.

REFERENCES

Acevedo-Whitehouse, K., F. Gulland, D. Greig, and W. Amos. 2003. "Inbreeding: Disease susceptibility in California sea lions." *Nature* 422:35.
Adams, E. S. 1990. "Boundary disputes in the territorial ant *Azteca trigona*: Effects of asymmetries in colony size." *Animal Behaviour* 39:321–28.
———. 2001. "Approaches to the study of territory size and shape." *Annual Review of Ecology and Systematics* 32:277–303.
Adams, W. M., and M. Mulligan, eds. 2003. *Decolonizing Nature: Strategies for Conservation in a Postcolonial Era*. London/Sterling, VA: Earthscan / James & James.
Aichele, S. R., M. Borgerhoff Mulder, S. James, and K. Grimm. 2014. "Attitudinal and behavioral characteristics predict high risk sexual activity in rural Tanzanian youth." *PLoS One* 9:e99987. https://doi.org/10.1371/journal.pone.0099987.
Anderson, J. A., R. A. Johnston, A. J. Lea, F. A. Campos, T. N. Voyles, M. Y. Akinyi, S. C. Alberts, E. A. Archie, and J. Tung. 2021. "High social status males experience accelerated epigenetic aging in wild baboons." *eLife* 10:e66128. https://doi.org/10.7554/eLife.66128.
Anderson, R. M., and R. M. May. 1991. *Infectious Diseases of Humans: Dynamics and Control*. Oxford: Oxford University Press.
Anderson, T. M., S. White, B. Davis, R. Erhardt, M. Palmer, M. Kosmala, A. Swanson, and C. Packer. 2016. "The spatial and temporal distribution of African savannah herbivores: Species associations and habitat occupancy in a landscape context." *Philosophical Transactions of the Royal Society B* 371:20150314. https://doi.org/10.1098/rstb.2015.0314.
Atkins, J. L., R. A. Long, J. Pansu, J. H. Daskin, A. B. Potter, M. Stalmans, C. E. Tarnita, and R. M. Pringle. 2019. "Cascading impacts of large-carnivore extirpation in an African ecosystem." *Science* 364:173–77.
Baghai, M., J. R. B. Miller, L. J. Blanken, H. T. Dublin, K. H. Fitzgerald, P. Gandiwa, K. Laurenson, et al. 2018. "Models for the collaborative management of Africa's protected areas." *Biological Conservation* 218:73–82.
Baldus, R. D. 2021. "The end of the game." *Sports Afield* 244:32–33.
Balme, G., M. Rogan, L. Thomas, R. Pitman, G. Mann, G. Whittington-Jones, N. Midlane, et al. 2019. "Big cats at large: Density, structure, and spatio-temporal patterns of a leopard population free of anthropogenic mortality." *Population Ecology* 61:256–67. https://doi.org/doi/10.1002/1438-390X.1023.
Balme, G., R. Slotow, and L. T. B. Hunter. 2010. "Edge effects and the impact of non-protected areas in carnivore conservation: Leopards in the Phinda-Mkhuze Complex, South Africa." *Animal Conservation* 13:315–23.
Balme, G. A., and L. T. B. Hunter. 2013. "Why leopards commit infanticide." *Animal Behaviour* 86:791–99. https://doi.org/10.1016/j.anbehav.2013.07.019.
Balme, G. A., J. R. B. Miller, R. T. Pitman, and L. T. B. Hunter. 2017. "Caching reduces kleptoparasitism in a solitary, large felid." *Journal of Animal Ecology* 86:634–44. https://doi.org/10.1111/1365-2656.12654.
Balme, G. A., R. T. Pitman, H. S. Robinson, J. R. B. Miller, P. J. Funston, and L. T. B. Hunter. 2017. "Leopard distribution and abundance is unaffected by interference competition with lions." *Behavioral Ecology* 28:1348–58. https://doi.org/10.1093/beheco/arx098.
Banerjee, K., Y. V. Jhala, K. S. Chauhan, and C. V. Dave. 2013. "Living with lions: The economics of coexistence in the Gir forests, India." *PLoS One* 8:e49457. https://doi.org/10.1371/journal.pone.0049457.
Barthold, J. A., A. J. Loveridge, D. W. Macdonald, C. Packer, and F. Colchero. 2016. "Bayesian estimates of male and female African lion mortality for future use in population management." *Journal of Applied Ecology* 53:295–304. https://doi.org/10.1111/1365-2664.12594.
Barua, M., S. A. Bhagwat, and S. Jadhav. 2013. "The hidden dimensions of human-wildlife conflict: Health impacts, opportunity and transaction costs." *Biological Conservation* 157:309–16.
Bates, J. H. 1984. "Transmission, pathogenesis, pathology and clinical manifestations of tuberculosis." In *The Mycobacteria: A Source Book*, 991–1005. New York: Marcel Dekker.
Bauer, H., K. Nowell, P. Henschel, P. Funston, G. Chapron, L. Hunter, D. W. Macdonald, and C. Packer. 2015. "Lion (*Panthera leo*) populations are declining rapidly across Africa, except in intensively managed areas." *Proceedings of the National Academy of Sciences* 112 (48): 14894–99. https://doi.org/10.1073/pnas.1500664112.
Beach, F. A. 1976. "Sexual attractivity, proceptivity, and receptivity in female mammals." *Hormones and Behavior* 7:105–38.

Becker, M., R. McRobb, F. Watson, E. Droge, B. Kanyembo, J. Murdoch, and C. Kakumbi. 2013. "Evaluating wire-snare poaching trends and the impacts of by-catch on elephants and large carnivores." *Biological Conservation* 158:26–36. https://doi.org/10.1016/j.biocon.2012.08.017

Bednekoff, P. A. 2015. "Sentinel behavior: A review and prospectus. *Advances in the Study of Behavior* 47:115–45. https://doi.org/10.1016/bs.asb.2015.02.001.

Begg, C. M., J. R. B. Miller, and K. S. Begg. 2018. "Effective implementation of age restrictions increases selectivity of sport hunting of the African lion." *Journal of Applied Ecology* 55:139–46. https://doi.org/10.1111/1365-2664.12951.

Berger, K. M., and E. M. Gese. 2007. "Does interference competition with wolves limit the distribution and abundance of coyotes?" *Journal of Animal Ecology* 76:1075–85. https://doi.org/10.1111/j.1365-2656.2007.01287.x.

Bertram, B. C. R. 1975a. "Social factors influencing reproduction in wild lions." *Journal of Zoology* 177 (4): 463–82. https://doi.org/10.1111/j.1469-7998.1975.tb02246.x.

———. 1975b. "Weights and measures of lions." *East African Wildlife Journal* 13:141–43.

———. 1976. "Kin selection in lions and in evolution." In *Growing Points in Ethology*, edited by P. P. G. Bateson and R. A. Hinde, 281–301. Cambridge: Cambridge University Press.

Bhalla, S. 2017. "Demography and ranging behaviour of lions (*Panthera leo*) within a human-occupied landscape in northern Kenya." PhD dissertation, University of Oxford.

Blackburn, S., J. G. C. Hopcraft, J. O. Ogutu, J. Matthiopoulos, and L. Frank. 2016. "Human-wildlife conflict, benefit sharing and the survival of lions in pastoralist community-based conservancies." *Journal of Applied Ecology* 53:1195–1205. https://doi.org/10.1111/1365-2664.12632.

Boehm, C. 2011. *Moral Origins: Social Selection and the Evolution of Virtue, Altruism, and Shame.* New York: Basic Books.

Boesch, C., and H. Boesch. 1989. "Hunting behavior of wild chimpanzees in the Tai National Park." *American Journal of Physical Anthropology* 78:547–73.

Boeskorov, G. G., V. V. Plotnikov, A. V. Protopopov, G. F. Baryshnikov, P. Fosse, L. Dalén, D. W. G. Stanton, et al. 2021. "The preliminary analysis of cave lion cubs *Panthera spelaea* (Goldfuss, 1810) from the permafrost of Siberia." *Quaternary* 4:24. https://doi.org/10.3390/quat4030024.

Boone, R. B., S. J. Thirgood, and J. G. C. Hopcraft. 2006. "Serengeti wildebeest migratory patterns modeled from rainfall and new vegetation growth." *Ecology* 87:1987–94.

Borrego, N., and M. Gaines. 2016. "Social carnivores outperform asocial carnivores on an innovative problem." *Animal Behaviour* 114:21–26. https://doi.org/10.1016/j.anbehav.2016.01.013.

Borrego, N., A. Ozgul, R. Slotow, and C. Packer. 2018. "Lion population dynamics: Do nomadic males matter?" *Behavioral Ecology* 29:660–66. https://doi.org/10.1093/beheco/ary018.

Boukal, D. S., L. Berec, and V. Křivan. 2008. "Does sex-selective predation stabilize or destabilize predator-prey dynamics?" *PLoS-One* 37:e2687. https://doi.org/10.1371/journal.pone.0002687.

Boulain, N., G. Simioni, and J. Gignoux. 2007. "Changing scale in ecological modelling: A bottom up approach with an individual based vegetation model." *Ecological Modelling* 203:257–69.

Bowles, S. 2009. "Did warfare among hunter-gatherer groups affect the evolution of human social behaviors?" *Science* 324:1293–98.

Bowles, S., and H. Gintis. 2011. *A Cooperative Species: Human Reciprocity and Its Evolution.* Princeton, NJ: Princeton University Press.

Bradbury, J. W., and S. L. Vehrencamp. 1976. "Social organization and foraging in emballonurid bats." *Behavioral Ecology and Sociobiology* 1:383–404.

Brandell, E. E., N. M. Fountain-Jones, M. L. J. Gilbertson, P. C. Cross, P. J. Hudson, D. W. Smith, D. R. Stahler, C. Packer, and M. E. Craft. 2020. "Group density, disease, and season shape territory size and overlap of social carnivores." *Journal of Animal Ecology* 90:87–101. https://doi.org/10.1111/1365-2656.13294.

Bray, J., I. C. Gilby. 2020. "Social relationships among adult male chimpanzees (*Pan troglodytes schweinfurthii*): Variation in the strength and quality of social bonds." *Behavioral Ecology and Sociobiology* 74:112. https://doi.org/10.1007/s00265-020-02892-3.

Broekhuis, F., G. Cozzi, M. Valeix, J. W. McNutt, and D. W. Macdonald. 2013. "Risk avoidance in sympatric large carnivores: Reactive or predictive?" *Journal of Animal Ecology* 82:1098–1105. https://doi.org/10.1111/1365-2656.12077.

Brown, E. W., N. Yuhki, C. Packer, and S. J. O'Brien. 1994. "A lion lentivirus related to feline immunodeficiency virus: Epidemiologic and phylogenetic aspects." *Journal of Virology* 68:5953–68.

Brown, J. L. 1970. "Cooperative breeding and altruistic behavior in the Mexican jay, *Aphelocoma ultramarina*." *Animal Behaviour* 18:366–78.

Brown, J. L., M. Bush, C. Packer, A. E. Pusey, S. L. Monfort, S. J. O'Brien, D. L Janssen, and D. E. Wildt. 1991. "Developmental changes in pituitary-gonadal function in free-ranging lions (*Panthera leo*) of the Serengeti and Ngorongoro Crater." *Journal of Reproduction and Fertility* 91:29–40.

Brown, J. S., J. W. Laundre, J., and M. Gurung. 1999. "The ecology of fear: Optimal foraging, game theory, and trophic interactions." *Journal of Mammalogy* 80:385–99.

Bruce, H. M. 1959. "Exteroceptive block to pregnancy in the mouse." *Nature* 184:105. https://doi.org/10.1038/184105a0.

Burger, K. M., E. M. Gese, and J. Berger. 2008. "Indirect effects and traditional trophic cascades: A test involving wolves, coyotes and pronghorns." *Ecology* 89:818–28.

Burkhart, J. C., S. Gupta, N. Borrego, S. R. Heilbronner, and C. Packer. 2022. "Oxytocin promotes social proximity and decreases vigilance in African lions." *iScience* 25:104049. https://doi.org/10.1016/j.isci.2022.104049.
Bygott, J. D., B. C. R. Bertram, and J. P. Hanby. 1979. "Male lions in large coalitions gain reproductive advantages." *Nature* 282:839–41.
Caputo, P. 2002. "Maneless in Tsavo." *National Geographic*. April 2002, 38–53.
Caraco, T., and L. L. Wolf. 1975. "Ecological determinants of group sizes in foraging lions." *American Naturalist* 109:343–52.
Carbone, C., T. Maddox, P. J. Funston, M. G. Mills, G. F. Grether, and B. Van Valkenburgh. 2009. "Parallels between playbacks and Pleistocene tar seeps suggest sociality in an extinct sabretooth cat, *Smilodon*." *Biology Letters* 5:81–85.
Carlson, A. 1986. "Group territoriality in the rattling cisticola, *Cisticola chiniana*." *Oikos* 47:181–89.
Carpenter, M. A., M. J. Appel, M. E. Roelke-Parker, L. Munson, H. Hofer, M. East, and S. J. O'Brien. 1998. "Genetic characterization of canine distemper virus in Serengeti carnivores." *Veterinary Immunology and Immunopathology* 65:259–66.
Chakrabarti, S., and Y. V. Jhala. 2017. "Selfish partners: Resource partitioning in male coalitions of Asiatic lions." *Behavioral Ecology* 28:1532–39. https://doi.org/10.1093/beheco/arx118.
———. 2019. "Battle of the sexes: A multi-male mating strategy helps lionesses win the gender war of fitness." *Behavioral Ecology* 30:1050–61. https://doi.org/10.1093/beheco/arz048.
Chakrabarti, S., J. K. Bump, Y. V. Jhala, and C. Packer. 2021. "Contrasting levels of social distancing between the sexes in lions." *iScience* 24:102406. https://doi.org/10.1016/j.isci.2021.102406.
Chapman, S., and G. Balme. 2010. "An estimate of leopard population density in a private reserve in KwaZulu-Natal, South Africa, using camera-traps and capture-recapture models." *South African Journal of Wildlife Research* 40:114–20.
Chardonnet, B. 2019. *Africa Is changing: Should Its Protected Areas Evolve? Reconfiguring the Protected Areas in Africa.* Gland, Switzerland: IUCN.
Chauvenet, A. L. M., S. M. Durant, R. Hilborn, and N. Pettorelli. 2011. "Unintended consequences of conservation actions: Managing disease in complex ecosystems." *PLoS One* 6:e28671. https://doi.org/10.1371/journal.pone.0028671.
Cheney, D. L. 1987. "Interactions and relationships between groups." In *Primate Societies*, edited by B. B. Smuts, D. L. Cheney, R. M. Seyfarth, R. W. Wrangham, and T. T. Struhsaker, 267–81. Chicago: University of Chicago Press.
Chetri, M., M. Odden, K. Sharma, Ø. Flagstad, and P. Wegge. 2019. "Estimating snow leopard density using fecal DNA in a large landscape in north-central Nepal." *Global Ecology and Conservation* 17:e00548. https://doi.org/10.1016/j.gecco.2019.e00548.
Clark, A. B. 1978. "Sex ratio and local resource competition in a prosimian primate." *Science* 201:163–65. https://doi.org/10.1126/science.201.4351.163.
Clark, C. W. 1987. "The lazy adaptable lions: A Markovian model of group foraging." *Animal Behaviour* 35:361–68.
Cleaveland, S., and K. Hampson. 2017. "Rabies elimination research: Juxtaposing optimism, pragmatism and realism." *Proceedings of the Royal Society* 284:20171880. https://doi.org/10.1098/rspb.2017.1880.
Cleaveland, S., T. Mlengeya, M. Kaare, D. Haydon, T. Lembo, K. Laurenson, and C. Packer. 2007. "The conservation relevance of epidemiological research into carnivore viral diseases in the Serengeti." *Conservation Biology* 21:613–22.
Cleaveland, S., T. Mlengeya, R. R. Kazwala, A. Michel, M. T. Kaare, S. L. Jones, E. Eblate, G. M. Shirima, and C. Packer. 2005. "Tuberculosis in Tanzanian wildlife." *Journal of Wildlife Diseases* 41:446–53.
Clifton, K. E. 1990. "The costs and benefits of territory sharing for the Caribbean coral reef fish, *Scarus iserti*." *Behavioral Ecology and Sociobiology* 26:139–47.
Clutton-Brock, T. H. 1991. *The Evolution of Parental Care*. Princeton, NJ: Princeton University.
Clutton-Brock, T. H., P. N. M. Brotherton, A. F. Russell, M. J. O'Riain, D. Gaynor, R. Kansky, A. Griffin, et al. 2001. "Cooperation, control, and concession in meerkat groups." *Science* 291:476–81. https://doi.org/10.1126/science.291.5503.478.
Clutton-Brock, T. H., S. J. Hodge, G. Spong, A. F. Russell, N. R. Jordan, N. C. Bennett, L. L. Sharpe, and M. B. Manser. 2006. "Intrasexual competition and sexual selection in cooperative mammals." *Nature* 444:1065–68. https://doi.org/10.1038/nature05386.
Clutton-Brock, T. H., and G. R. Iason. 1986. "Sex ratio variation in mammals." *Quarterly Review of Biology* 61:339–74.
Coals, P., A. Dickman, J. Hunt, A. Grau, R. Mandisodza-Chikerema, D. Ikanda, D. W. Macdonald, and A. Loveridge. 2020. "Commercially driven lion part removal: What is the evidence from mortality records?" *Global Ecology and Conservation* 24:e01327. https://doi.org/10.1016/j.gecco.2020.e01327.
Cooper, S. M. 1991. "Optimal hunting group size: The need for lions to defend their kills against loss to spotted hyaenas." *African Journal of Ecology* 29:130–36.
Craft, M., P. Hawthorne, C. Packer, and A. P. Dobson. 2008. "Dynamics of a multihost pathogen in a carnivore community." *Journal of Animal Ecology* 77:1257–64.
Craft, M., F. Vial, E. Miguel, S. Cleaveland, A. Ferdinands, and C. Packer. 2016. "Interactions between domestic and wild carnivores around the greater Serengeti ecosystem." *Animal Conservation* 20:193–204. https://doi.org/10.1111/acv.12305.
Craft, M., E. Volz, C. Packer, and L. A. Meyers. 2009. "Distinguishing epidemic waves from disease spillover in a wildlife population." *Proceedings of the Royal Society B* 276:1777–85.
———. 2010. "Disease transmission in territorial populations: The small-world network of Serengeti lions." *Royal Society Interface* 8:776–86. https://doi.org/10.1098/rsif.2010.0511.

Creel, S., N. Creel, D. E. Wildt, and S. L. Monfort. 1992. "Behavioral and endocrine mechanisms of reproductive suppression in Serengeti dwarf mongooses." *Animal Behaviour* 43:231–45. https://doi.org/10.1016/S0003-3472(05)80219-2.

Creel, S., J. M'soka, E. Droge, E. Rosenblatt, M. S. Becker, W. Matandiko, and T. Simpamba. 2016. "Assessing the sustainability of African lion trophy hunting, with recommendations for policy." *Ecological Applications* 26:2347–57. pmid:27755732.

Crofoot, M. C., and I. C. Gilby. 2012. "Cheating monkeys undermine group strength in enemy territory." *Proceedings of the National Academy of Sciences* 109:501–5. https://doi.org/10.1073/pnas.1115937109.

Crook, J. H. 1964. "The evolution of social organization and visual communication in the weaver birds (*Ploceinae*)." *Behaviour* Suppl. 10:1–178.

Cui, Z., H. Zhang, X. Chen, C. Zhang, W. Ma, C. Huang, W. Zhang, et al. 2018. "Pursuing sustainable productivity with millions of smallholder farmers." *Nature* 555:363–66. https://doi.org/10.1038/nature25785.

Dagg, A. I. 1998. "Infanticide by male lions hypothesis: A fallacy influencing research into human behavior. *American Anthropologist* 100 (4): 940–50. http://www.jstor.org/stable/681818

Darwin, C. 1871. *The Descent of Man, and Selection in Relation to Sex.* London: John Murray.

Davies, A. B., C. J. Tambling, D. G. Marneweck, N. Ranc, D. J. Druce, J. P. G. M Cromsigt, E. le Roux, and G. P. Asner. 2021. "Spatial heterogeneity facilitates carnivore coexistence." *Ecology* 102:e03319. https://doi.org/10.1002/ecy.3319.

Davies, N. B., M. de L. Brooke, and A. Kacelnik, A. 1996. "Recognition errors and probability of parasitism determine whether reed warblers should accept or reject mimetic cuckoo eggs." *Proceedings of the Royal Society B* 263:925–31.

Davies, N. B., and A. I. Houston. 1981. "Owners and satellites: The economics of territory defence in the pied wagtail, *Motacilla alba*." *Journal of Animal Ecology* 50:157–80.

Davies, N. G., and C. I. Jarvis, CMMID COVID-19 Working Group. 2021. "Increased mortality in community-tested cases of SARS-CoV-2 lineage B.1.1.7." *Nature* 593:270–74. https://doi.org/10.1038/s41586-021-03426-1.

Dias, P. C. 1996. "Sources and sinks in population biology." *Trends in Ecology and Evolution* 11:326–30.

Dickman, A. J. 2010. "Complexities of conflict: The importance of considering social factors for effectively resolving human–wildlife conflict." *Animal Conservation* 13:458–66. https://doi.org/10.1111/j.1469-1795.2010.00368.x.

Dickman, A., L. Hazzah, C. Carbone, and S. Durant. 2014. "Carnivores, culture and 'contagious conflict': Multiple factors influence perceived problems with carnivores in Tanzania's Ruaha landscape." *Biological Conservation* 178:19–27. https://doi.org/10.1016/j.biocon.2014.07.011.

Dickman, A. J., E. A. Macdonald, and D. W. Macdonald. 2011. "A review of financial instruments to pay for predator conservation and encourage human–carnivore coexistence." *Proceedings of the National Academy of Sciences* 108:13937–44. https://doi.org/10.1073/pnas.1012972108.

DiMinin, E., R. Slotow, C. Fink, H. Bauer, and C. Packer. 2021. "A pan-African spatial assessment of human conflicts with lions and elephants." *Nature Communications* 12:2978. https://doi.org/10.1038/s41467-021-23283-w.

Dixson, A., B. Dixson, and M. Anderson. 2005. "Sexual selection and the evolution of visually conspicuous sexually dimorphic traits in male monkeys, apes, and human beings." *Annual Review of Sex Research* 16:1–19. https://doi.org/10.1080/10532528.2005.10559826.

Dorman, S. E., C. L. Hatem, S. Tyagi, K. Aird, J. Javier Lopez-Molina, M. Louise, M. Pitt, et al. 2004. "Susceptibility to tuberculosis: Clues from studies with inbred and outbred New Zealand white rabbits." *Infection and Immunity* 72:1700–05.

Dublin, H. T., and J. O. Ogutu. 2015. "Population regulation of African buffalo in the Mara-Serengeti ecosystem." *Wildlife Research* 42:382–93. https://doi.org/10.1071/WR14205.

Dublin, H. T., A. R. E. Sinclair, S. Boutin, E. Anderson, M. Jago, and P. Arcese. 1990. "Does competition regulate ungulate populations? Further evidence from Serengeti, Tanzania." *Oecologia* 82:283–88.

Elliot, N. B., M. Valeix, D. W. Macdonald, and A. J. Loveridge. 2014. "Social relationships affect dispersal timing revealing a delayed infanticide in African lions." *Oikos* 123:1049–56. https://doi.org/10.1111/oik.01266.

Elwood, R. W., and H. F. Kennedy. 1994. "Selective allocation of parental and infanticidal responses in rodents: A review of mechanisms. In *Infanticide and Parental Care*, edited by S. Parmigiani and F. vom Saal, 397–425. Chur, Switzerland: Harwood.

Emlen, S. T. 1994. "Benefits, constraints and the evolution of the family." *Trends in Ecology and Evolution* 9:282–85.

Estes, R. D., J. L. Atwood, and A. B. Estes. 2006. "Downward trends in Ngorongoro Crater ungulate populations, 1986–2005: Conservation concerns and the need for ecological research." *Biological Conservation* 131:106–20.

Estes, R. D., and R. Small. 1981. "The large herbivore populations of Ngorongoro Crater." *African Journal of Ecology* 19:175–85. https://doi.org/10.1111/j.1365-2028.1981.tb00661.x.

Everatt, K. T., R. Kokes, and C. Lopez Pereira. 2019. "Evidence of a further emerging threat to lion conservation: Targeted poaching for body parts." *Biodiversity and Conservation* 28:4099–114. https://doi.org/10.1007/s10531-019-01866-w.

Fattebert, J., G. Balme, T. Dickerson, R. Slotow, and L. Hunter. 2015. "Density-dependent natal dispersal patterns in a leopard population recovering from over-harvest." *PLoS One* 10 (4): e0122355. https://doi.org/10.1371/journal.pone.0122355.

Ferreira, S. M., and P. J. Funston. 2010. "Estimating lion population variables: Prey and disease effects in Kruger National Park, South Africa." *Wildlife Research* 37:194–206.

———. 2020. "African lion and spotted hyaena changes in Kruger National Park, South Africa." *African Studies Quarterly* 19:38–50.

Fisher, R. A. 1930. *The Genetical Theory of Natural Selection*. Oxford, UK: Clarendon Press.

Fosbrooke, H. 1963. "The stomxys plague in Ngorongoro 1962." *African Journal of Ecology* 1:124–26.

———. 1972. *Ngorongoro: The Eighth Wonder*. London: Andre Deutsch.

Frank, L. G. 1986. "Social organization of the spotted hyaena *Crocuta crocuta*. II. Dominance and reproduction." *Animal Behaviour* 34:1510–27.

Fryxell, J., A. Mosser, A. R. E. Sinclair, and C. Packer. 2007. "Group formation stabilizes predator-prey dynamics." *Nature* 449:1041–44.

Fryxell, J. M., J. Greever, and A. R. E. Sinclair. 1988. "Why are migratory ungulates so abundant?" *American Naturalist* 131:781–98.

Fryxell, J.M., S. Mduma, J. Masoy, A.R.E. Sinclair, G.J.C. Hopcraft, and C. Packer. 2022. "Stabilizing effects of group formation by Serengeti herbivores on predator-prey dynamics." *Frontiers in Ecology and Evolution* 10:981842. https://doi.org/10.3389/fevo.2022.981842.

Funston, P. J., M. G. L. Mills, and H. C. Biggs. 2001. "Factors affecting the hunting success of male and female lions in the Kruger National Park." *Journal of Zoology* 253:419–31.

Gaston, A. J. 1978. "The evolution of group territorial behavior and cooperative breeding." *American Naturalist* 112:1091–1100.

Gazda, S. K., R. C. Connor, R. K. Edgar, and F. A. Cox. 2005. "Division of labour with role specialization in group hunting bottlenose dolphins (*Tursiops truncatus*) off Cedar Key, Florida." *Proceedings of the Royal Society B* 272:135–40. https://doi.org/10.1098/rspb.2004.2937.

Geffroy, B., and M. Douhard. 2019. "The adaptive sex in stressful environments. *Trends in Ecology and Evolution* 34: 628–40. https://doi.org/10.1016/j.tree.2019.02.012.

Gilbert, D., C. Packer, A. E. Pusey, J. C. Stephens, and S. J. O'Brien. 1991. "Analytical DNA fingerprinting in lions: Parentage, genetic diversity and kinship." *Journal of Heredity* 82:378–86.

Gilby, I. C., Z. P. Machanda, D. C. Mjungu, J. Rosen, M. N. Muller, A. E. Pusey, and R. W. Wrangham. 2015. "'Impact hunters' catalyse cooperative hunting in two wild chimpanzee communities." *Philosophical Transactions of the Royal Society B* 370:20150005. http://dx.doi.org/10.1098/rstb.2015.0005.

Giraldeau, L. A., and D. Gillis. 1988. "Do lions hunt in group sizes that maximize hunters' daily food returns?" *Animal Behaviour* 36:611–13.

Gittleman, J. L., and B. Van Valkenburgh. 1997. "Sexual dimorphism in the canines and skulls of carnivores: Effects of size, phylogeny and behavioural ecology." *Journal of Zoology* 242:97–117.

Goldman, M. J., J. R. De Pinho, and J. Perry. 2013. "Beyond ritual and economics: Maasai lion hunting and conservation politics." *Oryx* 47:490–500.

Goller, K. V., R. D. Fyumagwa, V. Nikolin, M. L. East, M. Kilewo, S. Speck, T. Müller, et al. 2010. "Fatal canine distemper infection in a pack of African wild dogs in the Serengeti ecosystem, Tanzania." *Veterinary Microbiology* 146:245–52. https://doi.org/10.1016/j.vetmic.2010.05.018.

Goodall, J. 1986. *The Chimpanzees of Gombe: Patterns of Behavior*. Boston: Belknap Press.

Goodrich, J. M., L. L. Kerley, E. N. Smirnov, D. G. Miquelle, L. McDonald, H. B. Quigley, M. G. Hornocker, and T. McDonald. 2008. "Survival rates and causes of mortality of Amur tigers on and near the Sikhote-Alin Biosphere Zapovednik. *Journal of Zoology* 276:323–29. https://doi.org/10.1111/j.1469-7998.2008.00458.x.

Grange, S., P. Duncan, J.-M. Gaillard, A. R. E. Sinclair, P. J. P. Gogan, C. Packer, H. Hofer, and M. East. 2004. "What limits the Serengeti zebra population?" *Oecologia* 140:523–32.

Green, D. S., L. Johnson-Ulrich, H. E. Couraudi, and K. E. Holekamp. 2018. "Anthropogenic disturbance induces opposing population trends in spotted hyenas and African lions." *Biodiversity Conservation* 27:871–89. https://doi.org/10.1007/s10531-017-1469-7.

Grey, J. N. C., V. T. Kent, and R. A. Hill. 2013. "Evidence of a high-density population of harvested leopards in a montane environment." *PloS One* 8:e82832. pmid:24349375.

Grinnell, J. 2002. "Modes of cooperation during territorial defense by African lions." *Human Nature* 13:85–104.

Grinnell, J., and K. McComb. 1996. "Maternal grouping as a defense against infanticide by males: Evidence from field playback experiments on African lions." *Behavioral Ecology* 7:55–59.

———. 2001. "Roaring and social communication in African lions: The limitations imposed by listeners." *Animal Behaviour* 62:93–98.

Grinnell, J., C. Packer, and A. E. Pusey. 1995. "Cooperation in male lions: Kinship, reciprocity or mutualism?" *Animal Behaviour* 49:95–105.

Grueter, C. C., K. Isler, and B. J. Dixson. 2015. "Are badges of status adaptive in large complex primate groups?" *Evolution and Human Behavior* 36:398–406.

Guiserix, M., N. Bahi-Jaber, D. Fouchet, F. Sauvage, and D. Pontier. 2008. "The canine distemper epidemic in Serengeti: Are lions victims of a new highly virulent canine distemper virus strain, or is pathogen circulation stochasticity to blame?" *Journal of the Royal Society, Interface* 4:1127–34. https://doi.org/10.1098/rsif.2007.0235.

de Haan, C., ed. 2016. *Prospects for Livestock-Based Livelihoods in Africa's Drylands. World Bank Studies.* Washington, DC: World Bank. http://hdl.handle.net/10986/24815.

Hager, R., and C. B. Jones. 2009. *Reproductive Skew in Vertebrate Species.* Cambridge, UK: Cambridge University Press.

Halton, T. L., and F. B. Hu. 2004. "The effects of high protein diets on thermogenesis, satiety and weight loss: A critical review." *Journal of the American College of Nutrition* 23:373–85. https://doi.org/10.1080/07315724.2004.10719381.

Hamel, S., J. M. Gaillard, N. G. Yoccoz, A. Loison, C. Bonenfant, and S. Descamps. 2010. "Fitness costs of reproduction depend on life speed: empirical evidence from mammalian populations." *Ecology Letters* 13:915–35. https://doi.org/10.1111/j.1461-0248.2010.01478.x.

Hammerstein, P. 1981. "The role of asymmetries in animal contests." *Animal Behaviour* 29:193–205. https://doi.org/10.1016/S0003-3472(81)80166-2.

Hammerstein, P., and S. E. Riechert. 1988. "Payoffs and strategies in spider territorial contests: ESS-analyses of two ecotypes." *Evolution and Ecology* 2:115–38.

Hampson, K., J. Dushoff, S. Cleaveland, D. Haydon, M. Kaare, C. Packer, and A. Dobson. 2009. "Transmission dynamics and prospects for the elimination of canine rabies." *PLoS-Biology* 7:e1000053. https://doi.org/10.1371/journal.pbio.1000053.

Hanby, J. P., and J. D. Bygott. 1979. "Population changes in lions and other predators." In *Serengeti: Dynamics of an Ecosystem*, edited by A. R. E. Sinclair and M. Norton-Griffiths, 249–62. Chicago: University of Chicago Press.

———. 1987. "Emigration of subadult lions." *Animal Behaviour* 35:161–69.

Hanby, J. P., J. D. Bygott, and C. Packer. 1995. "Ecology, demography and behavior of lions in two contrasting habitats: Ngorongoro Crater and the Serengeti Plains." In *Serengeti II: Research, Management and Conservation of an Ecosystem*, edited by P. Arcese and A. R. E. Sinclair, 315–31. Chicago: University of Chicago Press.

Harihar, A., P. Chanchani, J. Borah, R. J. Crouthers, Y. Darman, T. N. E. Gray, S. Mohamad, et al. 2018. "Recovery planning towards doubling wild tiger *Panthera tigris* numbers: Detailing 18 recovery sites from across the range." *PLoS One* 13:e0207114. https://doi.org/10.1371/journal.pone.0207114.

Havmøller, R. W., S. Tenan, N. Scharff, and F. Rovero. 2019. "Reserve size and anthropogenic disturbance affect the density of an African leopard (*Panthera pardus*) meta-population." *PLoS One* 14:e0209541. https://doi.org/10.1371/journal.pone.0209541.

Hayward, M. W. 2006. "Prey preferences of the spotted hyaena (*Crocuta crocuta*) and degree of dietary overlap with the lion (*Panthera leo*)." *Journal of Zoology* 270:606–14. https://doi.org/10.1111/j.1469-7998.2006.00183.x.

Hayward, M. W., and G. I. H. Kerley. 2005. "Prey preferences of the lion (*Panthera leo*)." *Journal of Zoology* 267:309–22. https://doi.org/10.1017/S0952836905007508.

———. 2009. "Fencing for conservation: Restriction of evolutionary potential or a riposte to threatening processes?" *Biological Conservation* 142:1–13. https://doi.org/10.1016/j.biocon.2008.09.022.

Hazzah, L., S. Dolrenry, L. Naughton, C. T. Edwards, O. Mwebi, F. Kearney, and L. Frank. 2014. "Efficacy of two lion conservation programs in Maasailand, Kenya." *Conservation Biology* 28:851–60. pmid:24527992.

Heald, F. 1989. "Injuries and diseases in *Smilodon californicus* Bovard, 1904 (*Mammalia, Felidae*) from Rancho La Brea, California." *Journal of Vertebrate Paleontology* 9: 24A.

Heinsohn, R., and C. Packer. 1995. "Complex cooperative strategies in group-territorial African lions." *Science* 269:1260–62.

Heinsohn, R., C. Packer, and A. E. Pusey. 1996. "Development of cooperative territoriality in juvenile lions." *Proceedings of the Royal Society B* 263:475–79.

Hilborn, R., P. Arcese, M. Borner, J. Hando, G. Hopcraft, M. Loibooki, S. Mduma, and A. R. E. Sinclair. 2006. "Effective enforcement in a conservation area." *Science* 314:1266.

Hilborn, A., N. Pettorelli, T. Caro, et al. 2018. "Cheetahs modify their prey handling behavior depending on risks from top predators." *Behavioral Ecology and Sociobiology* 72:74. https://doi.org/10.1007/s00265-018-2481-y.

Hofer, H., and M. L. East. 1993a. "The commuting system of Serengeti spotted hyenas: How a predator copes with migratory prey. II. Intrusion pressure and commuters' space use." *Animal Behaviour* 46:559–74.

———. 1993b. "The commuting system of Serengeti spotted hyaenas: How a predator copes with migratory prey. III. Attendance and maternal care." *Animal Behaviour* 46:575–89.

Hohmann, G., K. Potts, A. N'Guessan, A. Fowler, R. Mundry, J. U. Ganzhorn, and S. Ortmann. 2010. "Plant foods consumed by *Pan*: Exploring the variation of nutritional ecology across Africa." *American Journal of Physical Anthropology* 141:476–85.

Holekamp, K. E., L. Smale, and M. Szykman. 1996. "Rank and reproduction in the female spotted hyaena." *Journal of Reproduction and Fertility* 108:229–37.

Holekamp, K. E., M. Szykman, E. E. Boydston, and L. Smale. 1999. "Association of seasonal reproductive patterns with changing food availability in an equatorial carnivore, the spotted hyaena (*Crocuta crocuta*). *Journal of Reproduction and Fertility* 116:87–93.

Höner, O. P., B. Wachter, M. L. East, and H. Hofer. 2002. "The response of spotted hyaenas to long-term changes in prey populations: Functional response and interspecific kleptoparasitism." *Journal of Animal Ecology* 71:236–46.

Höner, O. P., B. Wachter, M. L. East, V. A. Runyoro, and H. Hofer. 2005. "The effect of prey abundance and foraging tactics on the population dynamics of a social, territorial carnivore, the spotted hyena." *Oikos* 108:544–54. https://doi.org/10.1111/j.0030-1299.2005.13533.x.

Höner, O. P., B. Wachter, K. V. Goller, H. Hofer, V. Runyoro, D. Thierer, R. D. Fyumagwa, T. Müller, and M. L. East. 2012. "The impact of a pathogenic bacterium on a social carnivore population." *Journal of Animal Ecology* 81:36–46. https://doi.org/10.1111/j.1365-2656.2011.01873.x.

Hopcraft, J. G. C., T. M. Anderson, S. Pérez-Vila, E. Mayemba, and H. Olff. 2012. "Body size and the division of niche space: Food and predation differentially shape the distribution of Serengeti grazers." *Journal of Animal Ecology* 81:201–13. https://doi.org/10.1111/j.1365-2656.2011.01885.x.

Hopcraft, J. G. C., J. M. Morales, H. L. Beyer, M. Borner, E. Mwangomo, A. R. E. Sinclair, H. Olff, and D. T. Haydon. 2014. "Competition, predation, and migration: Individual choice patterns of Serengeti migrants captured by hierarchical models." *Ecological Monographs* 84:355–72.

Hopcraft, J. G. C., A. R. E. Sinclair, and C. Packer. 2005. "Planning for success: Serengeti lions seek prey accessibility rather than abundance." *Journal of Animal Ecology* 74:559–66.

Hrdy, S. B. 1974. "Male-male competition and infanticide among the langurs (*Presbytis entellus*) of Abu, Rajasthan." *Folia Primatologica* 22:19–58.

———. 1979. "Infanticide among animals: A review, classification, and examination of the implications for reproductive strategies of females." *Ethology and Sociobiology* 1:13–40.

Hubel, T., J. Myatt, N. Jordan, O. P. Dewhirst, J. W. McNutt, and A. M. Wilson. 2016. "Additive opportunistic capture explains group hunting benefits in African wild dogs." *Nature Communications* 7:11033. https://doi.org/10.1038/ncomms11033.

Huerta, D., dir. 2015. *Brothers in Blood: The Lions of Sabi Sand*. Aquavision Productions.

Hunter, L. T. B., K. Pretorius, L. C. Carlisle, M. Rickelton, C. Walker, R. Slotow, and J. D. Skinner. 2007. "Restoring lions *Panthera leo* to northern KwaZulu-Natal, South Africa: Short-term biological and technical success but equivocal long-term conservation." *Oryx* 41:196–204.

Hurley, W. L., and P. K. Theil. 2011. "Perspectives on immunoglobulins in colostrum and milk." *Nutrients* 3:442–74. https://doi.org/10.3390/nu3040442.

Ikanda, D., and C. Packer. 2008. "Ritual vs. retaliatory killing of African lions in the Ngorongoro Conservation Area, Tanzania." *Endangered Species Research* 6:67–74.

Israel, P. 2009. "The war of lions: Witch-hunts, occult idioms and post-socialism in northern Mozambique." *Journal of Southern African Studies* 35:155–74. http://www.jstor.org/stable/40283220.

Jackson, C. R., E. H. Masenga, E. E. Mjingo, A. B. Davies, F. Fossøy, R. D. Fyumagwa, E. Røskaft, and R. F. May. 2018. "No evidence of handling-induced mortality in Serengeti's African wild dog population." *Ecology and Evolution* 9:1110–18. https://doi.org/10.1002/ece3.4798.

Jędrzejewski W., H. S. Robinson, M. Abarca, K. A. Zeller, G. Velasquez, E. A. D. Paemelaere, J. F. Goldberg, et al. 2018. "Estimating large carnivore populations at global scale based on spatial predictions of density and distribution: Application to the jaguar (*Panthera onca*)." *PLoS One* 13:e0194719. https://doi.org/10.1371/journal.pone.0194719.

Jevons, W. S. 1865. *The Coal Question: An inquiry concerning the progress of the nation, and the probable exhaustion of our coal-mines*. London: Macmillan.

Jhala, Y. V., K. Banerjee, S. Chakrabarti, P. Basu, K. Singh, C. Dave, and K. Gogoi. 2019. "Asiatic lion: Ecology, economics, and politics of conservation." *Frontiers in Ecology and Evolution* 7:312. https://doi.org/10.3389/fevo.2019.00312.

Johnson, D. D. P., and D. W. Macdonald. 2003. "Sentenced without trial: Reviling and revamping the resource dispersion hypothesis." *Oikos* 101:433–40.

Jones, B. T. B. 2010. "The evolution of Namibia's communal conservancies." In *Community Rights, Conservation and Contested Land*, edited by F. Nelson, 106–20. London: Earthscan.

Joppa, L. N, S. R. Loarie, and S. L. Pimm. 2009. "On population growth near protected areas." *PLoS One* 4:e4279. https://doi.org/10.1371/journal.pone.0004279.

Keet, D. F., N. P. J. Kriek, M. Penrith, and A. Michel.1997. "Tuberculosis in lions and cheetahs." *Proceedings of a Symposium on Lions and Leopards as Game Ranch Animals*, 151–56. Onderstepoort, South Africa.

Keet, D. F., A. L. Michel, R. G. Bengis, P. Becker, D. S. van Dyk, M. van Vuuren, V. P. M. G. Rutten, and B. L. Penzhorn. 2010. "Intradermal tuberculin testing of wild African lions (*Panthera leo*) naturally exposed to infection with *Mycobacterium bovis*." *Veterinary Microbiology* 144:384–91.

Kissui, B. M. 2008. "Livestock predation by lions, leopards, spotted hyenas, and their vulnerability to retaliatory killing in the Maasai steppe, Tanzania." *Animal Conservation* 11:422–32.

Kissui, B. M., and C. Packer (2004). "Top-down regulation of a top predator: Lions in the Ngorongoro Crater." *Proceedings of the Royal Society B* 271:1867–74.

Kleiman, D. G., and J. F. Eisenberg. 1973. "Comparisons of canid and felid social systems from an evolutionary perspective." *Animal Behaviour* 21:637–59.

Klingel, H. 1969. "The social organization and population ecology of the plains zebra, *Equus quagga*." *Zoologica Africana* 4:249–63.

Koenig, W. D., F. A. Pitelka, W. J. Carmen, R. L. Mumme, and M. T. Stanback. 1992. "The evolution of delayed dispersal in cooperative breeders." *Quarterly Review of Biology* 67:111–50.

Kokko, H., and R. A. Johnstone. 1999. "Social queuing in animal societies: A dynamic model of reproductive skew." *Proceedings of the Royal Society B* 266:571–78. https://doi.org/10.1098/rspb.1999.0674.

Kortlandt, A. 1972. *New Perspectives on Ape and Human Evolution*. Amsterdam: Stichting voor Psychobiologie.

Kosmala, M., P. Miller, S. Ferreira, P. Funston, D. Keet, and C. Packer. 2016. "Estimating wildlife disease dynamics in complex systems using approximate Bayesian computation models." *Ecological Applications* 26:295–308. http://dx.doi.org/10.1890/14-1808.1.

Kotze, R., M. Keith, C. W. Winterbach, H. E. K. Winterbach, and J. P. Marshal. 2018. "The influence of social and environmental factors on organization of African lion (*Panthera leo*) prides in the Okavango Delta." *Journal of Mammalogy* 99:845–58. https://doi.org/10.1093/jmammal/gyy076.

Krebs, C. J., R. Boonstra, and S. Boutin. 2018. "Using experimentation to understand the 10-year snowshoe hare cycle in the boreal forest of North America." *Journal of Animal Ecology* 87:87–100. https://doi.org/10.1111/1365-2656.12720.

Kress, M., and I. Talmor. 1999. "A new look at the 3:1 rule of combat through Markov Stochastic Lanchester models." *Journal of the Operational Research Society* 50:733–44. https://doi.org/10.1057/palgrave.jors.2600758.

Kruuk, H. 1972. *The Spotted Hyena*. Chicago: University of Chicago Press.

Kruuk, H., and D. Macdonald. 1985. "Group territories of carnivores: Empires and enclaves." In *Behavioural Ecology*, edited by R. M. Sibly and R. H. Smith, 521–26. Oxford: Blackwell Scientific Publications.

Kulldorff, M. 1997. "A spatial scan statistic." *Communications in Statistics: Theory and Methods* 26:1481–96. https://doi.org/10.1080/03610929708831995.

Kumar, U., N. Awasthi, Q. Qureshi, and Y. Jhala. 2019. "Do conservation strategies that increase tiger populations have consequences for other wild carnivores like leopards?" *Scientific Reports* 9:14673. https://doi.org/10.1038/s41598-019-51213-w.

Kushnir, H., E. Olson, T. Juntunen, D. Ikanda, and C. Packer. 2014. "Using landscape characteristics to predict risk of lion attacks in southeastern Tanzania." *African Journal of Ecology* 52:524–32. https://doi.org/10.1111/aje.12157.

Kushnir, H., and C. Packer. 2019. "Perceptions of risk from man-eating lions in southeastern Tanzania." *Frontiers in Ecology and Evolution: Conservation and Restoration Ecology* 7:47. https://doi.org/10.3389/fevo.2019.00047.

Langergraber, K., J. C. Mitani, and L. Vigilant. 2007. "The limited impact of kinship on cooperation in wild chimpanzees." *Proceedings of the National Academy of Sciences* 104:7786–90. https://doi.org/10.1073/pnas.0611449104.

Larom, D., M. Garstang, K. Payne, R. Raspet, and M. Lindeque. 1997. "The influence of surface atmospheric conditions on the range and area reached by animal vocalizations." *Journal of Experimental Biology* 200:421e431. https://jeb.biologists.org/content/200/3/421.

Laundré, J. W., L. Hernández, and K. B. Altendorf. 2001. "Wolves, elk, and bison: Reestablishing the 'landscape of fear' in Yellowstone National Park, U.S.A." *Canadian Journal of Zoology* 79:1401–09.

Laurenson, M. K. 1994. "High juvenile mortality in cheetahs (*Acinonyx jubatus*) and its consequences for maternal care." *Journal of Zoology* 234:387–408.

Lawson, D. W., M. Borgerhoff Mulder, M. E. Ghiselli, E. Ngadaya, B. Ngowi, S. G. M. Mfinaga, K. Hartwig, and S. James. 2014. "Ethnicity and child health in northern Tanzania: Maasai pastoralists are disadvantaged compared to neighbouring ethnic groups." *PLoS One* 9:e110447. https://doi.org/10.1371/journal.pone.0110447.

Ledger, H. P. 1968. "Body composition as a basis for a comparative study of some East African mammals." *Symposium of the. Zoological Society of London* 21:289–310.

Lehmann, K. D. S., T. M. Montgomery, S. M. Maclachlan, J. M. Parker, O. S. Spagnuolo, K. J. Vandewetering, P. S. Bills, and K. E. Holekamp. 2017. "Lions, hyenas and mobs (Oh my!)." *Current Zoology* 63:313–22. https://doi.org/10.1093/cz/zow073.

Lembo, T., K. Hampson, D. Haydon, M. Craft, A. Dobson, et al. 2008. "Exploring reservoirs dynamics: A case study of rabies in the Serengeti ecosystem." *Journal of Applied Ecology* 45:1246–57.

Lembo, T., K. Hampson, M. T. Kaare, D. Knobel, R. R. Kazwala, D. T Haydon, and S. Cleaveland. 2010. "The feasibility of canine rabies elimination in Africa: Dispelling doubts with data." *PLoS-Neglected Tropical Diseases* 4:e626. https://doi.org/10.1371/journal.pntd.0000626.

Levick, S. R., and K. H. Rogers. 2008. "Structural biodiversity monitoring in savanna ecosystems: Integrating LiDAR and high resolution imagery through object-based image analysis." In *Object-Based Image Analysis: Spatial Concepts for Knowledge-Driven Remote Sensing Applications*, edited by T. Blaschke, S. Lang, and G. J. Hay. 477–92. Berlin: Springer.

Li, C., E. Hoffland, T. W. Kuyper, Y. Yu, C. Zhang, H. Li, F. Zhang, and W. van der Werf. 2020. "Syndromes of production in intercropping impact yield gains." *Nature Plants* 6:653–60. https://doi.org/10.1038/s41477-020-0680-9.

Lindsey, P., J. Allan, P. Brehony, A. Dickman, A. Robson, C. Begg, H. Bhammar, et al. 2020. "Conserving Africa's wildlife and wildlands through the COVID-19 crisis and beyond." *Nature Ecology and Evolution* 4:1300–10. https://doi.org/10.1038/s41559-020-1275-6.

Lindsey, P., G. Balme, V. Booth, and N. Midlane. 2012. "The significance of African lions for the financial viability of trophy hunting and the maintenance of wild land." *PLoS One* 7:e29332. https://doi.org/10.1371/journal.pone.0029332.

Lindsey, P. A., G. A. Balme, P. Funston, P. Henschel, L. Hunter, H. Madzikanda, et al. 2013. "The trophy hunting of African lions: Scale, current management practices and factors undermining sustainability." *PLoS One* 8:e73808. https://doi.org/10.1371/journal.pone.0073808.

Lindsey, P. A., J. R. B. Miller, L. S. Petracca, L. Coad, A. J. Dickman, K. H. Fitzgerald, M. V. Flyman, et al. 2018. "More than $1 billion needed annually to secure Africa's protected areas with lions." *Proceedings of the National Academy of Sciences* 115:E10788–96. https://doi.org/10.1073/pnas.1805048115.

Lindsey, P. A., L. S. Petracca, P. J. Funston, H. Bauer, A. Dickman, K. Everatt, M. Flyman, et al. 2017. "The performance of African protected areas for lions and their prey." *Biological Conservation* 209:137–49. https://doi.org/10.1016/j.biocon.2017.01.011.

Lindstrom, E. 1986. "Territory inheritance and the evolution of group-living in carnivores." *Animal Behaviour* 34:1825–35.

Loibooki, M., H. Hofer, K. L. I. Campbell, and M. L. East. 2002. "Bushmeat hunting by communities adjacent to the Serengeti National Park, Tanzania: The importance of livestock ownership and alternative sources of protein and income." *Environmental Conservation* 29:391–98.

Loveridge, A. J., and S. Canney. 2009. *African Lion Distribution Modeling Project, Final Report.* Horsham, UK: Born Free Foundation.

Loveridge, A. J., G. Hemson, Z. Davidson, and D. W. Macdonald. 2010. "African lions on the edge: Reserve boundaries as 'attractive sinks.'" In *Biology and Conservation of Wild Felids*, edited by D. W. Macdonald and A. J. Loveridge, 283–304. Oxford: Oxford University Press.

Loveridge A. J., M. Valeix, G. Chapron, Z. Davidson, G. Mtare, and D. W. Macdonald. 2016. "Conservation of large predator populations: Demographic and spatial responses of African lions to the intensity of trophy hunting." *Biological Conservation* 204B:247–54. https://doi.org/10.1016/j.biocon.2016.10.024.

Lyke, M. A., J. Dubach, and M. B. Briggs. 2013. "A molecular analysis of African lion (*Panthera leo*) mating structure and extra-group paternity in Etosha National Park." *Molecular Ecology* 22:2787–96. https://doi:10.1111/mec.12279.

Macdonald, D. W. 1983. "The ecology of carnivore social behaviour." *Nature* 301:379–84.

MacDonald, D. W., and P. D. Moehlman.1982. "Cooperation, altruism, and restraint in the reproduction of carnivores." *Perspectives in Ethology* 5:433–67

MacNulty, D. R., D. W. Smith, J. A. Vucetich, L. D. Mech, D. R. Stahler, and C. Packer. 2009. "Predatory senescence in ageing wolves." *Ecology Letters* 12:1347–56. https://doi:10.1111/j.1461-0248.2009.01385.x.

Maher, M. C., I. Bartha, S. Weaver, J. di Iulio, E. Ferri, L. Soriaga, F. A. Lempp, et al. 2022. "Predicting the mutational drivers of future SARS-CoV-2 variants of concern." *Science Translational Medicine* 14:eabk3445.

Malo, A. F., F. Martinez-Pastor, F. Garcia-Gonzalez, J. Garde, J. D. Ballou, and R. C. Lacy. 2017. "A father effect explains sex-ratio bias." *Proceedings of the Royal Society. B* 284:20171159. http://dx.doi.org/10.1098/rspb.2017.1159.

Massaro, A. P., E. E. Wroblewski, D. C. Mjungu, J. Feldblum, E Boehm, N. Desai, S. Kamenya, et al. "The monopolizability of mating opportunities promotes within-community killing in chimpanzees," *American Journal of Primatology*. In review.

May, R. M. 1972. "Limit cycles in predator-prey communities." *Science* 177:900–02.

Maynard Smith, J. 1982. *Evolution and the Theory of Games.* Cambridge, UK: Cambridge University Press.

Maynard Smith, J., and D. Harper. 2003. *Animal Signals.* Oxford: Oxford University Press.

Mbaiwa, J. O., and O. I. Mbaiwa. 2006. "The effects of veterinary fences on wildlife populations in Okavango Delta, Botswana." *International Journal of Wilderness* 12:17–23.

Mbizah, M. M., M. Valeix, D. W. Macdonald, and A. J. Loveridge. 2019. "Applying the resource dispersion hypothesis to a fission-fusion society: A case study of the African lion (*Panthera leo*)." *Ecology and Evolution* 9:9111–19. https://doi.org/10.1002/ece3.5456.

McCarthy, K. P., T. D. Fuller, M. Ming, T. M. McCarthy, L. Waits, and K. Jumabaev. 2010. "Assessing estimators of snow leopard abundance." *Journal of Wildlife Management* 72:1826–33. https://doi.org/10.2193/2008-040.

McComb, K. E., C. Packer, and A. E. Pusey. 1994. "Roaring and numerical assessment in contests between groups of female lions *Panthera leo*." *Animal Behaviour* 47:379–87.

McComb, K. E., A. E. Pusey, C. Packer, and J. Grinnell. 1993. "Female lions can identify potentially infanticidal males from their roars." *Proceedings of the Royal Society B* 252:59–64.

McEvoy, O. 2019. "The management of lions (*Panthera leo*) in small, fenced wildlife reserves." PhD dissertation, Rhodes University.

McEvoy, O. K., S. M. Miller, T. Bodasing, W. Beets, N. Borrego, A. Burger, S. Ferreira, B. Courtenay, C. Hanekom, M. Hofmeyr, C. Packer, et al. 2019. "The use of contraceptive techniques in managed wild African lion (*Panthera leo*) populations to mimic open system cub recruitment." *Wildlife Research* 46:398-408. https://doi.org/10.1071/WR18079.

Mduma, S. A. R., A. R. E. Sinclair, and R. Hilborn. 1999. "Food regulates the Serengeti wildebeest: A 40-year record." *Journal of Animal Ecology* 68:1101–22.

Mech, L. David. 1999. "Alpha status, dominance, and division of labor in wolf packs." *Canadian Journal of Zoology* 77:1196–1203.

Metzger, K. L., A. R. E. Sinclair, R. Hilborn, J. G. C. Hopcraft, and S. A. R. Mduma. 2010. "Evaluating the protection of wildlife in parks: The case of African buffalo in Serengeti." *Biodiversity and Conservation* 19:3431–44.

Miller, J. R. B., G. Balme, P. A. Lindsey, A. J. Loveridge, M. S. Becker, C. Begg, H. Brink, et al. 2016. "Aging traits and sustainable trophy hunting of African lions." *Biological Conservation* 201:160-68. http://dx.doi.org/10.1016/j.biocon.2016.07.003.

Miller, J. R. B., R. T. Pitman, G. K. H. Mann, A. K. Fuller, and G. A. Balme. 2018. "Lions and leopards coexist without spatial, temporal or demographic effects of interspecific competition." *Journal of Animal Ecology* 87:1709–26. https://doi.org/10.1111/1365-2656.12883.

Miller, S. M., C. Bissett, A. Burger, B. Courtenay, T. Dickerson, D. J. Druce, S. Ferreira, et al. 2013. "Management of reintroduced lions in small, fenced reserves in South Africa: An assessment and guidelines." *South African Journal of Wildlife Research* 43:138–54.

Miller, S. M., D. J. Druce, D. L. Dalton, C. K. Harper, A. Kotze, C. Packer, R. Slotow, and P. Bloomer. 2019. "Genetic rescue of an isolated African lion population." *Conservation Genetics* 21:41–53. https://doi.org/10.1007/s10592-019-01231-y.

Miller, S. M., and P. J. Funston. 2014. "Rapid growth rates of lion (*Panthera leo*) populations in small, fenced reserves in South Africa: A management dilemma." *South African Journal of Wildlife Research* 44:43–55.

Mkonyi, F. J., A. B. Estes, M. J. Msuha, L. L. Lichtenfeld, and S. M. Durant. 2017a. "Socio-economic correlates and management implications of livestock depredation by large carnivores in the Tarangire ecosystem, northern Tanzania." *International Journal of Biodiversity Science, Ecosystem Services & Management* 13:248–63. https://doi.org/10.1080/21513732.2017.1339724.

———. 2017b. "Fortified bomas and vigilant herding are perceived to reduce livestock depredation by large carnivores in the Tarangire-Simanjiro Ecosystem, Tanzania." *Human Ecology* 45:513–23. https://doi.org/10.1007/s10745-017-9923-4.

Moehlman, P. D., J. O. Ogutu, H.-P. Piepho, V. Runyoro, M. Coughenour, and R. Boone. 2020. "Long-term historical and projected herbivore population dynamics in Ngorongoro Crater, Tanzania." *PLoS One* 15:e0212530. https://doi.org/10.1371/journal.pone.0212530.

Moll, R. J., K. M. Redilla, T. Mudumba, A. B. Muneza, S. M. Gray, L. Abade, M. W. Hayward, et al. 2017. "The many faces of fear: A synthesis of the methodological variation in characterizing predation risk." *Journal of Animal Ecology* 86:749–65. https://doi.org/10.1111/1365-2656.12680.

Moss, C. J., H. Croze, and P. C. Lee, eds. 2010. *The Amboseli Elephants: A Long-Term Perspective on a Long-Lived Mammal*. Chicago: University of Chicago Press.

Mosser, A. 2008. "Group territoriality of the African lion: Behavioral adaptation in a heterogeneous landscape." PhD dissertation, University of Minnesota.

Mosser, A., J. Fryxell, L. Eberly, and C. Packer. 2009. "Serengeti real estate: Density versus fitness-based indicators of lion habitat quality." *Ecology Letters* 12:1050–60.

Mosser, A., M. Kosmala, and C. Packer. 2015. "The evolution of group territoriality." *Behavioral Ecology* 26:1051–59. https://doi.org/10.1093/beheco/arv046.

Mosser, A., and C. Packer. 2009. "Group territoriality and the benefits of sociality in the African lion, *Panthera leo*." *Animal Behaviour* 78:359–70.

Munson, L., K. A. Terio, R. Kock, T. Mlengeya, M. E. Roelke, B. Dubovi, B. Summers, A. R. E. Sinclair, and C. Packer. 2008. "Climate extremes and co-infections determine mortality during epidemics in African lions." *PLoS One* 3:e2545.

Mweetwa, T., D. Christianson, M. Becker, S. Creel, E. Rosenblatt, J. Merkle, E. Droge, H. Mwape, J. Masonde, and T. Simpamba. 2018. "Quantifying lion (*Panthera leo*) demographic response following a three-year moratorium on trophy hunting." *PLoS One* 13:e0197030. https://doi.org/10.1371/journal.pone.0197030.

Nagel, D., S. Hilsberg, A. Benesch, and J. Scholz. 2003. "Functional morphology and fur patterns in recent and fossil *Panthera* species." *Scripta Geologica* 126:227–40.

Naude, V. N., G. A. Balme, J. O'Riain, L. T. B. Hunter, J. Fattebert, T. Dickerson, and J. M. Bishop. 2020. "Unsustainable anthropogenic mortality disrupts natal dispersal and promotes inbreeding in leopards." *Ecology and Evolution* 10:3605–19. https://doi.org/10.1002/ece3.6089.

Navara, K. J. 2013. "Hormone-mediated adjustment of sex ratio in vertebrates." *Integrative and Comparative Biology* 53:877–87. https://doi.org/10.1093/icb/ict081.

Niang, I., O. C. Ruppel, M. A. Abdrabo, A. Essel, C. Lennard, J. Padgham, and P. Urquhart. 2014. Chapter 22, "Africa," in *Climate Change 2014: Impacts, Adaptation, and Vulnerability. Part B: Regional Aspects. Contribution of Working Group II to the Fifth Assessment Report of the Intergovernmental Panel on Climate Change*, edited by V. R. Barros, C. B. Field, D. J. Dokken, M. D. Mastrandrea, K. J. Mach, T. E. Bilir, M. Chatterjee, et al., 1199–1265. Cambridge, UK and NY: Cambridge University Press.

Nikolin, V. M., X. A. Olarte-Castillo, N. Osterrieder, H. Hofer, E. Dubovi, C. J. Mazzoni, E. Brunner, et al. 2017. "Canine distemper virus in the Serengeti ecosystem: Molecular adaptation to different carnivore species." *Molecular Ecology* 26: 2111–30. https://doi.org/10.1111/mec.13902.

Noë, R. 2006. "Cooperation experiments: Coordination through communication versus acting apart together." *Animal Behaviour* 71:1–18. https://doi.org/10.1016/j.anbehav.2005.03.037.

O'Brien, S. J., J. S. Martenson, C. Packer, L. Herbst, V. de Voss, P. Jocelyn, J. Ott-Jocelyn, D. E. Wildt, and M. Bush. Biochemical genetic variation in geographically isolated populations of African and Asiatic lions. *National Geographic Research* 3:114–24.

Owen-Smith, N., D. R. Mason, and J. O. Ogutu. 2005. "Correlates of survival rates for 10 African ungulate populations: Density, rainfall and predation." *Journal of Animal Ecology* 74:774–88.

Owen-Smith, N., and M. G. L. Mills. 2008. "Shifting prey selection generates contrasting herbivore dynamics within a large-mammal predator-prey web." *Ecology* 89:1120–33.

Owens, M., and D. Owens. 1984. *Cry of the Kalahari*. Boston: Houghton-Mifflin.

Packer, C. 1986. "The ecology of sociality in felids." In *Ecological Aspects of Social Evolution*, edited by D. I. Rubenstein and R. W. Wrangham, 429–51. Princeton, NJ: Princeton University Press.

———. 1994. *Into Africa*. Chicago: University of Chicago Press.

———. 2001. "Infanticide is no fantasy." *American Anthropologist* 102:829–31.

———. 2015. *Lions in the Balance: Man-Eaters, Manes, and Men with Guns*. Chicago: University of Chicago Press.

Packer, C., S. Altizer, M. Appel, E. Brown, J. Martenson, S. J. O'Brien, M. Roelke-Parker, et al. 1999. "Viruses of the Serengeti: Patterns of infection and mortality in African lions." *Journal of Animal Ecology* 68:1161–78.

Packer, C., H. Brink, B. M. Kissui, H. Maliti, H. Kushnir, and T. Caro. 2011. "The effects of trophy hunting on lion and leopard populations in Tanzania." *Conservation Biology* 25:142–53.

Packer, C., and J. Clottes. 2000. "When lions ruled France." *Natural History* 109 (9):52–57.

Packer, C., D. Gilbert, A. E. Pusey, and S. J. O'Brien. 1991. "A molecular genetic analysis of kinship and cooperation in African lions." *Nature* 351:562–65.

Packer, C., R. Hilborn, A. Mosser, B. Kissui, J. Wilmshurst, M. Borner, G. Hopcraft, and A. R. E. Sinclair. 2005. "Ecological change, group territoriality and non-linear population dynamics in Serengeti lions." *Science* 307:390–93.

Packer, C., R. Holt, P. Hudson, K. Lafferty, and A. Dobson. 2003. "Keeping the herds healthy and alert: Implications of predator control for infectious disease." *Ecology Letters* 6:797–802.

Packer, C., D. Ikanda, B. Kissui, and H. Kushnir. 2005. "Ecology: Lion attacks on humans in Tanzania." *Nature* 436:927–28.

Packer, C., M. Kosmala, H. S. Cooley, H. Brink, L. Pintea, D. Garshelis, G. Purchase, et al. 2009. "Sport hunting, predator control and conservation of large carnivores." *PLoS One* 4:e5941. https://doi.org/10.1371/journal.pone.0005941.

Packer, C., S. Lewis, and A. E. Pusey. 1992. "A comparative analysis of non-offspring nursing." *Animal Behaviour* 43:265–81.

Packer, C., A. Loveridge, S. Canney, T. Caro, S. T. Garnett, M. Pfeifer, K. K. Zander, et al. 2013. "Conserving large carnivores: Dollars and fence." *Ecology Letters* 16:635–41. https://doi.org/10.1111/ele.12091.

Packer, C., and S. Polasky. 2018. "Reconciling corruption with conservation triage: Should investments shift from the last best places?" *PLoS Biology* 16:e2005620. https://doi.org/10.1371/journal.pbio.2005620.

Packer, C., and A. E. Pusey. 1982. "Cooperation and competition in coalitions of male lions: Kin selection or game theory?" *Nature* 296:740–42.

———. 1983a. "Male takeovers and female reproductive parameters: A simulation of oestrus synchrony in lions (*Panthera leo*)." *Animal Behaviour* 31:334–40.

———. 1983b. "Adaptations of female lions to infanticide by incoming males." *American Naturalist* 121:716–28.

———. 1984. "Infanticide in carnivores." In *Infanticide: Comparative and Evolutionary Perspectives*, edited by G. Hausfater and S. B. Hrdy, 31–42. New York: Taylor and Francis.

———. 1985. "Asymmetric contests in social mammals: Respect, manipulation and age-specific aspects." In *Evolution*, edited by P. J. Greenwood and M. Slatkin, 173–86. Cambridge: Cambridge University Press.

———. 1987. Intrasexual cooperation and the sex ratio in African lions. *American Naturalist* 130:636–42.

———. 1995. "The lack clutch in a communal breeder: Lion litter size is a mixed evolutionarily stable strategy." *American Naturalist* 145:833–41.

Packer, C., A. E. Pusey, and L. Eberly. 2001. "Egalitarianism in female African lions." *Science* 293:690–93.

Packer, C., A. E. Pusey, H. Rowley, D. A. Gilbert, J. Martenson, and S. J. O'Brien. 1991. "Case study of a population bottleneck: Lions of Ngorongoro Crater." *Conservation Biology* 5:219–30.

Packer, C., and L. M. Ruttan. 1988. "The evolution of cooperative hunting." *American Naturalist* 132:159–98.

Packer, C., D. Scheel, and A. E. Pusey. 1990. "Why lions form groups: Food is not enough." *American Naturalist* 136:1–19.

Packer, C., S. Shivakumar, M. E. Craft, H. Dhanwatey, P. Dhanwatey, B. Gurung, A. Joshi, et al. 2019. "Species-specific spatiotemporal patterns of leopard, lion and tiger attacks on humans." *Journal of Applied Ecology* 56:585–93. https://doi.org/10.1111/1365-2664.13311.

Packer, C., A. Swanson, D. Ikanda, and H. Kushnir. 2011. "Fear of darkness, the full moon and the lunar ecology of African lions." *PLoS One* 6:e22285. https://doi.org/10.1371/journal.pone.0022285.

Packer, C., M. Tatar, and D. A. Collins. 1998. "Reproductive cessation in female mammals." *Nature* 392:807–11.

Palmer, M. S., J. Fieberg, A. Swanson, M. Kosmala, and C. Packer. 2017. "A 'dynamic' landscape of fear: Prey responses to spatiotemporal variations in predation risk across the lunar cycle." *Ecology Letters* 20:1364–73. https://doi.org/10.1111/ele.12832.

Palomares, F., and T. M. Caro. 1999. "Interspecific killing among mammalian carnivores." *American Naturalist* 153:492–508.

Palombit, R. A. 2015. "Infanticide as sexual conflict: Coevolution of male strategies and female counterstrategies." *Cold Spring Harbor Perspectives in Biology* 7:a017640. https://doi.org/10.1101/cshperspect.a017640.

Patterson, B., R. Kays, S. Kasiki, and V. Sebestyen. 2006. "Developmental effects of climate on the lion's mane (*Panthera leo*)." *Journal of Mammalogy* 87:193–200.

Pennycuick, C. J., and J. Rudnai. 1970. "A method of identifying individual lions *Panthera leo* with an analysis of the reliability of identification." *Journal of Zoology* 160:497–508. https://doi.org/10.1111/j.1469-7998.1970.tb03093.x.

Périquet, S., H. Fritz, E. Revilla. 2014. "The Lion King and the Hyaena Queen: Large carnivore interactions and coexistence." *Biological Reviews* 90:1197–1214. https://doi.org/10.1111/brv.12152.

Perrigo, G., W. C. Bryant, and F. S. vom Saal. 1990. "A unique neural timing system prevents male mice from harming their own offspring." *Animal Behaviour* 39:535–39.

Peterhans, J. C. K., and T. P. Gnoske. 2001. "The science of 'man-eating' among lions *Panthera Leo* with a reconstruction of the natural history of the 'Man-eaters of Tsavo.'" *Journal of East African Natural History* 90:1–40.

Pickett, S. T. A., M. L. Cadenasso, and T. L. Benning. 2003. "Biotic and abiotic variability as key determinants of savanna heterogeneity at multiple spatiotemporal scales." In *The Kruger experience: Ecology and management of Savanna heterogeneity*, edited by J. T. du Toit, K. H. Rogers, H. C. Biggs, 22–40. Washington, DC: Island Press.

Pluhácek, J., and L. Bartos. 2000. "Male infanticide in captive plains zebra, *Equus burchelli*." *Animal Behaviour* 59: 689–94. https://doi.org/10.1006/anbe.1999.1371.

Polasky, S., E. Nelson, D. Tilman, J. Gerber, J. A. Johnson, F. Isbell, J. Hill, and C. Packer. "Halting the great degradation of nature through economic development." Proceedings of the National Academy of Sciences. In review.

Polis, G. A., and R. D. Holt. 1992. "Intraguild predation: The dynamics of complex trophic interactions." *Trends in Ecology and Evolution* 7:151–54.

Pringle, R. M. 2017. "Upgrading protected areas to conserve wild biodiversity." *Nature* 546:91–99. https://doi.org/10.1038/nature22902.

Pulliam, H. R. 1988. "Sources, sinks, and population regulation." *American Naturalist* 132:652–61.

Pusey, A. E., and C. Packer. 1994a. "Non-offspring nursing in social carnivores: Minimizing the costs." *Behavioral Ecology* 5:362–74.

———. 1994b. "Infanticide in lions." In *Protection and Abuse of Young in Animals and Man*, edited by S. Parmigiani, B. Svare, and F. vom Saal, 277–99. London: Harwood.

Rees, A. 2009. *The Infanticide Controversy*. Chicago: University of Chicago Press.

Reeve, H. K., S. T. Emlen, and L. Keller. 1998. "Reproductive sharing in animal societies: Reproductive incentives or incomplete control by dominant breeders?" *Behavioral Ecology* 9:267–78. https://doi.org/10.1093/beheco/9.3.267.

Reid, J. M., P. Arcese, and L. F. Keller. 2003. "Inbreeding depresses immune response in song sparrows (*Melospiza melodia*): Direct and intergenerational effects." *Proceedings of the Royal Society of London B* 270:2151–57.

Rentsch, D., and C. Packer. 2015. "Bushmeat consumption and the impact on migratory wildlife in the Serengeti ecosystem." *Oryx* 49:287–94.

Riggio, J., A. Jacobson, L. Dollar, H. Bauer, M. Becker, A. Dickman, P. Funston, et al. 2013. "The size of savannah Africa: A lion's (*Panthera leo*) view." *Biodiversity Conservation* 22:17–35. https://doi.org/10.1007/s10531-012-0381-4.

Ripa, J., and A. R. Ives. 2003. "Food web dynamics in correlated and autocorrelated environments." *Theoretical Population Biology* 64:369–84.

Roelke-Parker, M. E., L. Munson, C. Packer, R. Kock, S. Cleaveland, M. Carpenter, S. J. O'Brien, et al. 1996. "A canine distemper virus epidemic in Serengeti lions (*Panthera leo*)." *Nature* 379:441–45.

Sachs, R. 1967. "Liveweights and measurements of Serengeti game animals." *East African Widlife Journal* 5:24–36.

Salerno, J., M. Borgerhoff Mulder, M. N. Grote, M. Ghiselli, and C. Packer. 2016. "Household livelihoods and conflict with wildlife in community-based conservation areas across northern Tanzania." *Oryx* 50:702–12.

Salerno, J., J. Mwalyoyo, T. Caro, E. Fitzherbert, and M. Borgerhoff Mulder. 2017. "The consequences of internal migration in Sub-Saharan Africa: A case study. *BioScience* 67:664–71. https://doi.org/10.1093/biosci/bix041.

Samuni, L., A. Preis, T. Deschner, C. Crockford, and R. M. Wittig. 2018. "Reward of labor coordination and hunting success in wild chimpanzees." *Communications Biology* 1:138. https://doi.org/10.1038/s42003-018-0142-3.

Schaller, G. B. 1972. *The Serengeti Lion*. Chicago: University of Chicago Press.

Scheel, D. 1993. "Profitability, encounter rates and the prey choice of African lions." *Behavioral Ecology* 4:90–97.

Scheel, D., and C. Packer. 1991. "Group hunting behaviour of lions: A search for cooperation." *Animal Behaviour* 41:697–710.

Searchinger, T., R. Waite, C. Hanson, J. Ranganathan, P. Dumas, and E. Matthews. 2018. *Creating a Sustainable Food Future: A Menu of Solutions to Feed Nearly 10 Billion People by 2050*. Washington, DC: World Resources Institute.

Sela, S., H. M. van Es, B. N. Moebius-Clune, R. Marjerison, J. J. Melkonian, D. Moebius-Clune, R. Schindelbeck, and S. Gomes. 2016. "Adapt-N outperforms grower-selected nitrogen rates in Northeast and Midwest USA strip trials." *Agronomy Journal* 108:1726–34. https://doi.org/10.2134/agronj2015.0606.

Selous, F. C. 1908. *African Nature Notes and Reminiscences*. London: Macmillan.

Senda, A., E. Hatakeyama, R. Kobayashi, K. Fukuda, Y. Uemura, T. Saito, C. Packer, O. T. Oftedal, and T. Urashima. 2010. "Chemical characterization of milk oligosaccharides of an African lion (*Panthera leo*) and a clouded leopard (*Neofelis nebulosa*)." *Animal Science Journal* 81:687–93.

Silk, J. B., J. C. Beehner, T. Bergman, C. Crockford, A. L. Engh, L. Moscovice, R. M. Wittig, R. M. Seyfarth, and D. L. Cheney. 2010. "Female chacma baboons form strong, equitable, and enduring social bonds." *Behavioral Ecology and Sociobiology* 64:1733–47.

Silk, J. B., and G. R. Brown. 2008. "Local resource competition and local resource enhancement shape primate birth sex ratios." *Proceedings of the Royal Society B* 275:1761–65. https://doi.org/10.1098/rspb.2008.0340.

Silk, J. B., E. Willoughby, and G. R. Brown. 2005. "Maternal rank and local resource competition do not predict birth sex ratios in wild baboons." *Proceedings of the Royal Society B* 272:859–64. https://doi.org/10.1098/rspb.2004.2994.

Sinclair, A. R. E. 1977. *The African Buffalo: A Study of Resource Limitation of Populations*. Chicago: University of Chicago Press.

———. 1979a. "The Serengeti environment." In *Serengeti: Dynamics of an Ecosystem*, edited by A. R. E. Sinclair, M. Norton-Griffiths, 31–45. Chicago: University of Chicago Press.

———. 1979b. "The eruption of the ruminants." In *Serengeti: Dynamics of an Ecosystem*, edited by A. R. E. Sinclair and M. Norton-Griffiths, 82–103. Chicago: University of Chicago Press.

Sinclair, A. R. E., and P. Arcese. 1995. "Population consequences of predation-sensitive foraging: The Serengeti wildebeest." *Ecology* 76:882–91.

Sinclair, A. R. E., and R. Beyers. 2021. *A Place Like No Other: Discovering the Secrets of Serengeti*. Princeton, NJ: Princeton University Press.

Sinclair, A. R. E., J. D. Keyyu, S. A. R. Mduma, M. Mtahika, E. Kisamo, J. G. C. Hopcraft, J. M. Fryxell, et al. 2015. "The role of research in conservation and the future of the Serengeti." In *Serengeti IV: Sustaining Biodiversity in a Coupled Human-Natural System*. Chicago: University of Chicago Press.

Sinclair, A. R. E., S. Mduma, and J. S. Brashares. 2003. "Patterns of predation in a diverse predator–prey system." *Nature* 425:288–90.

Sinclair, A. R. E., K. L. Metzger, J. M. Fryxell, C. Packer, A. E. Byrom, M. E. Craft, K. Hampson, et al. 2013. "Asynchronous food web pathways could buffer the response of Serengeti predators to El Niño Southern Oscillation." *Ecology* 94:1123–30.

Sinclair, A. R. E., and M. Norton Griffiths. 1979. *Serengeti: Dynamics of an Ecosystem*. Chicago: University of Chicago Press.

Smith, J. E., R. C. Van Horn, K. S. Powning, A. R. Cole, K. E. Graham, S. K. Memenis, and K. Holekamp. 2010. "Evolutionary forces favoring intragroup coalitions among spotted hyenas and other animals." *Behavioral Ecology* 21:284–303.

Smuts, G. L. 1975. "Why we killed those lions." *African Wildlife* 29 (3): 30–33.

———. 1978. "Interrelations between predators, prey and their environment." *Bioscience* 28:316–20.

Somerville, K. 2020. *Humans and Lions: Conflict, Conservation and Coexistence*. London: Routledge.

Stacey, P. B., and J. D. Ligon. 1987. "Territory quality and dispersal options in the acorn woodpecker, and a challenge to the habitat-saturation model of cooperative breeding." *American Naturalist* 130:654–76.

Staerk, J., F. Colchero, M. Tidiére, D. A. Conde, K. S. Simonsen, and C. Packer. 2020. "Reproductive patterns in the genus *Panthera*." *European Association of Zoos and Aquaria*, poster.

Stander, P., W. Steenkamp, and L. Steenkamp. 2018. *Vanishing Kings: Lions of the Namib Desert*. Johannesburg, South Africa: HPH Publishing.

Stander, P. E. 1992. "Cooperative hunting in lions: The role of the individual." *Behavioral Ecology and Sociobiology* 29:445–54.

Stephens, D. W., and E. L. Charnov. 1982. "Optimal foraging: Some simple stochastic models." *Behavioral Ecology and Sociobiology* 10:251–63.

Strampelli, P., L. Andresen, K. Everatt, M. Somers, and J. Rowcliffe. 2019. "Leopard (*Panthera pardus*) density in southern Mozambique: Evidence from spatially explicit capture-recapture in Xonghile Game Reserve." *Oryx* 55:405–11. https://doi.org/10.1017/S0030605318000121.

Strauss, M., M. Kilewo, D. Rentsch, and C. Packer. 2015. "Food supply and poaching limit giraffe abundance in the Serengeti." *Population Ecology* 57:505–16. https://doi.org/10.1007/s10144-015-0499-9.

Strauss, M., and C. Packer. 2015. "Did elephants and giraffes mediate change in the prevalence of palatable species in an East African *Acacia* woodland?" *Journal of Tropical Ecology* 31:1–12. https://doi.org/10.1017/S0266467414000625.

Strauss, M. K. L., and C. Packer. 2012. "Using claw marks to study lion predation on giraffes of the Serengeti." *Journal of Zoology* 289:134–42. https://doi.org/10.1111/j.1469-7998.2012.00972.x.

Swanson, A., T. Caro, H. Davies-Mostert, M. G. L. Mills, D. W. Macdonald, M. Borner, E. Masenga, and C. Packer. 2014. "Cheetahs and wild dogs show contrasting patterns of suppression by lions." *Journal of Animal Ecology* 83:1418–27. https://doi.org/10.1111/1365-2656.12231.

Swanson, A., J. Forester, T. Arnold, M. Kosmala, and C. Packer. 2016. "In the absence of a landscape of fear: How lions, hyenas, and cheetahs coexist. *Ecology and Evolution* 6:8534–45. https://doi.org/10.1002/ece3.2569.

Swanson, A., M. Kosmala, C. Lintott, and C. Packer. 2016. "A generalized approach for producing, quantifying, and validating citizen science data from wildlife images." *Conservation Biology* 30:520–31. https://doi.org/10.1111/cobi.12695.

Swanson, A., M. Kosmala, C. Lintott, R. Simpson, A. Smith, and C. Packer. 2015. "Snapshot Serengeti, high-frequency, fine-scale annotated camera trap images of 40 mammalian species in an African savanna. *Scientific Data* 2:150026. https://doi.org/10.1038/sdata.2015.26.

Thondhlana, G., S. M. Redpath, P. O. Vedeld, L. van Eeden, U. Pascual, K. Sherren, and C. Murata. 2020. "Non-material costs of wildlife conservation to local people and their implications for conservation interventions." *Biological Conservation* 246:108578.

Tilman, D., M. Clark, D. R. Williams, K. Kimmel, S. Polasky, and C. Packer. 2017. "Future threats to biodiversity and pathways to their prevention." *Nature* 546:73–81. https://doi.org/10.1038/nature22900.

Tinbergen, N. 1948. "Social releasers and the experimental method required for their study." *Wilson Bulletin* 60:6–51.

Traulsen, A., and M. A. Nowak. 2006. "Evolution of cooperation by multilevel selection." *Proceedings of the National Academy of Sciences* 103:10952–55. https://doi.org/10.1073/pnas.0602530103.

Trethowan, P., A. Fuller, A. Haw, T. Hart, A. Markham, A. Loveridge, R. Hetem, et al. 2017. "Getting to the core: Internal body temperatures help reveal the ecological function and thermal implications of the lions' mane." *Ecology and Evolution* 7:253–62. https://doi.org/10.1002/ece3.2556.

Trinkel, M., D. Cooper, C. Packer, and R. Slotow. 2011. "Inbreeding depression increases susceptibility to bovine tuberculosis in lions: An experimental test using an inbred–outbred contrast through translocation." *Journal of Wildlife Diseases* 47:494–500.

Trinkel, M., N. Ferguson, A. Reid, C. Reid, M. Somers, L. Turelli, J. Graf, M. Szykman, D. Cooper, P. Haverman, G. Kastberger, C. Packer, and R. Slotow. 2008. "Translocating lions into an inbred lion population in the Hluhluwe-iMfolozi Park, South Africa." *Animal Conservation* 11:138–43.

Trinkel, M., P. Funston, M. Hofmeyr, D. Hofmeyr, S. Dell, C. Packer, and R. Slotow. 2010. "Inbreeding and density dependent population growth in a small, isolated lion population." *Animal Conservation* 13:374–82.

Troyer, J. L., M. E. Roelke, J. M. Jespersen, N. Baggett, V. Buckley-Beason, D. MacNulty, M. Craft, C. Packer, et al. 2011. "FIV diversity: FIV_{Ple} subtype composition may influence disease outcome in African lions." *Veterinary Immunology and Immunopathology* 143:338–46.

Tucker, M. A., et al. 2018. "Moving in the Anthropocene: Global reductions in terrestrial mammalian movements." *Science* 469:466–69. https://doi.org/10.1126/science.aam9712.

Tulogdi, A., L. Biro, B. Barsvari, M. Stankovic, J. Haller, and M. Toth. 2015. "Neural mechanisms of predatory aggression in rats: Implications for abnormal intraspecific aggression." *Behavior and Brain Research* 283:108–115.

Tulogdi, A., M. Toth, J. Halasz, E. Mikics, T. Fuzesi, and J. Haller. 2010. "Brain mechanisms involved in predatory aggression are activated in a laboratory model of violent intra-specific aggression." *European Journal of Neuroscience* 32:1744–53.

UNDESA. 2013. *World Population Prospects: The 2012 Revision, Key Findings and Advance Tables.* ESA/P/WP.227. New York: United Nations Department of Economic and Social Affairs.

Valeix, M., A. J. Loveridge, and D. W. Macdonald. 2012. "Influence of prey dispersion on territory and group size of African lions: A test of the resource dispersion hypothesis." *Ecology* 93:2490–96.

VanderWaal, K., A. Mosser, and C. Packer. 2009. "Optimal group size, dispersal decisions and post-dispersal relationships in female African lions." *Animal Behaviour* 77:949–54.

Van Orsdol, K. G. 1984. "Foraging behaviour and hunting success of lions in Queen Elizabeth National Park, Uganda." *African Journal of Ecology* 22:79–99. https://doi.org/10.1111/j.1365-2028.1984.tb00682.x.

Van Valkenburgh, B., and P. A. White. 2021. "Naturally occurring tooth wear, tooth fracture, and cranial injuries in large carnivores from Zambia." *PeerJ* 9:e11313. https://doi.org/10.7717/peerj.11313.

Veldhuis, M. P., E. S. Kihwele, J. P. G. M. Cromsigt, J. O. Ogutu, J. G. C. Hopcraft, N. Owen-Smith, and H. Olff. 2019. "Large herbivore assemblages in a changing climate: Incorporating water dependence and thermoregulation." *Ecology Letters* 22:1536–46. https://doi.org/10.1111/ele.13350.

Veldhuis, M. P., M. E. Ritchie, J. O., Ogutu, T. A. Morrison, C. M. Beale, A. B. Estes, W. Mwakilema, et al. 2019. "Cross-boundary human impacts compromise the Serengeti-Mara ecosystem." *Science* 363:1424–28.

Viana, M., S. Cleaveland, J. Matthiopoulos, Jo Halliday, C. Packer, M. E. Craft, K. Hampson, et al. 2015. "Dynamics of a morbillivirus at the domestic-wildlife interface: Canine distemper virus in domestic dogs and lions." *Proceedings of the National Academy of Sciences* 112:1464–69. https://doi.org/10.1073/pnas.1411623112.

Vollset, S. E., E. Goren, C.-W. Yuan, J. Cao, A. E. Smith, T. Hsiao, et al. 2020. "Fertility, mortality, migration, and population scenarios for 195 countries and territories from 2017 to 2100: A forecasting analysis for the Global Burden of Disease Study." *The Lancet* 396:1285–1306. https://doi.org/10.1016/S0140-6736(20)30677-2.

Volterra, V. 1926. "Fluctuations in the abundance of a species considered mathematically." *Nature* 118:558–60.

vom Saal, F. S. 1985. "Time-contingent change in infanticide and parental behavior induced by ejaculation in male mice." *Physiology and Behavior* 34:7–15.

Walters, J. R., and R. M. Seyfarth. 1987. "Conflict and cooperation." In *Primate Societies*, edited by B. B. Smuts, D. L. Cheney, R. M. Seyfarth, R. W. Wrangham, and T. T. Struhsaker, 306–17. Chicago: University of Chicago Press.

Waser, P. M. 1981. "Sociality or territorial defense? The influence of resource renewal." *Behavioral Ecology and Sociobiology* 8:231–37.

Watts, H. E., and K. E. Holekamp. 2008. "Interspecific competition influences reproduction in spotted hyenas." *Journal of Zoology* 276:402–10.

———. 2009. "Ecological determinants of survival and reproduction in the spotted hyena." *Journal of Mammalogy* 90:461–71.

Weckworth, J., B. Davis, E. Dubovi, N. Fountain-Jones, C. Packer, S. Cleaveland, M. Craft, et al. 2020. "Cross-species transmission and evolutionary dynamics of canine distemper virus during a spillover in African lions of Serengeti National Park." *Molecular Ecology* 29:4308–21. https://doi.org/10.1111/mec.15449.

Weckworth, J., B. W. Davis, M. E. Roelke-Parker, R. P. Wilkes, C. Packer, E. Eblate, M. K. Schwartz, and L. S. Mills. 2020. "Identifying candidate genetic markers of CDV cross-species pathogenicity in African lions." *Pathogens* 9:872. https://doi.org/10.3390/pathogens9110872.

West, P. 2005. "The lion's mane." *American Scientist* 93:226–35.

West, P. M., H. MacCormick, G. Hopcraft, K. Whitman, M. Ericson, M. Hordinsky, and C. Packer. 2006. "Wounding, mortality and mane morphology in African lions, *Panthera leo*." *Animal Behaviour* 71:609–19.

West, P. M., and C. Packer. 2002. "Sexual selection, temperature and the lion's mane." *Science* 297:1339–43.

Whitman, K., A. Starfield, H. Quadling, and C. Packer. 2004. "Sustainable trophy hunting in African lions." *Nature* 428:175–78.

Wijers, M., P. Trethowan, B. du Preez, S. Chamaillé-Jammes, A. J. Loveridge, D. W. Macdonald, and A. Markham. 2021. "The influence of spatial features and atmospheric conditions on African lion vocal behaviour." *Animal Behaviour* 174:63–76. https://doi.org/10.1016/j.anbehav.2021.01.027.

Wildt, D. E., M. Bush, K. L. Goodrowe, C. Packer, A. E. Pusey, J. L. Brown, P. Joslin, and S. J. O'Brien. 1987. "Reproductive and genetic consequences of founding isolated lion populations." *Nature* 329:328–31.

Williams, J. M., G. W. Oehlert, J. V. Carlis, and A. E. Pusey. 2004. "Why do male chimpanzees defend a group range?" *Animal Behaviour* 68:523–32.

Williams, V. L., A. J. Loveridge, D. J. Newton, and D. W. Macdonald. 2017. "A roaring trade? The legal trade in *Panthera leo* bones from Africa to East-Southeast Asia." *PloS One* 12:e0185996.

Wilson, E. O. 1975. *Sociobiology: The New Synthesis*. Cambridge: The Belknap Press of Harvard University Press.

Wilson, M., C. Boesch, B. Fruth, T. Furuichi, I. C. Gilby, C. Hashimoto, C. L. Hobaiter, et al. (2014). "Lethal aggression in *Pan* is better explained by adaptive strategies than human impacts." *Nature* 513:414–17. https://doi.org/10.1038/nature13727.

Wittemyer, G., P. Elsen, W. Bean, A. C. O. Burton, and J. Brashares. 2008. "Accelerated human population growth at protected area edges." *Science* 321:123–26.

Woodroffe, R., and L. G. Frank. 2006. "Lethal control of African lions (*Panthera leo*): Local and regional population impacts." *Animal Conservation* 8:91–98. https://doi-org.ezp2.lib.umn.edu/10.1017/S1367943004001829.

Woolfenden, G. E., and J. W. Fitzpatrick. 1984. *The Florida Scrub Jay*. Princeton, NJ: Princeton University Press.

Wrangham, R. W. 2018. "Two types of aggression in human evolution." *Proceedings of the National Academy of Sciences* 115:245–53. https://doi.org/10.1073/pnas.1713611115.

Zimova, M., K. Hackländer, J. M. Good, J. Melo-Ferreira, P. C. Alves, and L. S. Mills. 2018. "Function and underlying mechanisms of seasonal colour moulting in mammals and birds: What keeps them changing in a warming world?" *Biology Reviews* 93:1478–98. https://doi.org/10.1111/brv.12405.

INDEX

Terms refer to lions unless otherwise indicated.

age: classification of, 28; and coalition composition, 109–10, 118–20; and conflict outcomes, 168; and fertility, 93, 95–96; and hunting, 312–14; and mating behavior, 110; and mortality, 24; of "trophy" lions, 314
aging, 28–30, 40–41, 93, 95–97
ala-mayo, 297–301
Approximate Bayesian Computation (ABC) modeling, 293
associations
  among lions: around kills, 155, 159; between adults and cubs, 34, 91–92, 94, 121, 129; between females and females, 25, 90 (see also crèches); between first-order kin, 35, 37; between males and females, 71, 214, 217; between males and males, 118, 120, 212, 214; between prides, 26, 183
  interspecific: between lions and herbivores, 262, 264, 266, 276; between lions and other carnivores, 267, 269–70, 273–74, 288

Babesia parasite, 237–40, 249, 251, 255, 291
babysitting, 88–94
Balme, Guy, 268, 271
behavior: activity patterns, 20, 23–24; leading/lagging, 171–76; ownership of consorts, 108–9, 137; ownership of food, 38–39, 178; peri-partum, 76–79, 99; maternal protective (see babysitting; takeovers); paternal protective, 70, 122–23
belly size, 16–24, 140–42, 157, 159, 180, 231, 277; intragroup variance in, 38–41; and mating, 113; and milk production, 21; of mothers and cubs, 81, 88, 90–93, 101, 103; in Ngorongoro Crater, 242–43
Bertram, Brian, xi, 4, 16, 20, 35, 43, 63, 73, 220
birth order effects, 97
birth synchrony, 68–69, 96–100, 104
body size: of lions (see belly size; chest girth); of prey species, 251–52, 335
Borrego, Natalia, 329n1
Botswana, 149, 212, 268, 318
bovine tuberculosis, 245, 291–94, 326–27
Bruce effect, 68
buffalo, 5, 11, 123, 150–64, 169, 195, 208, 214, 224–30, 237–38, 240, 243, 251, 253, 255, 260, 263, 265, 275, 278, 303, 326, 328, 330, 335; as a disease vector, 291–94
Burkhart, Jessica, 221, 223
bushmeat hunting, 315, 317
bush pig, 303, 305–8
Bygott, David, xi, 4, 53n5, 134n1, 154n7, 263, 265n4

camera traps: in study area, 261–65, 268, 270, 274; in South African reserves, 268, 273
canine distemper virus (CDV), 226, 232–41, 251, 255, 282–91, 326–28; genome of, 290; impact of on host population genetics, 248; transmission of, 287, 289
Canney, Susan, 316
captive lions, 98–99, 140, 221–23
cave lions, 145, 147
cheetah, 20, 23, 88, 89, 154, 196, 230, 250–53, 265, 267, 268, 270
chest girth, 20, 22, 29
chimpanzees, 114, 116, 164, 184, 221
citizen science. See Snapshot Serengeti project
Cleaveland, Sarah, 283, 327
climate change (global), 325
coalitions, 32, 49–58, 60–75, 107–31, 233; competition within, 40, 108–16, 137, 138; composition and size of, 54, 112–20, 192–93, 213; evolutionary context for, 212–22; extinction risk for, 180–81; father-son, 131; longevity of, 53, 56; loss of members from, 114, 116; in Ngorongoro Crater, 245–47; and pride residence, 51, 117, 123–28, 182, 246; and reproductive performance, 52, 74, 110, 112–13, 115, 193, 212–14; size effects of, 124–25, 129; in study population versus other populations, 213–18
communication. See roaring
competition, 166–92, 195–208, 219; between cubs, 38–39, 80–87; between coalition members, 40, 108–16, 137–38; between lions and other carnivores, 265–81; intra versus intergroup, 221
conservation, 282–326; via community engagement, 247, 300, 302; via culling, 250, 293, 294; and economic development, 321, 324–26, 330; via fencing, 265, 267–68, 316, 318–22; funding of, 320–21; via predator reintroduction, 265, 267–68, 318; via regulation of hunting, 312–14; via translocation, 245, 247, 293–94; via vaccination, 251, 284–91; via wildlife corridor, 247, 302
consortships, 63–71, 107–16, 122–23, 128–30, 137, 144–47, 329; and ownership behavior, 35, 109–11, 219–21; preferences for, 115, 146
controversy, regarding infanticide, 73–74
cooperation, xi, 1, 114–16, 195–222; in cub rearing, 43n3, 47, 76–105; in fighting, 53, 167–76; in hunting, 149, 154–64, 329
Craft, Meggan, 283, 286

crèches, 32, 38, 42, 68, 70, 74, 76–105; benefits of, 96, 166–67, 195, 220; fathers and, 122; takeover effects on, 74
cubs, 59–104; abandoned, 79, 99; and fathers, 118, 121–25, 129, 131; nursing behavior of, 38–39, 80–87; orphaned, 95–96; survival factors for, 21, 30, 40, 62, 68, 73, 81, 88, 96–98, 100–105, 126, 143–44, 230, 232–33, 236, 245–46, 314

Darwin, Charles, xi, 133, 219, 221
die-offs. See canine distemper virus (CDV)
diet, 5, 8–12, 16, 50, 123, 140, 149, 150–64, 169, 178, 180, 195, 208, 214, 216–17, 224–31, 237, 238, 240–44, 250–66, 274–78, 300, 303, 315–17, 335
DiMinin, Enrico, 320
disease, xi, 57, 80, 88, 224, 232–45, 248, 251, 255, 257, 260, 275, 282–94, 316, 325; varieties of (see Babesia parasite; bovine tuberculosis; canine distemper virus (CDV); feline immunodeficiency virus; rabies; rinderpest)
dispersal, 32–36, 62, 102–4, 130, 144, 184–92, 247; female, 43, 47–48, 53–56, 60, 63, 64, 128, 129, 184, 187–92, 196–98, 206, 230; male, 29, 32, 34, 60, 104, 119, 192, 212, 245–46
DNA. See genetics
dogs: domestic, 282–91, 296, 319, 326–28; wild, 149, 154, 250–51, 265–69, 278, 281
dominance, 35–40, 42, 57, 156

ecosystem. See food web; study area
eland, 10–12, 260, 317, 335
elephant, 35, 195, 251–52, 255n1, 264, 318–22, 326
epizootic. See rinderpest
Etosha National Park, 149, 164, 208, 214–16
evolution, xi, 1, 40, 73, 75, 166–67, 204–10, 330; and group selection, 219–22; and hunting behavior, 160–64; and interspecific competition, 265–78; and sexual selection, 140–45
extinction: localized, 250; of prides and coalitions, 180–81

feline immunodeficiency virus, 237, 240
fencing, 101, 208, 265, 267, 268, 281, 316, 318–22, 328
fertility: female, 16, 19–20, 28–29, 68–73, 93–96; male, 141
fighting, 107–9, 133–45; between groups, 168, 170, 204; between individuals, 138–39; risks of, 38, 169. See also wounding
fission-fusion patterns: in coalitions, 118, 212; in prides, 25–26, 80, 150, 156, 160, 164, 216, 281
food intake: 11, 38–40, 101–2; and birth rate, 16, 19, 21; and body size, 20, 22; composition of (see diet); and cub survival, 21, 80; and group size, 154–60; and mane characteristics, 141–42; patterns of, 24, 142, 209, 229; types of (see foraging, nursing)
food object trials, 223
food web, 250–81
foraging, 20–24, 149–64, 250–81; hunting, 154–63, 179, 197, 199–200, 209, 217; scavenging, 23, 279–80
Fosbrooke, Henry, 53n5, 236, 237n4
Fountain-Jones, Nick, 307
Fryxell, John, 257, 281

game theory, 99–101, 161–64
gazelle, 5, 9–12, 16, 150–51, 154–58, 164, 224–25, 230–31, 253–54, 260, 263–65, 274–78, 335

genetics: of canine distemper virus, 234, 290; of lions, 28, 35, 36, 248; genetic rescue, 293–94
giraffe, 5, 11, 160, 250–55, 335
Gir Forest, 40, 208, 214–18, 271, 311
GIS analysis, 263, 266
grandmothers, 93–96
Grinnell, Jon, 53, 70, 171–72
grouping patterns. See associations
group selection, 219–22
group territoriality, 47, 163, 167–76, 195, 204–10, 222, 330

habitat. See study area
Hanby, Jeannette, xi, 53n5, 134n1, 154n7, 265n4
hartebeest, 5, 10, 12, 260, 317, 335
heat stress, 140–42
Heinsohn, Rob, 171, 174
hippopotamus, 251–52
Hluhluwe iMfolozi Park (HIP), 245, 268, 292–93
Hopcraft, Grant, 4, 196, 197
human–lion interactions: attacks on humans by lions, 302–9; retaliatory killing of lions by pastoralists, 294–302; sport hunting of lions, 309–14. See also conservation
hunting: by humans (see human–lion relations); by lions (see foraging)
hyenas, x, 23, 35, 42, 50, 144, 234, 253; competition with lions, 154, 178, 180, 196, 251, 253, 261, 265–81; as disease vectors, 283–91, 328; and leopards, 271; human–wildlife conflict, 294–97; nursing habits of, 80, 82, 83, 86; predation on lion cubs, 88, 230

Ikanda, Dennis, 297, 302, 328
impala, 5, 250–54, 263–65
inbreeding, 102–4; and conservation efforts, 302; in Ngorongoro Crater, 2, 35, 57, 128, 141, 245–49, 257, 260, 278; in South African lions, 292–93, 326
infanticide, 59–75, 88, 104, 123–24, 230; and fertility, 61, 65, 69; in other Panthera, 68n2, 75, 166; and subadult survival, 62; and trophy hunting, 309, 312, 314; in zebra, 253
interbirth intervals, 20–21, 42, 52, 61, 65, 67, 96–97, 101–3
interpride encounters, 167–80
intersex lion, 122–23, 128
interspecies proximities. See associations
interspecific disease transmission, 282–91

jackals, 23, 251, 253, 283–91, 328
jaguars, 98, 145, 167, 204, 206, 212
Jansson, Ingela, 17, 169–70, 247, 300, 302

Kenya: historic lion attacks in, 307, 309; Maasai Mara Reserve in, 232, 274, 297; pride size in, 208, 331; Tsavo National Park in, 140
kinship: and infanticide, 122; kin selection, 1, 109–12, 220; and mating, 112, 128–29; among pridemates, 35–37, 82, 85, 86; and recruitment, 43, 131. See also nursing; paternity
Kissui, Bernard, 294, 302, 328
Kock, Richard, 236
Kosmala, Margaret, 204, 261, 293

Kruger National Park, 208, 250, 255, 260, 275, 278, 291–94
Kushnir, Hadas, 302–6, 310–11

leopards, 1–2, 47, 68n2, 75, 88, 98, 137, 144–45, 166–67, 184–85, 189, 204, 206, 208, 212, 219, 251, 253, 265, 281, 329; attacking humans, 294–97, 307, 309, 311, 326; competing with lions, 268–73; predation on lion cubs by, 88, 230
Lindi District, Tanzania, 302–10
Lion Guardians project, 247, 300, 302
lions compared to: baboons, 111n3; cuckoos, 128; hyenas, 35, 42–43, 80, 82–83, 86; extinct species, 145, 147, 208; humans, 221; other group-hunting animals, 161–64; other group territorial species, 202; other litter-bearing animals, 86–88; other matrilineal animals, 35; other nursing mammals, 86–88; other *Panthera* species, 68n2, 75, 98–99, 106, 133, 137, 145, 166–67, 184, 202–7, 212, 215, 219, 222, 307–11; primates, 35, 73, 145; rodents, 68, 86, 122
litters, 96–104; kinship within, 36; "lost," 77, 99, 245, 246; sex ratio of, 103, 105; size of, 95, 98, 100; spacing of, 61, 65, 67, 97, 103. *See also* crèches; nursing; pregnancy and birth
livestock predation by lions, 294–302
local resource competition/enhancement, 102–4
Loveridge, Andrew, 316
lunar cycle. *See* nocturnal activity

Maasai, 247, 275, 285, 294–302, 316n9
Maasai Mara Reserve and Conservancies, 232, 253–54, 271, 274–75, 286, 321, 324
Macdonald, David, 202
man-eating lions. *See* human–lion relations: attacks on humans by lions
manes, 133–48, 325; characteristics of, 134, 136; growth of, 135; and fitness, 143; and heat stress, 142; lions' preferences for, 146
mating. *See* consortships
McComb, Karen, 70, 171
meerkats, 35, 43
methods: of age estimation, 312n6; using camera traps, 261–63; using dummies, 88, 145, 329n1; of estimating maternity, 42; of estimating per capita food intake, 156; of individual identification, 2–4; of quantifying mane quality, 133–34; of radio tracking, 11; using thermal cameras, 140–42. *See also* modeling; playback experiments
milk production, 20–21, 80–82, 86–87
modeling: of big-cat attacks on humans, 307–9; of disease transmission, 283–93; of lion carrying capacity, 316; of population dynamics, 257–63; of sociality, 204–8; of sport hunting, 309–14
morbillivirus. *See* canine distemper virus (CDV); rinderpest
mortality, 21, 28–29, 95, 241; for cubs (*see* cubs: survival factors for); from disease, 230–49, 291–93; for dispersed subadult males, 60; for females, 48, 63, 100–102, 180–83; and inbreeding, 245–46; in population modeling, 204–6, 309; by predation, 230, 232; and predator-prey dynamics, 259; for solitaries, 49, 176; and wounds, 139, 177, 183. *See also* human–lion relations; infanticide

Mosser, Anna, 180–84, 196, 204
Mozambique, 302, 306n3, 312–13, 315, 321
Munson, Linda, 232, 236–37, 327–28

Namibia, 149, 164, 208, 214–16, 321
National Cancer Institute, 327
National Science Foundation (NSF), xii, 283, 328
naturalistic experiments: lunar cycle observations, 263, 266; predator removal, 250–55
natural selection. *See* evolution
network modeling, 283–91
Ngorongoro Conservation Area (NCA), 5, 240, 247, 297–98, 300–301, 313, 323
Ngorongoro Crater, data unique to, 9, 240–49, 255–56, 260, 275–77
nocturnal activity, 16, 20–25, 27–28, 79, 149, 154, 156, 166, 169, 263–66; attacks on humans, 303–9; ranging, 16, 79; roaring, 25, 28, 172, 218
nomadic males, 11, 15–16, 32–35, 49–53, 68–73, 107–8, 130, 167–72, 221, 232–33, 300; radio tracking of, 309; versus resident males, 172–73. *See also* dispersal; takeovers
Normalized Difference Vegetation Index (NDVI), 229, 263
numeracy (ability to count), 25, 171–75
nursing, 38–39, 42, 77, 80–88, 102, 220

oribi, 252–54
orphans, 93–96, 143
oxytocin, 20n4, 221–23

Palmer, Meredith, 263
parasites. *See* Babesia parasite
paternity, 36, 62, 73, 109–14, 118–30, 214–16, 219n5, 245, 327; and father-cub interactions, 118, 121, 122–23, 125, 129, 131; skew in, 110, 112–13, 329; uncertainty of, 66, 68
philopatry: in leopards, 185; in lions, 184, 186, 206, 220
playback experiments, 53, 54, 70, 72, 144n4, 146, 171–76, 221–23
poaching. *See* bushmeat hunting
population dynamics: density dependence, 240–43, 255–56, 275–76; population cycling (for predators and prey), 255–61; population decline (for lions), xi, 232, 240, 247, 260, 275, 282, 297, 312–13; source-sink dynamics, 144, 196, 198–99, 294. *See also* study population
predator removal experiment, 250–55
preferences: in consorts, 68, 111, 114–15, 128; in manes, 146; in prey types, 158
pregnancy and birth, 16, 18, 21, 61–69, 76–78, 96–97, 101–3
prey, 150–60, 224–43, 250–66, 275–77; accessibility of, 224–30, 244; behavior of, 261–65; body size of, 159, 252; as a disease vector, 238, 292; impact of bushmeat hunting on, 315, 317; quantities of, 9, 12, 225, 228, 256, 258, 266; ranging patterns of, 8
prides, 25, 55, 88–96, 154–60, 180–93, 195–213, 233; composition of, 36, 48, 190; encounters between, 167–80; evolutionary considerations regarding, 88–96, 202–8, 219–22; longevity of, 53, 56, 181; proximity of pride-mates within, 26, 77, 218; and reproductive success, 44–46, 183, 193 (*see also* crèches); size and size effects of, 125, 158–59, 193, 210–11, 213; spatial distribution of, 26, 183, 189–91, 201
Pusey, Anne, xi, 4, 21, 50, 80, 150

rabies, 265n4, 286, 291
radio collar tracking, 76–78
ranging patterns: of cheetahs, 268, 270; of hyenas, 286, 288; of leopards, 185; of lions, 13–18, 77, 79, 184–91, 198, 201, 203, 215, 234, 257, 263; of wild dogs, 265, 268–69
recruitment: female, 32–35, 43, 47, 53, 63, 202–6, 210n4, 220; during conflicts, 166, 171, 182; male, 58, 114, 130–31, 212. *See also* dispersal
reedbuck, 161
Rentsch, Dennis, 315
reproduction, 40–56, 59–75, 101–4, 192; cycle of, 32–33, 61, 65; in lions compared to other felids, 215; reproductive behavior (*see* consortships); reproductive success factors, 29–30, 44–46, 52, 64, 193, 199, 259 (*see also* cubs: survival factors for); reproductive synchrony, 69; sperm abnormality, 141. *See also* inbreeding; paternity
Resource Dispersion Hypothesis, 202–6
rhinoceros, 251–52
rinderpest, 224–30, 243, 251, 309
roaring, 25–27, 166–76, 217–18. *See also* playback experiments
Roelke, Melody, 232, 234, 327
Rosengren, Daniel, xii, 17, 76–77, 142, 182
Rufiji District, Tanzania, 303–11

sabretooth cats, 208
SaTScan modeling, 307, 309
Savannas Forever-Tanzania, 295, 324
Schaller, George, ix–xii, 4, 53, 58, 66, 88, 133, 154, 182, 250, 268
Scheel, Dave, 154n2, 156–57, 180
scientific responsibility, 327–28
Selous, Frederick, 140
senescence. *See* aging
Serengeti Ecological Monitoring Program, 278
Serengeti Research Institute, 250
sexual selection, 140–45
sex differences: in body size, 22; in cub sex ratio, 103, 105, 193; in heat tolerance, 140–42; in hunting style, 150, 154–55; in mortality and reproduction, 28–30, 101–3; in roaring, 25, 27, 171, 173–74, 216–18; in social life cycle, 32, 33, 53; in wounding rates, 177. *See also* intersex lion
SimSimba population model, 309–14
Sinclair, Tony, 227
Snapshot Serengeti project, 261–66, 271
sociality, 32, 163–67, 195, 202–12, 219–22, 249, 257–61, 265, 277–81, 325, 329; female (*see* crèches; prides); male (*see* coalitions); and population stability, 258–59, 281
solitary lions, 26, 47–50, 57–58, 99–100, 192, 331; feeding by, 151, 156, 162, 277, 279–80; roaring by, 172, 178; solitary males, 35, 53, 109, 114, 116, 119; solitary mothers and cubs, 91, 100; vulnerability of, 176–180, 181
South Africa: conservation efforts in, 265, 267, 268, 277, 316, 318, 328; lions in, 123, 185, 208, 221, 245, 250, 273, 281, 291–93, 316, 319, 329–31
Southern Oscillation Index, 227–30
stochastic modeling, 286, 289
Stomoxys flies, 236–37
study area, 5–20; available biomass in, 12, 209, 335; carrying capacity of, 319; human impact on, 322; human populations around, 295; hydrography of, 199–200; maps of, 6, 79, 201, 255; rainfall in, 7, 229, 231; wildfire in, 228
study population: coalitions in, 58; prides in, 55; size of over time, 224–30, 240–44; versus other lion populations, 66, 101, 111n2, 123, 140, 149, 164, 208, 213–18, 268, 271, 275, 291–94, 300, 313
subadults, 32, 43, 47, 60–64, 97, 122, 144; in conflicts, 167–68, 174–75; feeding by, 38–39, 155–56; ranging by, 15–16, 184; survival of, 62, 192, 233; wounding of, 137, 139. *See also* dispersal
Sukuma, 295, 306
survival. *See* mortality
Sustainable Development Goals, 321, 324n12, 330
Swanson, Ali, 261, 263

takeovers, 48, 60–75, 77, 80; and cub/subadult survival, 62, 143, 192, 232–33, 242; effects of trophy hunting on, 309, 312, 314; and the reproductive cycle, 61, 65, 67, 104–6, 122, 128; risks and consequences of, 74, 219, 240, 318
Tanzania: hunting of lions in, 309–13, 326, 328, 330; lion attacks in, 304–5, 309, 311, 326–28; National Parks authority in, 250, 312–13
Tarangire National Park in, 294, 296–97
temperature sensitivity. *See* heat stress
Terio, Karen, 237, 328
territory, 11, 166–92, 195–208, 210n1, 212, 214, 219, 222, 234n2, 278, 331; maps of, 190–91, 199–201, 203, 205, 207. *See also* group territoriality; ranging patterns
testosterone, 122, 134–35, 141, 145
tick-borne parasite. *See Babesia* parasite
tigers, 1–2, 98, 133, 137, 145, 166–67, 204, 206, 212, 219, 221n6, 271, 282, 307–9, 311, 316n9
topi, 10, 12, 250–55, 260, 263–64, 278, 317, 335
trophic cascades, 275–77
Tsavo, 140, 144, 307, 309, 327

U.S. Endangered Species Act, 313

vaccination, 251, 265, 283–91, 328
vibrissae. *See* whisker spots
vocalization. *See* roaring

warthogs, 5, 10, 12, 150–64, 224n1, 230–31, 253–60, 278, 317, 335
waterbuck, 260, 317
West, Peyton, 88, 133, 140, 142, 144
whisker spots, 2–4
Whitman, Karyl, 309, 312
wildebeest, 5, 8–12, 16, 50, 140, 149, 150–58, 162, 178, 180, 208, 214, 216, 224–31, 240, 243–44, 250–66, 274–75, 278, 294, 300, 303, 315–17, 326, 335
wolves, 40, 43, 265, 275
wounding, 116, 133–34, 137, 139, 142–44, 169, 176–78, 182–83

Zambia, 157n3, 312–13, 315, 321, 324, 330
zebra, 5, 8–12, 16, 140, 150–62, 178, 180, 184, 214, 216, 224–25, 230–31, 240–44, 250–66, 274–78, 294, 303, 317, 326, 335
Zimbabwe, 60, 141, 312–13, 316, 321, 330